ADVANCES IN AUTOCLAVED AERATED CONCRETE

PROCEEDINGS OF THE 3RD RILEM INTERNATIONAL SYMPOSIUM ON AUTOCLAVED AERATED CONCRETE / ZÜRICH / SWITZERLAND / 14-16 OCTOBER 1992

Advances in Autoclaved Aerated Concrete

Edited by
FOLKER H. WITTMANN
Swiss Federal Institute of Technology, Zürich

Taylor & Francis
Taylor & Francis Group

LONDON AND NEW YORK

The texts of the various papers in this volume were set individually by typists under the supervision of each of the authors concerned.

Authorization to photocopy items for internal or personal use, or the internal or personal use of specific clients, is granted by Taylor & Francis, provided that the base fee of US$1.00 per copy, plus US$0.10 per page is paid directly to Copyright Clearance Center, 27 Congress Street, Salem, MA 01970, USA. For those organizations that have been granted a photocopy license by CCC, a separate system of payment has been arranged. The fee code for users of the Transactional Reporting Service is: 90 5410 086 9/92 US$1.00 + US$0.10.

Published by Taylor & Francis
2 Park Square, Milton Park, Abingdon, Oxon, OX14 4RN
270 Madison Ave, New York NY 10016

Transferred to Digital Printing 2006

ISBN 90 5410 0869
© 1992 Taylor & Francis

Advances in Autoclaved Aerated Concrete, Wittmann (ed.) © 1992 Taylor & Francis. ISBN 90 5410 086 9

Table of contents

3 Heat and mass transfer

4 Crack formation and durability

5 Reinforced components

6 Masonry

7 Ecology and new developments

Advances in Autoclaved Aerated Concrete, Wittmann (ed.) © 1992 Taylor & Francis. ISBN 90 5410 086 9

Preface

A first RILEM International Symposium on lightweight concrete was held in Göteborg, Sweden, in 1960. In 1982 the second RILEM International Symposium was organized in Lausanne, Switzerland, and the third RILEM International Symposium took place in Zürich after a period of ten years, in October 1992.

This conference marked the end of two RILEM Technical Committees, 51-ALC and 78-MCA, whose task was the preparation of 'Recommendations for Test Methods' (51-ALC) and a 'Recommended Practice for the Construction with Autoclaved Aerated Concrete' (78-MCA), respectively.

The 'Recommended Practice' has been presented by members of the Technical Committee on a one-day-seminar at the Swiss Federal Institute of Technology, Zürich, on October 14th. The full text will be published by Chapman and Hall, England. All Recommended RILEM Test Methods will be included in this volume.

The seminar was followed by a RILEM Symposium. All papers presented on this occasion are printed in this volume. Recent advances in research and technology of autoclaved aerated concrete are presented.

At the end of this volume a comprehensive bibliography which covers the period following the second RILEM Symposium on autoclaved aerated concrete until now is included. In order to facilitate a literature review an author index has also been prepared and added.

It is hoped that this volume contributes to a better understanding of an interesting building material and that in this way a still wider application may be enhanced.

Zürich, August 1992   F. H. Wittmann

Advances in Autoclaved Aerated Concrete, Wittmann (ed.) © 1992 Taylor & Francis. ISBN 90 5410 086 9

Organization

3rd RILEM International Symposium on Autoclaved Aerated Concrete

AAC-3

Swiss Federal Institute of Technology (ETH), Zürich, October 14th-16th, 1992

Scientific Organizing Committee:

Folker H.Wittmann – *Chairman*
Samuel Aroni
Norman J.Bright
Dieter Hums
Yoshikazu Kanoh
Poul Nerenst
Mike Robinson
Reinhard Schramm
Göte Svanholm

Local Organizing Committee:

Andreas Gerdes
Peter Lunk
Thomas Müller

Sponsors:

RILEM Réunion Internationale des Laboratoires d'Essais et de Recherches sur les Matériaux et les
        Constructions
EAACA European Autoclaved Aerated Concrete Association
ETHZ   Eidgenössische Technische Hochschule Zürich
WTA   Wissenschaftlich-Technische Arbeitsgemeinschaft für Bauwerkserhaltung und Denk-
        malpflege

1 Production, technology and properties

Advances in Autoclaved Aerated Concrete, Wittmann (ed.) © 1992 Taylor & Francis. ISBN 90 5410 086 9

On production and application of AAC worldwide

W. Dubral
YTONG AG, Munich, Germany

ABSTRACT: The large number of AAC plants which are either operated or under construction worldwide, is a proof of the increasing popularity and esteem of AAC and of the variety in the application of its products. AAC plants are located in more than 40 countries, on all continents, in the main climatic regions of the earth and in seismic areas such as Japan, Iran, Italy and Mexico.

As a result, AAC is, in contrast to almost all other building materials, available in almost the same quality worldwide.

1 HISTORICAL REVIEW

AAC is the result of a systematic development, which began approximately 100 years ago. The presently known AAC-type was developed at the end of the twenties in Sweden by combining the processes a) to porose concrete by means of metal powder and b) to harden concrete through autoclaving. Thereby, a new building material was created which united a wealth of characteristics such as low weight, high material strength, very good heat insulation, excellent fire protection and good sound insulation. From then it took another 10 to 20 years until reinforced AAC elements were developed which were first used mainly in Scandinavia as roof and floor units and wall panels.

The Second World War halted temporarily the quick spread of AAC. Only afterwards - since roughly 1950 - did its successful expansion in Europe and many parts of the world begin. The fast development of the worldwide production of AAC products in the years from 1929 to 1975 is made clear in Figure 1: in 1955 there were only 13 YTONG plants or licensed plants; 20 years later, there were already 40 plants worldwide.

The global success was first of all established by the companies YTONG, Siporex and Durox which are still well-known. Further companies followed with their own

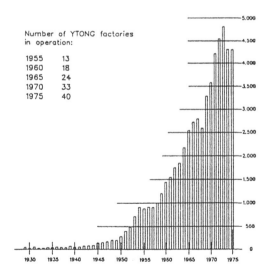

Number of YTONG factories in operation:

Year	Number
1955	13
1960	18
1965	24
1970	33
1975	40

Fig. 1: Worldwide production of YTONG-products 1929-1975

differing technologies such as Hebel in Germany and Hendriksen og Hendriksen (H+H) in Denmark.

Neither did the government-controlled enterprises of the communist countries of Eastern Europe remain idle. On the basis of the western technologies own processes were developed among which the Polish Unipol method surely is the most well-known. Similar new developments occured for instance in the CSFR (Armaporit) and Romania (Forexim).

In the meantime there are AAC plants working on almost all continents with roughly 200 plants in Western and Eastern Europe, including the Asian part of the former Soviet Union, approximately 50 plants in the rest of Asia, (although we have no details on Chinese production) as well as 16 plants in the rest of the world.

Fifteen AAC plants are presently under construction in the world.

2 ON AAC PRODUCTION WORLD-WIDE

The results of this study were taken from surveys as well as ascertained from evaluating statistics issued by AAC producers and associations. They might not be complete, since it was particularly difficult to receive the desired information within a very limited period of time. This is particulary valid for the former Soviet Union and China, but also for other nations outside Europe.

Verifiable and precise production figures for 1991 could only be determined for Western Europe. These included the "national delivery figures" of the E.A.A.C.A, the European Autoclaved Aerated Concrete Association, which comprises associations and firms from more than 10 Western European nations at present.

In the following pages the differing AAC technologies such as YTONG, Hebel, Siporex, Durox, H+H, Unipol and others are handled separately. The individual technologies differ above all in the cutting process as well as in

Table 1: AAC factories in Western Europe, production figures 1991

| Country | Number of factories | AAC-Technoloy | | | | | | Production 10^3m^3 | | |
		YTONG	Hebel	Siporex	Durox	H+H	Others	Total	unrein-forced	rein-forced (%)
Germany	32	8	13	-	1	-	10	4.470	3.610	860 (19%)
U.K.	13	6	1	1	1	4	-	2.400	2.400	- (0%)
France	5	1	1	2	1	-	-	390	310	80 (20%)
Netherlands	3	-	-	-	3	-	-	380	250	130 (34%)
Belgium	2	1	-	1	-	-	-	360	230	130 (36%)
Denmark	2	-	-	-	-	2	-	110	100	10 (9%)
Italy	2	-	1	1	-	-	-	110	90	20 (18%)
Portugal	1	1	-	-	-	-	-	45	45	- (0%)
Spain	1	-	-	-	-	1[1]	-	-	-	- (-%)
Sweden	2	1	-	1	-	-	-	150	40	110 (73%)
Austria	1	1	-	-	-	-	-	160	140	20 (13%)
Finland	1	-	-	1	-	-	-	40	20	20 (50%)
Norway	1	1	-	-	-	-	-	20	10	10 (50%)
Switzerland	1	-	-	1	-	-	-	15	15	- (0%)
Total	67	20	16	8	6	7	10	8.650	7.260	1.390 (16%)

1) Start of production end of 1991

the pre- and post-process steps. These differences are not however further discussed here.

2.1 Western Europe

In Western Europe, 8.65 mio m³ AAC were produced in 1991, 84 % of which was non-reinforced material (see Table 1). Germany is the leading producer of AAC in Western Europe with 4.47 mio m³ coming from 32 plants; more than 50 % of all reinforced and non-reinforced products are produced in Germany.

Next in the league is the United Kingdom with approximately 2.4 mio m³ exclusively block products from 13 plants; reinforced elements are imported to a low extent from other countries. Taking both countries together they have with their 45 plants a share of almost 80 % of the 1991 AAC production in Western Europe.

Most of the remaining 12 countries, listed in Table 1, produce with only 1 or 2 plants, exceptions being France (5) and the Netherlands (3).

There are no AAC plants in Greece, Ireland or Luxembourg. These markets are supplied via their neighbouring countries. Greece however, produces foamed concrete using Hebel technology.

Here it should be remarked that Western Europe is presently confronted with a serious recession in the construction industry, which has so far only spared the former West Germany. This means that many Western European countries have to produce far below their normal productive capacity.

This must be taken into account when considering the production figures in Table 1.

2.2 Eastern Europe

Table 2 shows that there are an estimated 120 AAC plants in the Eastern European countries with a total capacity of 22 - 24 mio m³.

The number and the planned productive capacities for plants in Poland, the CSFR, Romania, Yugoslavia, Hungary and Bulgaria were known in sufficient detail. It was however necessary to proceed from the following estimations in the case of the former Soviet Union: the total capacity of plants for AAC and foamed concrete probably amount to approximately 20 - 40 mio m³, the AAC share could range roughly around 10 - 12 mio m³, which would correspond to a capacity of 60 plants. Approximately 20 % will be reinforced products, above all prefabricated room-sized facade elements with integrated windows and

Table 2. AAC factories in Eastern Europe

Country	Number of factories	AAC-Technology						Capacity $10^3 m^3$
		Unipol	Hebel	Siporex	Durox	YTONG	Others	
Soviet Union [1]	60 [2]			- n/a -				10.-12.000
Poland	27	23	-	3.	1	-	-	5.000
ČSFR	13	6	-	1	4	-	2	2.900
Romania	12	-	8	-	-	1	3	2.000
Yugoslavia	5	-	-	3	-	1	1 [3]	1.000
Hungary	2	1	1	-	-	-	-	900
Bulgaria	1	-	1	-	-	-	-	200

Total approx. 120   Capacity approx. 22.-24.000

1) Asian S.U. included   2) Estimated   3) H+H

doors. Some Russian products have very high densities (800 kg/m³) and do not share the quality characteristics of Western European products.

The brochure "Polish Cellular Concrete

Process and Application" issued in 1981 by CEBET, the Polish Research and Development Centre of Concrete Industry in Warsaw, describes the development of Polish AAC production as shown in Fig.2.

Years	1951-52	1953	1955	1960	1965	1979	1975	1980
Millions m³	-	0.2	0.3	0.7	1.8	3.2	3.2	6.0

Fig. 2:   Development of manufacturing cellular concrete in Poland

By 1981, 27 AAC plants had been built in Poland with a planned capacity of more than 5.5 mio m³. 30 plants had been exported into 11 different countries, mainly the former states of the Eastern bloc. In 1988, 4.6 mio m³ of AAC were produced in Poland, 15 % of which were reinforced products, i.e. roof and floor units and wall panels, but also prefabricated facade elements. The capacity of the presently existing 27 plants amounts to 5 mio m³.

The CSFR runs 13 plants, 3 of which are smaller plants with pure block production. Four of the remaining 10 plants use the Calsilox technique (Durox). The share of the reinforced products amounts to roughly 20 %. The products are similar to the ones in Poland except for roof units and wall panels, which are only produced in negligible quantities. The total capacity amounts to approximately 2.9 mio m³. Romania's 12 plants have a technical capacity of

approximately 2 mio m³. There are 8 plants that work with Hebel technology - 2 original ones and 6 imitations.

Hungary has 2 large plants with 450.000 m³ each; both produce only non-reinforced products.

All Eastern European countries tried during the times of the communist state-controlled economy to run the plants with a 3 or 4 shift operation (i.e. also on Sundays and public holidays), since there were no sales problems at that time. Presently however, we assume that there is a considerably lower utilization of the plants, probably between 40 - 60 %.

2.3   Asia

Table 3 lists the AAC plants in Asia which are known to us.

For China, there is unfortunately no evaluable information. Apart from the 1966 Siporex plant in Beijing, there are supposed to be more than 30 AAC plants with a relatively

Table 3: AAC factories in Asia

Country	Number of factories	AAC-Technology						
		Siporex	YTONG	Hebel	Durox	Unipol	H+H	Others
Japan	20	6	4	7	2	-	-	1
South Korea	8	1	1	1	1	-	-	4
India	3	1	1	1	-	-	-	-
Iraq	3	2	-	-	-	1	-	-
Iran	3	1	1	-	-	1	-	-
Turkey	3	-	1	1	-	-	-	1[1]
Indonesia	2	-	-	-	-	-	2	-
Israel	2	-	2	-	-	-	-	-
Hongkong	1	-	1	-	-	-	-	-
Kuwait	1	-	-	-	1	-	-	-
Saudi Arabia	1	1	-	-	-	-	-	-
China	n/a (1)	1	-	-	-	-	-	-
Total	48	13	11	9	4	3	2	6

1) Wehrhahn Group

6

low degree of mechanization.

In recent years the main acquisition effort of the popular European AAC producers has been directed towards Asia. Within the last 5 years, new AAC plants in Japan, Korea, Hongkong, India, Israel and Iran have started production, further plants are under construction and are planned.

Japan with its 20 plants is the country with the highest production of AAC in Asia. They produce almost exclusively reinforced construction components. The annual production quantities amount to roughly 2.5 mio m³.

In South Korea 3 plants are operating at present; by Autumn 1992 there will be 8 plants. The total technical capacity will then amount to 1.0 mio m³.

For the past 3 years only 3 of the 4 Turkish plants have been in operation. Their total production in 1991 amounted to 400.000 m³, roughly 12 % of which was reinforced products.

We can say that, on the basis of about 50 Asian plants, each with an average capacity of 125.000 m³ per year, it is possible with good plant utilization to produce 6 mio m³ AAC per year.

2.4 Africa, America, Australia

These three continents together presently only contain 16 AAC plants: 7 in Africa, 7 in Central and South America and 3 in Australia (see Table 4).

In the USA and Canada it has so far not been possible to establish AAC. In Canada a Siporex plant was operating near Montreal during an eighteen year period from 1955 to 1972, but this remained for North America the first and only attempt.

The reasons for the lack of success so far could be the low price construction methods

Table 4. AAC factories in Africa, America and Australia

Area	Country	Number of factories	Siporex	YTONG	Hebel	Others
Africa	Egypt	4	1	3	-	-
	Algeria	2	2	-	-	-
	South Africa	1	-	-	-	-
America	Brazil	2	1	-	-	1
	Cuba	2	1	-	1	-
	Argentina	1	1	-	-	-
	Mexico	1	1	-	-	-
Australia	Australia	3	1[2]	-	1	1[1]
Total		16	7	3	2	3

1) Wehrhahn-Group   2) Operated by YTONG-licensee

for wall, roof and floor components in residential and non-residential buildings, and perhaps a perceived low profitability level for investors in the industry. Construction materials made of AAC can hardly compete in residential building constructions (single and multi-family houses) with cheap wood stud walls, wood decks and truss roofs and in non-residential constructions (industrial and commercial buildings) with metal decks, curtain walls and concrete block walls.

3   AAC IN CLIMATIC AND SEISMIC ZONES

As we have seen, AAC products are employed in both of the most important climatic regions of the earth, the temperate and the tropical climatic zones. But there is a preponderance in two of the temperate zones: the warm-humid regions, e.g. Central and Southern Europe, South Japan, Southern and Eastern China, the East and Southwest coast of Australia and cold-humid regions such as e.g. Scandinavia and large parts of the former Soviet Union.

There are few AAC plants in the two zones of the tropical climatic belt: the hot-arid zone, with countries such as Egypt, Israel, Kuwait, Iran, South Africa and the tropical-humid zone with countries such as Brazil, Cuba, India and Indonesia.

AAC has also been employed successfully for many years now in seismic areas, e.g. in Japan, Indonesia, Iran, Iraq, Turkey, Italy, Greece, Algeria and Mexico. Here, particular design and construction principles have to be met with in accordance with the national requirements. A document on "Seismic design" of RILEM TC 78 MCA dated March 1991 informs that "buildings which consist in full or partly of AAC have in general shown good resistance to earthquake forces in practice. The light weight of AAC reduces the seismic forces in common with some other materials. The fact that AAC can be used in most structural parts of low rise buildings makes a high degree of symmetry possible and provides deformational compatibility. In addition, the non-combustible and fire resisting nature of AAC material is an advantage against fires commonly associated with earthquakes".

4   ON AAC APPLICATIONS WORLD-WIDE

Non-reinforced and reinforced AAC products are employed adapted to the local construction traditions in all fields of residential and non-residential building construction.

4.1   Non-reinforced products

The main use of AAC blocks is for loadbearing internal and external walls, for basement and foundation walls, for partition walls, filler walls, linings and coverings, as well as for fillers in flooring constructions. Examples of special uses are linings of furnaces in material testing, heat-insulating coatings of fermentation towers and subconstructions of low-inclined residential building roofs such as in Spain.

Masonry in thin-joint processes (joint thickness 1 - 3 mm) is very common in Germany and other Central and Southern European states, but not in the United Kingdom and the Eastern European countries.

Dry masonry from AAC blocks is known

in Sweden, in the Netherlands and Germany, but could not so far establish a good market position. Basement walls of AAC have to be protected against water impact. Foundation walls below d.p.c. can, as done in the United Kingdom where there are normal soil conditions, also be erected without moisture protection.

The erection of reinforced masonry with AAC blocks is also possible.

A highly heat-insulating external wall version consists of masonry with prefabricated three layer blocks, whose central layer is made from heat-insulating materials, e.g. polyurethane. Several countries utilize large blocks of AAC rather than the smaller blocks which are put into place by hand, as they find this cheaper. These are erected on the building-site with minicranes into thin joint mortar; the usual dimensions of these large blocks in Central Europe are 1000x625 mm.

AAC blocks can be used with concrete T-beams to produce a simple and economical suspended flooring system; they can also be used as infill blocks in solid floors and in situ concrete floors. Such applications are very common in Israel where 1/3 of the total production of approximately 300.000 m³ is composed of such infill blocks. They are also employed in the United Kingdom, Austria and Portugal.

4.2   Reinforced products

Reinforced AAC products are used as roof and floor slabs, wall panels, loadbearing or non-loadbearing, as well as for lintels. The production of beams and columns is also possible. The reinforcements have to be protected against corrosion, exceptions from this rule can be seen in the Netherlands where wall panels are reinforced with longitudinal iron bars for protection during transportation; they are not linked by cross iron bars and do not have corrosion protection.

The maximum production length is 6 m , in some cases also 7,5 m. The maximum widths are 0.60 and 0.625 m, in some cases also 0.75 m. The thickness of reinforced AAC units usually ranges between 100 - 300 mm, in seismic regions however even less, like in

Japan where partition panels with expanded metal fabric reinforcement are produced at thicknesses of 30 - 50 mm.

In some countries, particularly in Eastern Europe, wall panels are assembled in the plant to large size wall elements with integrated windows. They are normally finished with an undercoat rendering in the plant, the finishing coat is applied on the site. Finished surface processing in the plant is possible e.g. by applying exposed concrete, rendering or acrylic resin coatings.

Moreover, Israel produces in its plants complete roomcells out of AAC elements.

Interesting finishing of facades in the construction of industrial buildings can also be achieved by employing AAC wall elements with specially formed surfaces. So Germany offers elements with trapezoidal cross sections, as well as coffer and chamfer elements. In Japan there are so-called "fashion panels" with a number of plant produced surface structures.

5   CONCLUSIONS

This description of the global production and use of AAC cannot be complete for the reasons we stated above. However it offers an interesting overview of the worldwide activities of the AAC industries and of most of the major product applications.

The success of AAC around the world cannot be fully appreciated without taking into account that it must compete in many countries against traditional building materials, such as laterite in the tropical regions, which are cheaper. It is indeed remarkable that in 40 countries, scattered over all continents and over all the most important climatic and seismic regions, AAC has been established as a universally applicable building material, and that this material is produced everywhere with an almost uniform quality.

Advances in Autoclaved Aerated Concrete, Wittmann (ed.) © 1992 Taylor & Francis. ISBN 90 5410 086 9

Influence of hydrothermal processing on the properties of autoclaved aerated concrete

T. Mitsuda & T. Kiribayashi
Ceramics Research Laboratory, Nagoya Institute of Technology, Tajimi, Japan

K. Sasaki
Ceramics Research and Development Division, INAX Corporation, Tokoname, Japan

H. Ishida
Ceramics Research Laboratory, Nagoya Institute of Technology, Tajimi & Ceramics Research and Development Division, INAX Corporation, Tokoname, Japan

ABSTRACT: The chemical reactions were investigated for the AAC block samples, which were treated at 180°C under saturated steam pressures for various times from 1 to 128 h. The hydrothermally formed materials consist of Ca-rich C-S-H and poorly crystalline tobermorite with varying Ca/Si ratios as an initial product, which react further with silica dissolved from quartz to form highly crystalline 1.1 nm tobermorite with increase of autoclaving time. As the reaction proceeds, the numbers of micropores less than 100 nm in diameter increases and shifts to more smaller size, and the strength also increases due to the binder effect of the tobermorite. However, the total pore volume does not change, remaining constant values in the AAC processing. For longer curing sample, some tobermorite decompose into fibrous xonotlite, causing a decrease of the strength.

1 INTRODUCTIION

Although a great deal of work has been published on the properties of autoclaved aerated concrete (AAC) products as a building material, e.g., thermal conductivity water absorption, shrinkage behavior, pore structure, and strength (Wittmann, 1983), very little is known about the hydrothermal chemistry involved during autoclaving process.

Mitsuda, Sasaki and Ishida (1992) reported the formation of calcium silicate hydrates during autoclaving process and the influence of these on the main properties for AAC; thermal behavior, micropore structure, and strength development. They used the block samples (550 kg/m³) made by an AAC factory, which were hardened by autoclaving at 180°C for various times from 1h to 64 h. The raw mixtures used were 20 wt % normal portland cement, 15% lime, 65% quartz sand. The results were: (1) AAC is hardened by hydrothermal processing at 180°C, in which the reaction proceeds by the quartz reacting with lime or C-S-H gels formed. (2) After 1 h curing, the hydrothermally formed materials consist of varied Ca-rich C-S-H and small amounts of poorly crystalline tobermorite.

With increasing times, all the C-S-H is exhausted and formed highly crystalline tobermorite. (3) Along with the formation of tobermorite, the highest frequency of pore size shifts from 6 um diameter to 50-100 nm, though the total pore volume remaining remains unchanged. The compressive strength increases with increase in the amounts of tobermorite formation. (4) After the completion of the reaction, long-term curing results in a change in the chemical composition of 1.1 nm tobermorite and some decompose into xonotlite. This causes a decrease of compressive strength.

The objective of the present work is to investigate another AAC using different contents of raw mixtures and to reconfirm the previous work.

2 EXPERIMENTAL PROCEDURE

2.1 Block-sample preparation

All block samples were provided to us by an AAC Research Laboratory after the following preparation procedure. The raw mixtures used were 36.7% normal portland cement, 5.9% lime, 5.9% gypsum, and 51.5% quartz sand (93.6% SiO_2, 1.2% Al_2O_3) hav-

Table I   Chemical compositions for the block samples before (green block) and after autoclaving (wt %).

Block sample	SiO_2	Al_2O_3	Fe_2O_3	CaO	MgO	K_2O	Na_2O	TiO_2	SO_3	Loss	Total
Green Block	55.74	2.77	2.28	28.97	0.76	0.16	0.00	0.61	2.15	6.56	100.00
8 h cured	53.23	2.52	2.07	27.38	0.70	0.13	0.00	0.61	1.98	11.37	99.99
128 h cured	51.78	2.50	2.16	27.50	0.73	0.16	0.00	0.61	2.06	12.51	100.01

Loss = loss on ignition

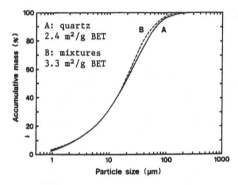

Fig.1 Integrated grain size distribution of raw materials.

ing almost the same specific surface area as the mixes (Fig. 1). These were treated with suspensions containing a trace of Al and poured into the mold (70 mm deep with 120 x 120 mm) to make a cellular green block. The density of the block was made to be 500 kg/m³ after autoclaving. The demolded block was then hardened by auto-claving at 180°C under saturated steam pressures for various times; 1, 2, 3, 4, 5, 8, 16, 32, 64 and 128 h. After auto-claving all the blocks were dried at 60°C before testing.

2.2   EXAMINATION

(A)   X-ray examination, thermal analysis and electron microscopy

All samples autoclaved for various times were examined by X-ray powder diffracto-metry, scanning electron microscopy and thermal analysis (TG-DTA) to detect the reaction phases.   Thermal shrinkage measurements were used for some block samples (20 mm length with 5 x 5 mm) to determine the relation between the heat resistance and the reaction products of the AAC samples.   The samples were also examined by analytical transmission electron microscopy (ATEM) to determine

the atomic ratios based on Si of the reaction products of calcium silicate hydrates.   The transmissions electron microscope (JEM-2000FX, JEOL) with a lithium-drifted silicon detector (TN-5500, Tracor Northern) was operated at 100 kV. The method used was reported by previous paper (Mitsuda et al., 1992).

(B)   Specific surface area and mercury intrusion porosity measurements

The specific surface area was measured by the N_2 gas absorption BET method, and pore distribution was determined using Autoscan-33 porosimeter, Quantachrome, by the mercury intrusion method.   For the both measurements, all block samples were crushed and sieved to a range of 710-1680 µm.

(C)   Compressive strength measurements

The compressive strength was measured by using a size of 15 x 15 x 15 mm for 6 samples of each block cured for desired time.   The loading was in a parallel direction with hydrogen gas generation and with a speed of 0.25 mm/min.
   In addition, the bulk density and porosity were determined for all block samples, using the pycnometer method with kerosine as the medium.

3   RESULTS AND DISCUSSION

Tables I and II show the chemical compo-sitions of some block samples and the characteristic properties of all the samples for various curing times.

3.1   Influence of curing time on the reaction products

(A)   Reaction products

In general, hydrothermal reaction of tobermorite formation proceeds through the sequence: Ca-rich   C-S-H ⟶ poorly

Table II   Properties of the block samples for various curing times

Samples curing time (h)	Bulk density (kg/m³)	Total pore vol (%)	Mercury pore vol (%)	BET surface (m²/g)	Compressive strength		Reaction rate of quartz (wt %)
					mean (MPa)	range (MPa)	
1	502	80.5	32.7	25.1	3.04	2.72 - 3.40	34 - 32
2	504	79.8	33.3	25.3	3.94	3.79 - 4.33	41 - 35
3	512	78.1	34.8	19.3	4.71	4.57 - 5.02	43 - 35
4	514	78.1	33.2	19.0	5.19	5.05 - 5.43	46 - 41
5	506	78.5	32.6	18.5	5.36	5.02 - 5.59	48 - 42
8	515	78.5	32.4	18.6	5.83	5.64 - 5.99	51
16	504	78.5	32.4	20.7	6.39	6.08 - 6.63	49
32	499	78.7	32.4	20.7	6.84	6.60 - 7.16	55
64	510	78.2	32.1	20.5	7.37	6.85 - 7.70	54
128	505	78.4	33.7	23.6	6.91	6.63 - 7.32	54

Compressive strength shows the values for 6 block samples. Reaction rate of quartz (reacted quartz / total quartz) was calculated by using the mean Ca/Si ratios (ATEM) of C-S-H and tobermorite in the samples and by estimating all of Ca and Si in the cement, and of Ca in the lime reacted to form C-S-H and tobermorite. Samples cured for 1 to 5 h show the values in the range of tobermorite and C-S-H respectively.

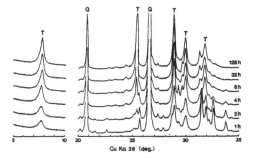

Fig. 2.   X-ray powder patterns of the samples cured for various time. T and Q indicate the peaks of tobermorite and quartz respectively.

crystalline tobermorite ⟶ highly crystalline 1.1 nm tobermorite (El-Hemaly et al., 1977; Sakiyama et al., 1977; Chan et al., 1978; Mitsuda et al., 1989). These products give distinctive X-ray powder patterns, differential thermal analysis (DTA) curves, chemical compositions, specific surface area and morphology under electron microscope (EM). Figures 2 and 3 show X-ray powder patterns and DTA curves.

For the sample autoclaved for 1 h, the products gave a weak basal spacing at 1.1 nm, using X-ray, a weak exothermic peak at 840°C, using DTA, and mixtures of fibrous aggregates and plate crystals under EM. This shows that the products consisted mainly of C-S-H and tobermorite. The results agree with those obtained by ATEM measurement, as will be

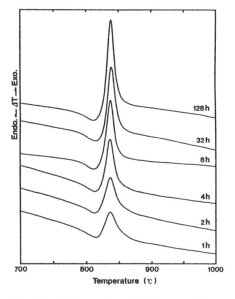

Fig. 3   DTA peaks of the samples cured for various times (20.0-22.7 mg for each sample).

discussed later. In addition, large amounts of quartz (about 67 wt %) remained in the sample, but no $Ca(OH)_2$.

With an increase of curing time, C-S-H reacted with dissolved silica from quartz and crystallized into tobermorite. For the samples cured more than 8 h, all the Ca and Si in the cement and lime in the raw mixtures reacted with quartz to give highly crystalline tobermorite having a

Table III   Analytical TEM results of calcium silicate hydrates for samples cured for various times

Sample curing time (h)	1	1	2	2	3
Phases analyzed	Tobermorite	C-S-H	Tobermorite	C-S-H	Tobermorite
No. of analyses	36	34	36	36	35
Ca/Si Mean	1.07	1.09	0.96	1.05	0.94
Range	0.93-1.39	0.93-1.50	0.89-1.07	0.94-1.19	0.85-1.09
Std	0.11	0.14	0.05	0.07	0.06
Al/Si Mean	0.03	0.04	0.03	0.04	0.04
Range	0.00-0.12	0.01-0.19	0.00-0.06	0.00-0.07	0.00-0.06
Std	0.02	0.04	0.01	0.01	0.02
Sample curing time (h)	3	4	4	5	5
Phases analyzed	C-S-H	Tobermorite	C-S-H	Tobermorite	C-S-H
No. of analyses	34	39	34	38	37
Ca/Si Mean	1.03	0.91	0.97	0.88	0.96
Range	0.91-1.23	0.81-1.03	0.86-1.14	0.79-0.98	0.87-1.13
Std	0.09	0.05	0.07	0.05	0.06
Al/Si Mean	0.06	0.04	0.04	0.04	0.04
Range	0.02-0.21	0.00-0.16	0.01-0.14	0.00-0.08	0.01-0.09
Std	0.04	0.03	0.02	0.02	0.02
Sample curing time (h)	8	16	32	64	128
Phases analyzed	Tobermorite	Tobermorite	Tobermorite	Tobermorite	Tobermorite
No. of analyses	37	36	33	32	34
Ca/Si Mean	0.86	0.88	0.82	0.83	0.83
Range	0.76-0.97	0.78-1.06	0.71-0.92	0.75-0.96	0.77-0.92
Std	0.05	0.06	0.05	0.05	0.04
Al/Si Mean	0.03	0.05	0.03	0.04	0.04
Range	0.00-0.10	0.00-0.16	0.00-0.10	0.00-0.08	0.02-0.08
Std	0.03	0.03	0.03	0.02	0.02

Std = standard deviation

fixed Ca/(Al+Si) ratio of crystal and then the reaction was retarded or almost stopped. The samples gave exotherm at 830-850°C by DTA, which is due to the crystallization of calcium silicate hydrates into wollastonite, and stronger peak for higher crystallinity with an increase of curing time.

Under the electron microscope, all samples included traces of round-grained particles to be hydrogarnets, containing Ca, Si, and Al as the main elements, together with Fe, Mg, S and K. Some amounts of gypsum were remained as an unhydrate in all samples cured for various times.

Although the samples cured for 1-2 h contained the products smallest amounts in all block samples, the BET surface gave the largest values of 25 m²/g. This is due to the specific surface area much larger in C-S-H than in tobermorite (Mitsuda et al., 1975; Mitsuda et al., 1986).

(B) Changing chemical composition of C-S-H and its crystallization into 1.1 nm tobermorite

Table III shows the atomic ratios based on Si for the major elements of calcium silicate hydrate obtained by ATEM measurements of the samples. The minor elements were also detected in the hydrates: Mg/Si, 0.00-0.02; Fe/Si, 0.00-0.04; K/Si, 0.03-0.04; S/Si, 0.01-0.04 (mean). Figure 4 was obtained by plotting the Ca/(Al+Si) and Al/(Al+Si) ratios, those of which are based on the tobermorite structure and on the Al substitution rate for Si, respectively. The results show that hydrothermally formed fibrous aggregates (C-S-H) and plate-like crystals (poorly crystalline tobermorite) initially give higher and more varied Ca/(Al+Si) ratio than does tobermorite in the samples cured for longer time. This is due to the hetero-

14

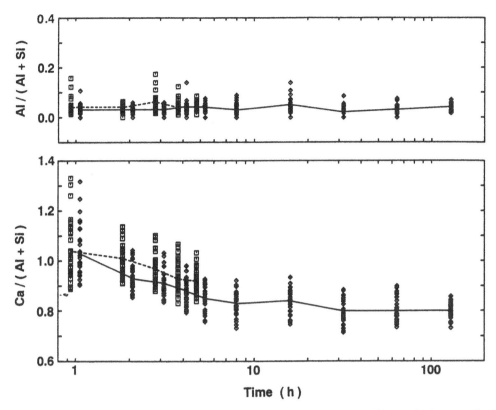

Fig. 4. Variations in the chemical composition of calcium silicate hydrates of the samples cured for various times. (□) and (◇) show the fibrous aggregates (C-S-H) and plate tobermorite respectively.

geneous reaction between the various components in the green block: Ca(OH)$_2$-quartz, cement-quartz, and C-S-H - quartz (Mitsuda et al., 1992). With an increase of curing time, the products formed initially reacts rapidly with the silica dissolved from quartz and crystallized into tobermorite with 0.8 Ca/(Al+Si), which is a stable ratio of highly cryst-alline one for hydrothermal preparation at 180°C.

For the substitution of Si by Al in C-S-H aggregates and tobermorites, both materials gave nearly the same Al/(Al+Si) ratio of 0.03-0.04 mean value for all samples cured various times. Some amounts of the Al in the raw materials partly formed hydrogarnets as mentioned early.

3.2 Influence of products on the physical properties

(A) Thermal shrinkage
For all samples, shrinkage increased

linearly with an increase of heating temperature and changed abruptly at around 800°C (Fig. 5). In this case, the amount of shrinkage slightly decrease with an increase of curing time and with the formation of highly crystallized tobermo-rite. This large shrinkage results the decomposition of the hydrates into wolla-stonite (CaSiO$_3$). The two steps of shrink-age at 100°-300°C and 300°-800°C agree with the TG results, losing water of the hydrates in the samples.

(B) Micropore distribution
In AAC processing the macropores generated by hydrogen remained unchanged in shape and size before and after autoclaving. However the micropores distribution is very sensitive to the products formed by hydrothermal reaction, being constant in the total volume of pores (Mitsuda et al., 1992). Fig. 6 shows the mercury intrusion porosity results. In the green block sample, the highest frequency of pores was 6 um in diameter, which shifted

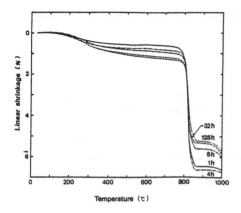

Fig. 5. Dilatometric variation of the block-samples cured for various times.

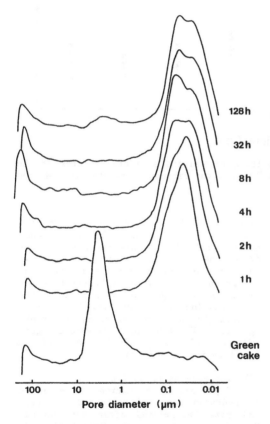

128 h

32 h

8 h

4 h

2 h

1 h

Green cake

Fig. 6. Pore size distribution of the block-samples before (green cake) and after autoclaving for various times.

to 50-100 nm for samples cured for 1-2 h, containing C-S-H and poorly crystalline tobermorite, and further spread to much

lower size for highly crystalline samples with an increase of curing time. This shows in Fig. 7, giving the pore volume ratios of 50-100 nm / less than 50 nm.

(C) Strength development
Table II shows the variation of bulk density and total pore volume including mercury intrusion results with curing time. The results gave nearly the same values for all samples. This indicates that the strength development results show no evidence of the strength being sensitive to density or porosity. The compressive strength of the samples (Fig. 8) increased with increasing time up to 64 h and tended to decrease for sample prolonged to 128 h. The increase in strength is directly proportional to increasing the reaction rate of quartz and the formation of micropores less than 100 nm in size, which results in increasing amounts of calcium silicate hydrates and their crystallinity obtained by autoclaving. After the formation of highly crystallized tobermorite having 0.8 Ca/(Al+Si) ratio, hydrothermal reaction is stopped for samples cured more than 32 h. The sample treated for the longer period (128 h) deteriorated in the strength with decrease of the BET value slightly. Under electron microscopy, the decomposition of some tobermorite (plates) into xonotlite (fibers) was observed. This causes a decrease of the the strength.

4   CONCLUSIONS

In the present paper the results are similar to those of previous paper (Mitsuda et al., 1992), in which the raw mixtures used were different contents.
    (1)   In AAC processing hydrothermal reaction initially produces C-S-H and poorly crystalline tobermorite having higher and various Ca/(Al+Si) ratios, those of which further react with dissolved silica from quartz to make highly crystalline 1.1 nm tobermorite of the 0.8 ratio. The analytical transmission electron microscopy is useful method for evaluation of reaction processing.
    (2)   The strength values increase with increasing amounts of calcium silicate hydrates, which causes the formation of micropores less than 100 nm in diameter size, showing total pore volume the same as the green block and all samples cured for various times.
(3)   After completion of the hydrothermal reaction, longer period of auto-

16

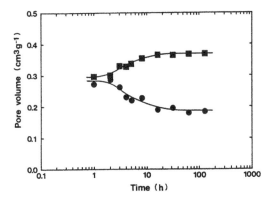

Fig. 7. Pore volume of the block-samples cured for various times, showing 50-100 nm (●) and less than 50 nm (■).

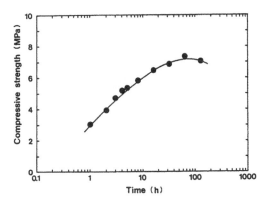

Fig. 8. Compressive strength of the block-samples as a function of curing time.

claving causes the some decomposition of plate tobermorite into fibrous xonotlite resulting a decrease of compressive strength.

REFERENCES

Chan. C.F. and T. Mitsuda 1978. Cem. Concr. Res., 8: 135-138.
El-Hemaly, S.A.S., T. Mitsuda and H.F.W. Taylor 1977. Cem. Concr. Res., 7: 429-438.
Mitsuda, T. and H.F.W. Taylor 1975. Cem. Concr. Res., 5: 203-210.
Mitsuda, T., S. Kobayakawa and H. Toraya 1986. Proceedings of the 8th Int. Cong. on Chemistry of Cements, Brazil: 173-178.
Mitsuda, T., H. Toraya, Y. Okada and M.
Shimoda 1989. Ceram. Trans., 5: 266-213.
Mitsuda, T., K. Sasaki and H. Ishida 1992. J. Am. Cerm. Soc., 75, 1858-1863.
Sakiyama, M. and T. Mitsuda 1978. Cem. Concr. Res,.7: 681-686.
Wittmann, F.H. edited, 1983. Autoclaved aerated concrete, moisture and properties. Amsterdam: Elsevier.

APPENDIX

Scanning electron microscopy

Freeze drying technique was applied in the preparation of specimens.
(A) Showing typical texture of calcium silicate hydrates formed in early stage (1 h).
(B) Showing mixtures of Ca-rich C-S-H and poorly crystalline tobermorite (4 h).
(C) & (D) Highly crystalline tobermorite gives plate crystals in the pore made by H_2 gas generation (5 and 8 h, respectively).
(E) Showing tobermorite growing near surrounding quartz (128 h).
(F) For longer curing sample, some tobermorite (plate) is decomposing into xonotlite (fiber) (128 h).
(G) Showing xonotlite growing after the decomposition of tobermorite (128 h).
(H) All samples include traces of round-grained particles considered to be hydrogarnets, containing Ca, Si, and Al as main elements, together with Fe, Mg, S, and K, (8 h).

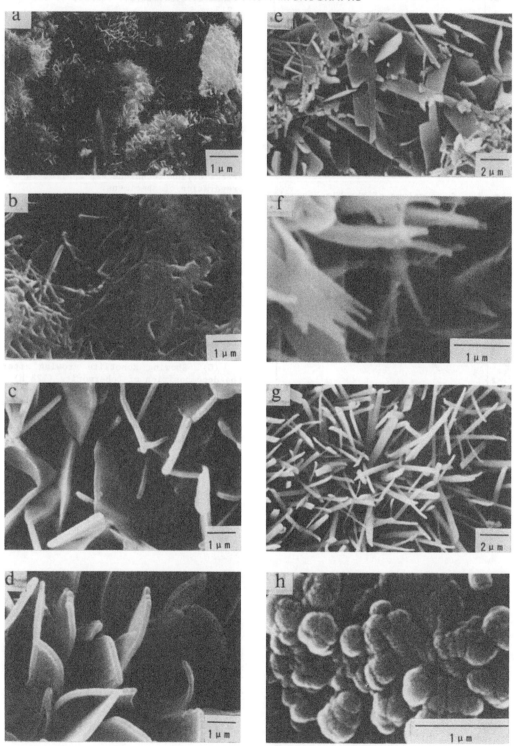

Advances in Autoclaved Aerated Concrete, Wittmann (ed.) © 1992 Taylor & Francis. ISBN 90 5410 086 9

Influence of particle size of quartz on the tobermorite formation

T. Mitsuda
Ceramics Research Laboratory, Nagoya Institute of Technology, Tajimi, Japan

K. Sasaki
Ceramics Research and Development Division, INAX Corporation, Tokoname, Japan

H. Ishida
Ceramics Research Laboratory, Nagoya Institute of Technology, Tajimi & Ceramics Research and Development Division, INAX Corporation, Tokoname, Japan

ABSTRACT: Autoclaved aerated concrete (AAC) constitutes tobermorite as a binder, which is formed in autoclaving process. The tobermorite formation was investigated by analytical electron microscopy using different particle sizes of quartz with cement and lime based on raw mix of AAC in hydrothermal suspensions. The mix with coarse quartz gives α-dicalcium silicate hydrate as an initial product, which reacts further with silica to form 1.1 nm tobermorite with increase of curing time. However, the mix with fine quartz gives 1.1 nm tobermorite with the mean ratio of Ca/(Al+Si) = 0.87 even after 1 h and 0.79 after 2 h. This suggests for the AAC processing that finer quartz reduces the autoclaving time.

1. Introduction

Tobermorite is the most important constituent of AAC as a binder, which is formed in the autoclaving process using raw materials of cement, lime and quartz sand. Taylor (1952) first found a crystalline 1.1 nm tobermorite in an aerated sand-lime block. The crystal structure of natural 1.1 nm tobermorite was determined by Megaw and Kelsy (1956) as a dreier single chain; theoretical composition of $Ca_5(Si_6O_{18}H_2)$ $4H_2O$. However, Wieker et al. (1982) first reported the NMR results for synthetic tobermorites, which gave a double-chain silicate structure. For this evidence, many workers have confirmed the double-chain like structure both for natural and synthetic 1.1 nm tobermorites. In thermal behavior, 1.1 nm tobermorite varies, notably in whether or not unidimensional lattice shrinkage occurs by about 300°C to give 0.93 nm form; specimens that do this are called normal, and ones that do not, anomalous (Mitsuda and Taylor, 1978). Mitsuda and Chan (1977) also reported anomalous type tobermorite formed in AAC product. Recently Taylor (1990) proposed the formula of anomalous type tobermorite with Al and Na substitution, $Ca_4(Si_{5.5}Al_{0.5}O_{17}H_2)Ca_{0.2}·Na_{0.1}·4H_2O$, which is based on double-chain structure

by the NMR results. This formula could represent 1.1 nm tobermorite in AAC products or other autoclaved ones using cement material.

For the hydrothermal chemistry involved during autoclaving process, very little is known, though a great deal of work has been published on the properties of AAC products as a building material (see papers and bibliography, edited by Wittmann, 1983). For the autoclaved calcium silicate products, analytical transmission electron microscopy (ATEM) is useful technique to determine chemical compositions of calcium silicate hydrate under electron microscope, even though products are crystals, poorly crystalline hydrates, gels, or mixtures of them (Mitsuda et al., 1989). The ATEM is normally used to obtain ratios of different elements and not absolute values. For any pair of elements, an empirical correction factor must be applied to convert X-ray intensity ratio. This factor is determined by analyzing substance of known composition. Cliff and Lorimer (1972) have experimentally shown the determination of K factor. The constant K varies with operating voltage but is independent of the sample thickness and composition as long as the thin film criterion is satisfied. The volume analyzed by ATEM can be small as few tenths of nm in each direction and if the sample is thin below 30 nm, fluorescence, absorption and other effects can be neglected. Simultaneously diffraction pattern can be obtained. In addition, autoclaved products for calcium silicate hydrates prepared in suspensions are thin specimen without grinding of the samples. However, the specimens hardened such as AAC product being analyzed cannot be identified of their exact location in the sample; for example tobermorite crystal in AAC is located either in pore, made by H_2 gas generation in the molding process, or around quartz in the raw mix.

Although having this disadvantage, the formation of calcium silicate hydrates during autoclaving process and the influence of these on the main properties for AAC were recently reported by us (Mitsuda, Sasaki et al., 1992; Mitsuda, Kiribayashi et al., 1992). They examined two cases using different contents of raw mixtures, which were autoclaved for various curing times at 180°C. For both cases, the tobermorite formation was examined by using ATEM. The results were: (1) The hydrothermal reaction initially produces Ca-rich C-S-H of fibrous aggregates with higher and various Ca/(Al+Si) ratios. (2) With an increase of curing time, C-S-H reacts further with dissolved silica from quartz, reducing their ratios, and crystallizes to 1.1 nm tobermorite of plate crystals with the ratio of about 0.8. (3) After the formation of highly crystalline tobermorite, the hydrothermal reaction is retarded or almost stopped, remaining quartz unreacted in the AAC product. (4) The long-term curing over 64 h or 128 h results the decomposition of tobermorite into xonotlite. (5) Tobermorite in AAC product is substituted by Al, resulting by using cement or quartz rock including Al-bearing minerals as an impurity.

Mitsuda (1990) examined chemical compositions of tobermorite (plate crystal) by using ATEM for 13 block-samples, which were made in various AAC factories with different contents of raw mixtures. The results showed that the main atomic ratios of Ca/(Al+Si) based on tobermorite are 0.84 in mean value for 442 crystals analyzed, though the Al/(Al+Si) ratios are variable from 0.03 to 0.15 for each AAC product.

In this work, the tobermorite formation was investigated by ATEM method using different particles of quartz with cement and lime based on raw mix of AAC in hydrothermal suspensions. The objective of the present work is to reduce the autoclaving time in the AAC processing.

2. EXPERIMENTAL PROCEDURE

2.1 Starting materials and preparations

The starting materials used were the mixtures of 20 wt % normal portland cement, 15% CaO (99.9% CaO) and 65% quartz (99.9% SiO_2), those of which are based on the raw materials of AAC product and their constituent ratios reported by the previous work (Mitsuda, Sasaki et al., 1992). The CaO is made by heating reagent grade $CaCO_3$ at 1050°C for 3 h. Three quartz samples were used, of different particle sizes; mean diameter 22.4 μm (0.17 m²/g BET surface area), 13.0 μm (0.41 m²/g) and 8.6 μm (0.91 m²/g). Fig. 1 shows their integrated

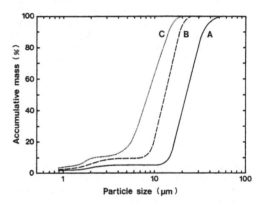

Fig. 1 Integrated particle size distribution of quartz. A, B and C show mean diameter 22.4 μm (0.17 m²/g BET surface area), 13.0 μm (0.41 m²/g) and 8.6 μm (0.91 m²/g) respectively.

grain size distributions.

All reactions were done using an autoclave (300 ml) at 180°C under saturated steam pressures. The mixtures with the different particle size of quartz were treated in stirred suspensions using a water / solids weight ratio of 15 and a stirring rate of 200 r.p.m. for the first 4 h, after which stirring was stopped. The heating time from 100°C to 180°C was 30 min. The curing times at 180°C were 1, 2, 4, 8 and 16 h. After autoclaving, all the products were filtered and dried in a vacuum at 80°C before testing.

2.2 Examination of products

All the samples autoclaved for various times were examined by X-ray powder diffraction and thermal analysis (DTA-TG) to detect the reaction phases, and by analytical transmission electron microscopy (ATEM) to determine the atomic ratios based on Si of the reaction products. For ATEM, autoclaved samples were first dispersed in an isoprophyl alcohol, and the suspended products were then allowed to settle on an electron microscope grid. The analytical transmission electron microscope (JEM-2000FX, JEOL, Tokyo, Japan) with a lithium-drifted silicon detector (TN-5500, Tracor Northern, USA) was operated at 100 kV. Determining the correction factor (K) (Cliff and Lorimer, 1975), natural minerals known the composition were used: xonotlite (Ca/Si), kaolinite (Al/Si), talc (Mg/Si), olivine (Fe/Si), sericite (K/Si) and gypsum (S/Si). These factors in particular, of the Ca/Si and Al/Si ratios checked frequently before and after the analysis of the samples. The counting time was 150 s per C-S-H aggregates (fibers) and tobermorite crystal (plate), using the

area analysis method. The number of analysis was 32-40 aggregates or crystals for each sample.

3. RESULTS AND DISCUSSION

Fig. 2 shows the X-ray powder patterns of all products using different particle size of quartz and curing for various times. Table I gives the ATEM results based on Si for the elements of calcium silicate hydrates, those of which were formed during autoclaving. Fig. 3 was obtained by plotting the major element ratios against the curing times. In this figure the Ca/(Al+Si) ratio is based on the tobermorite structure (Mitsuda, Sasaki et al., 1992; Mitsuda, Kiribayashi et al., 1992).

3.1 Effect of quartz particle size

(A) Products made using coarser quartz: mean diameter, 22.4 μm

X-ray and ATEM results show that reaction proceeds through the sequence; α-dicalcium silicate hydrate, Ca-rich C-S-H ⟶ C-S-H ⟶ poorly crystalline tobermorite ⟶ highly crystalline 1.1 nm tobermorite. The sample gives α-dicalcium silicate hydrate, $Ca_2(HSiO_4)(OH)$, as an initial and major product without lime or cement unreacted after 1 h treatment, which reacts further with dissolved silica from quartz and decomposes then into C-S-H (fibrous aggregates) for the sample cured above 2 h. With an increase of curing time the C-S-H slowly reacts with silica, resulting to crystallize plate tobermorite. The results agree to those reported by many workers. In general, by using portland cement, portland cement – coarse quartz, or lime – coarse quartz, hydrothermal reactions favor the formation of α-dicalcium silicate hydrate (plate crystal) at 160°-180°C as an initial product, which is unstable form in saturated silica solution and decomposes then into C-S-H for short curing time.

(B) Products made using median (13.0 μm) or fine (8.6 μm) quartz

The results confirm the previous works (Mitsuda et al., 1989; Mitsuda, Sasaki et al., 1992; Mitsuda, Kiribayashi et al., 1992) that Ca-rich C-S-H formed initially without α-dicalcium silicate hydrate reacts further with silica and finally crystallizes to 1.1 nm tobermorite. However, the reactions are faster for the fine quartz than for the median one. For

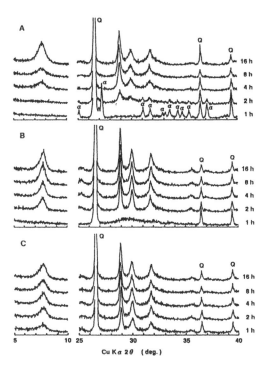

Fig. 2. X-ray powder patterns of the samples cured for various times, using coarse (A), median (B) and fine (C) quartz as for Fig. 1. α = α- dicalcium silicate hydrate; $Ca_2(HSiO_4)(OH)$. Q = quartz

Table I   Analytical TEM results of calcium silicate hydrates for samples cured various times, using coarse quartz as for Fig. 1 (A)

Curing time(h)	1	1	2	2	4
Phases analyzed	C-S-H	α-C$_2$SH	C-S-H	α-C$_2$SH	Tobermorite
No. of analyses	40	40	40	33	40
Ca/Si Mean	1.59	2.10	1.13	1.98	0.86
Range	1.44-1.81	1.93-2.27	1.00-1.22	1.81-2.14	0.79-0.93
Std	0.07	0.07	0.05	0.09	0.03
Al/Si Mean	0.04	0.00	0.00	0.00	0.00
Range	0.01-0.08	0.00-0.01	0.00-0.05	0.00-0.00	0.00-0.01
Std	0.01	0.00	0.01	0.00	0.00
Mg/Si Mean	0.01	0.01	0.01	0.00	0.00
Range	0.00-0.03	0.00-0.04	0.00-0.03	0.00-0.02	0.00-0.02
Std	0.01	0.01	0.01	0.01	0.01
Fe/Si Mean	0.02	0.00	0.00	0.00	0.00
Range	0.00-0.03	0.00-0.01	0.00-0.01	0.00-0.01	0.00-0.01
Std	0.01	0.00	0.00	0.00	0.00
K/Si Mean	0.03	0.03	0.02	0.03	0.02
Range	0.02-0.04	0.02-0.04	0.01-0.03	0.02-0.04	0.01-0.03
Std	0.01	0.00	0.00	0.01	0.00
S/Si Mean	0.03	0.01	0.01	0.01	0.01
Range	0.01-0.09	0.00-0.02	0.00-0.03	0.00-0.02	0.00-0.01
Std	0.01	0.01	0.01	0.01	0.00

Curing time(h)	8	16
Phases analyzed	Tobermorite	Tobermorite
No. of analyses	39	32
Ca/Si Mean	0.77	0.76
Range	0.70-0.88	0.71-0.80
Std	0.04	0.02
Al/Si Mean	0.00	0.00
Range	0.00-0.02	0.00-0.02
Std	0.00	0.01
Mg/Si Mean	0.00	0.01
Range	0.00-0.04	0.00-0.07
Std	0.01	0.01
Fe/Si Mean	0.00	0.00
Range	0.00-0.02	0.00-0.01
Std	0.00	0.00
K/Si Mean	0.02	0.02
Range	0.01-0.03	0.01-0.02
Std	0.01	0.01
S/Si Mean	0.00	0.01
Range	0.00-0.02	0.00-0.02
Std	0.01	0.01

st = standard deviation, α-C$_2$SH = α-dicalsium silicate hydrate

Table I   Analytical TEM results of calcium silicate hydrates for samples cured various times, using median quartz as for Fig. 1 (B) (continued)

Curing time(h)	1	2	4	8	16
Phases analyzed	C-S-H	C-S-H	Tobermorite	Tobermorite	Tobermorite
No. of analyses	40	40	39	32	40
Ca/Si Mean	1.38	0.88	0.79	0.76	0.76
Range	1.22-1.61	0.77-1.04	0.72-0.94	0.70-0.82	0.70-0.84
Std	0.08	0.05	0.05	0.03	0.04
Al/Si Mean	0.00	0.00	0.01	0.01	0.01
Range	0.00-0.04	0.00-0.03	0.00-0.09	0.00-0.03	0.00-0.02
Std	0.01	0.01	0.02	0.01	0.01
Mg/Si Mean	0.01	0.00	0.00	0.00	0.00
Range	0.00-0.02	0.00-0.07	0.00-0.03	0.00-0.02	0.00-0.02
Std	0.01	0.01	0.01	0.01	0.01
Fe/Si Mean	0.01	0.00	0.00	0.00	0.01
Range	0.00-0.02	0.00-0.03	0.00-0.01	0.00-0.01	0.00-0.01
Std	0.01	0.01	0.00	0.01	0.00
K/Si Mean	0.02	0.02	0.02	0.02	0.02
Range	0.00-0.05	0.00-0.03	0.00-0.03	0.00-0.04	0.00-0.03
Std	0.01	0.01	0.01	0.01	0.01
S/Si Mean	0.01	0.00	0.00	0.00	0.00
Range	0.00-0.04	0.00-0.03	0.00-0.01	0.00-0.02	0.00-0.02
Std	0.01	0.01	0.00	0.01	0.01

Table I   Analytical TEM results of calcium silicate hydrates for samples cured various times, using fine quartz as for Fig. 1 (C) (continued)

Curing time(h)	1	2	4	8	16
Phases analyzed	Tobermorite	Tobermorite	Tobermorite	Tobermorite	Tobermorite
No. of analyses	40	40	39	37	32
Ca/Si Mean	0.87	0.79	0.77	0.76	0.74
Range	0.77-1.01	0.72-0.87	0.72-0.84	0.73-0.84	0.71-0.83
Std	0.04	0.04	0.03	0.02	0.02
Al/Si Mean	0.00	0.01	0.01	0.00	0.01
Range	0.00-0.02	0.00-0.04	0.00-0.03	0.00-0.02	0.00-0.03
Std	0.01	0.01	0.01	0.01	0.01
Mg/Si Mean	0.01	0.00	0.00	0.00	0.01
Range	0.00-0.02	0.00-0.04	0.00-0.04	0.00-0.04	0.00-0.07
Std	0.01	0.01	0.01	0.01	0.02
Fe/Si Mean	0.00	0.01	0.00	0.01	0.01
Range	0.00-0.01	0.00-0.02	0.00-0.01	0.00-0.09	0.00-0.01
Std	0.00	0.01	0.00	0.00	0.00
K/Si Mean	0.02	0.02	0.02	0.02	0.02
Range	0.01-0.03	0.01-0.03	0.01-0.03	0.01-0.03	0.01-0.03
Std	0.00	0.01	0.01	0.00	0.00
S/Si Mean	0.00	0.00	0.00	0.00	0.00
Range	0.00-0.02	0.00-0.01	0.00-0.02	0.00-0.01	0.00-0.02
Std	0.00	0.00	0.01	0.00	0.00

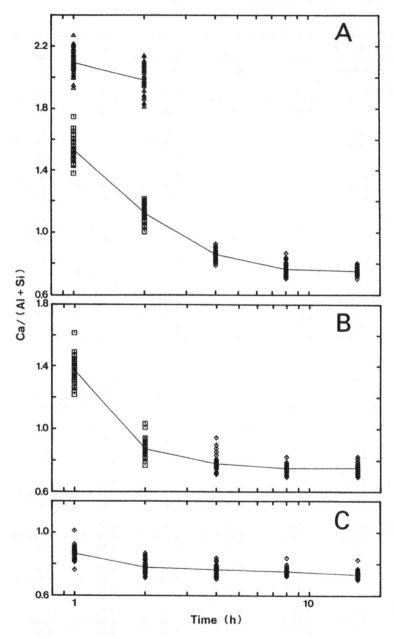

Fig. 3. Variations in the chemical composition of calcium silicate hydrates
of the samples cured for various times, using coarse (A), median (B) and fine
(C) quartz as for Fig. 1. (□), (◇) and (△) show the fibrous aggregates (C-
S-H), plate tobermorite and α -dicalcium silicate hydrate respectively.

X-ray patterns, the basal spacing of 1.1 nm was observed for the sample using the fine quartz after 1 h treatment, but not for the samples using the median quartz.

The kinetic results of the CaO – quartz – H_2O reaction at 180°C in suspensions showed that the reaction of quartz was mainly controlled by dissolution and gave a decrease in radius of the quartz at 0.85 μm/h (Chan et al., 1978). This indicates that the reaction rate of the mixtures to make tobermorite depends on the specific surface area of quartz; faster for finer. This evidence could be applied to AAC processing; how to reduce the autoclaving time.

(C) Changing chemical composition of calcium silicate hydrates during autoclaving

The ATEM results of the products using the coarse quartz show that the runs after 1 and 2 h gave α-dicalcium silicate hydrate with the Ca/Si ratios of 2.10 and 1.98 mean values respectively, those of which similar to the ideal ratios of 2.0. Regarding the change of major elements ratios, it is obvious that hydrothermally formed C-S-H aggregates (fibers) initially gives higher and more varied Ca/(Al+Si) ratios for the mixtures using coarser quartz. With increase of curing time, all the C-S-H decreases the ratios and crystallizes to 1.1 nm tobermorite in the range of 0.74 – 0.87 mean values, those of which similar to the ideal ratio of 0.8. For the runs using the fine quartz, the products gave plate crystals of 1.1 nm tobermorite with the 0.87 ratio after 1 h and decrease further to the 0.74 after 16 h. This suggests that C-S-H was initially formed for the samples cured less than 1h.

For minor elements in the products, Mg, Fe, K and S were in range of 0.00-0.03 mean values based on Si. The K and S cations could be balanced by substitution of Si by Al in the interlayer and caused anomalous behavior on heating (Mitsuda, 1977).

3.2 Substitution of Si by Al in C-S-H and tobermorite

In the present work, the mixtures have small amount of Al content and give the 0.016 ratio, if all the Al in cement materials made the tobermorite with Ca /(Al+Si) = 0.8. The results show the Al /(Al+Si) ratios = 0.00 – 0.02 mean values for all products. In the preparations, the Al content of the mixtures is sufficient to make Al-substituted tobermorite,

which has a ratio of up to 0.14 (Diamond et al., 1966; Mitsuda, 1970; Mitsuda et al., 1989). However, under the electron microscope, all samples include traces of round-grained particles considered to be hydrogarnets, containing Ca, Si and Al as the main elements, together with Fe, Mg, S and K. This evidence suggests that the aluminate phase constituting 5-10% of portland cement favor the formation of hydrogarnet under hydrothermal treatments. The results agree to those for AAC processing reported by previous papers (Mitsuda, Sasaki et al., 1992; Mitsuda, Kiribayashi et al., 1992).

3.3 Reaction rate of quartz

Under autoclaving treatments, all the Ca and Si in the cement and lime in the starting mixtures react with quartz to give highly crystalline tobermorite with Ca/(Al+Si) ratio of about 0.8, and then the reaction is retarded or almost stopped, remaining quartz unreacted in the samples. Fig. 4 gives the reaction rate of quartz (reacted quartz / total quartz)

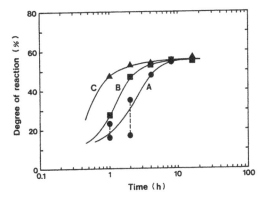

Fig. 4. Reaction rate showing reacted quartz / total quartz using coarse (A), median (B) and fine (C) quartz as for Fig. 1.

for all samples against a curing time, which is calculated by the ATEM results, assuming that the products have the ratios of Ca/(Al+Si) mean value for each sample cured for desired time.

4 CONCLUSIONS

In the present paper, the ATEM results are similar to those of previous works for AAC

(Mitsuda, Sasaki et al., 1992; Mitsuda, Kiribayashi et al., 1992) and suggest the application to the autoclaving time for the AAC processing.

(1) Reaction sequence is almost the same as for all samples using different quartz particles: Ca-rich C-S-H $\longrightarrow$ C-S-H $\longrightarrow$ 1.1 nm tobermorite. However, using the coarse quartz with lime and cement, the hydrothermal reactions give α-dicalcium silicate hydrate as an initial product, which reacts further with silica and decomposes into C-S-H.

(2) For the mixtures, the reaction rate is faster for finer quartz and the Ca/(Al+Si) ratios of C-S-H are higher and more varied for coarser quartz.

(3) The Al in cement partly forms hydrogarnets together with Mg, Fe, K and S, even though the content is sufficient to make Al-substituted tobermorite.

(4) Using the fine quartz, the products give 1.1 nm tobermorite with the ratio of 0.89 mean value even after 1 h and 0.79 after 2 h. This results suggest for the AAC processing that finer quartz reduce the autoclaving time.

REFERENCES

Chan, C.H., M. Sakiyama and T. Mitsuda 1978. Cem. Concr. Res., 8: 1-6.

Cliff, G. and G.W. Lorimer 1972. Proc. 5th European Congress on Electron Microscopy, Institute of Physics. Bristol: 140-145.

Cliff, G. and G.W. Lorimer 1975. J. Microscopy, 103: 203-206.

Diamond, S., J.L. White and W.L. Dolch 1966. Am. Mineral., 51: 388-401.

Megaw, H.D. and C.H. Kelsy 1956. Nature, 177: 390-391; 1959. Proc. 3rd Int. Symp. on Reactivity of Solids. Madrid. 1956, 3: 355-365.

Mitsuda, T. 1970. Mineral. J., 6: 143-158.

Mitsuda, T. and C.F. Chan 1977. Cem. Concr. Res., 7: 191-194.

Mitsuda, T. and H.F.W. Taylor 1978. Mineral. Mag., 42: 229-235.

Mitsuda, T., H. Toraya, Y. Okada and M. Shimoda 1989. Ceram. Trans., 5: 206-213.

Mitsuda, T. 1990. unpublished data.

Mitsuda, T. K. Sasaki and H. Ishida 1992. J. Am. Ceram. Soc., 75: 1858-1863.

Mitsuda, T., T. Kiribayashi, K. Sasaki and H. Ishida 1992. RILEM Int. Sympo. on Autoclaved Aerated Concrete, Zürich, in press.

Taylor, H.F.W. 1952. J. appl. Chem., 2: 3-5.

Taylor, H.F.W. 1990. p. 368, Cement Chemistry. London: Academic Press.

Wieker, W., A.-R. Grimmer, A. Wieker, M. Mägi, M. Tarmak and E. Lippmaa 1982. Cem. Concre. Res., 12: 333-339.

Wittmann, F.H. edited, 1983. Autoclaved Aerated Concrete, Moisture and Properties. Amsterdam: Elsevier.

Advances in Autoclaved Aerated Concrete, Wittmann (ed.) © 1992 Taylor & Francis. ISBN 90 5410 086 9

Influence of quartz particle size on the chemical and mechanical properties of autoclaved lightweight concrete

N. Isu
Ceramics Research Laboratory, Nagoya Institute of Technology, Tajimi & Research and Development Laboratory, Onoda A.L.C. Co., Ltd, Owariasahi, Japan

S. Teramura & K. Ido
Research and Development Laboratory, Onoda A.L.C. Co., Ltd, Owariasahi, Japan

T. Mitsuda
Ceramics Research Laboratory, Nagoya Institute of Technology, Tajimi, Japan

ABSTRACT: The ALC block-samples were prepared using a highly reactive quartz-rock with coarse (0.59 m²/g, BET), median (0.82 m²/g) and fine (2.62 m²/g) at 180°C under saturated steam pressures for 1 to 16 h. After 1h autoclaving, all the samples gave the product of 1.1 nm tobermorite and the compressive strength of 4.8 - 5.8 MPa, showing no effect of the quartz particle size. The reactions proceed to give higher crystallinity of tobermorite for coarser quartz-samples. For toughness, fracture energy (Gf) shows higher values for coarser quartz-samples cured for various times.

1 INTRODUCTION

Autoclaved Lightweight Concrete (ALC : using in Japan, synonymous with AAC) is produced by hydrothermal processing of various raw materials containing mainly quartz (50-70 wt%), portland cement (10-40 %) and lime (10-20 %) with gypsum (2-10 %) or waste materials. Therefore, ALC has lower Ca/Si ratio as a bulk composition than that of tobermorite as a binder, which is formed by autoclaving process, and is then composed of 13-20 vol % tobermorite, 5-10 % unreacted quartz and about 80 % pores for the product having 550 kg/m³.

For the hydrothermal reactions in ALC processing, Mitsuda, Sasaki et al. (1992-a) and Mitsuda, Kiribayashi et al. (1992) examined tobermorite formation by analytical electron microscopy. The results were : (1) ALC initially produces Ca-rich C-S-H of fibrous aggregates with higher and various Ca/(Al+Si) ratios. (2) With an increase of curing time, C-S-H reacts further with dissolved silica from quartz, reducing their ratios, and crystallized to 1.1 nm tobermorite of plate crystals with the ratio of about 0.8. (3) After the formation of highly crystalline tobermorite, the hydrothermal reaction is retarded or almost stopped, remaining quartz unreacted in the ALC product. Recently, Mitsuda, Sasaki

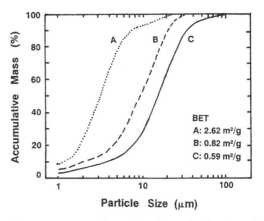

Fig. 1 Integrated particle sizes distribution of quartz samples A, B and C.

and Ishida (1992-b) reported the influence of quartz particle sizes on the tobermorite formation using cement and lime based on ALC constituent and in suspensions at 180 °C. The results showed that the mix with finer quartz gives 1.1 nm tobermorite with the mean ratio of Ca/(Al + Si) = 0.87 even after 1 h and 0.79 after 2 h. They also suggest the ALC processing that the finer quartz reduces the autoclaving time.

Wittmann and Gheorghita (1984) first reported the application of fracture toughness

Sample	SiO2	TiO2	Al2O3	Fe2O3	CaO	MgO	K2O	Na2O	Ig-loss	Total
A	91.82	0.20	4.82	0.50	0.05	0.26	1.27	-	1.10	100.02
B	98.55	0.07	0.72	0.11	0.03	0.06	0.23	-	0.23	100.00
C	98.18	0.08	0.90	0.18	0.03	0.07	0.28	-	0.28	100.00

Table I   Chemical compositions of quartz samples

mechanics to study crack propagation in ALC blocks. For the unreacted quartz related to the mechanical property of ALC, Teramura et al. (1988) reported that the coarser quartz particles contribute to develop the fracture toughness.

The objective of the present work is to investigate the tobermorite formation in ALC processing by using quartz with different particle sizes, which are used for autoclaved calcium silicate (xonotlite) product not for ALC as a highly reactive silica source, and to clarify the influence of these on the mechanical properties, comparing to the ALC Factory products.

2 EXPERIMENTAL PROCEDURE

2.1 *Quartz samples*

Quartz used were of three different particle sizes (quartz sample A, B, C), which are originally the same source rock using for autoclaved calcium silicate products except of ALC ones, and of quartz (quartz sample ALC) using for the ALC Factory product. The quartz sample A, B and C were of particle sizes: (A) mean diameter 3.1 μm (2.62 m^2/g, BET surface area), (B) 9.4 μm (0.82 m^2/g) and (C) 15.4 μm (0.59 m^2/g). All the A, B, C samples were prepared by grinding and hydraulic elutriation from the same source of silicious rock, which is composed of quartz varying in size from 5 μm to 20 μm together with a trace of sericite and kaolinite. During the seizing process, both materials were concentrated into the finer quartz particles and result the increase of Al2O3 content for the sample A. Fig. 1 and Table I show their integrated grain size distribution and chemical compositions respectively. Comparing to the mean diameter, even the coarsest sample C is finer than the particles of quartz sample ALC, which are prepared by ball-milling method.

2.2 *Block-sample (A, B, C) preparation using by quartz samples A, B, and C*

All block-samples A, B, and C were made to be 500 kg/m^3 by using 60.4 wt % quartz samples (A, B, C), respectively, 20.0 % normal portland cement, 15.6 % lime and 4 % gypsum with a trace of Al powder. After 4 h molding (300 x 300 x 300 mm), demolded green blocks were autoclaved at 180 °C under the saturated steam pressure for various times; 1, 2, 4, 8 and 16 h.

Before testing the samples, all the block samples were cut into four blocks to prepare the block-specimen (100 x 100 x 100 mm), and then dried at room temperature or at 60 °C for the fracture toughness test and for other examinations, respectively.

2.3 *Block-sample (ALC) preparation using by quartz samples ALC*

As for the referring sample, the block-samples (ALC) (300 x 300 x 300 mm) were prepared by using the quartz samples (ALC) with other raw mix of the ALC Factory, and were measured for some chemical and mechanical properties. The quartz particle size for the ALC sample has larger mean diameter than those of the coarse quartz sample (C) used for the block-sample (C).

2.4 *X-ray examination and analytical electron microscopy*

All the samples were examined by X-ray diffraction to detect the reaction phases and analytical transmission electron microscopy (ATEM: JEM-2000FX, JEOL ; with TN-5500, Tracor Northern) to determine the atomic ratios based on Si of the reaction products. The method used was described by Mitsuda, Sasaki et al. (1992-a).

2.5 Reaction rate of quartz

The reaction rate of quartz (reacted quartz / total quartz) was examined by the wet-chemical analysis (Ishii et al., 1978) to determine the amount of unreacted quartz of the block-samples.

2.6 Compressive strength and Young's modulus measurements

The compressive strength and Young's modulus were measured by using a size of 100 x 100 x 100 mm containing of 10 ± 2 wt % water for two block-specimens and one block-specimen of each block-sample cured for various times, respectively. The loading was in a vertical direction with H_2 gas generation. In addition, the bulk density was measured after the strength test.

2.7 Fracture toughness measurements

RILEM recommends using fracture energy (Gf) to evaluate mechanical property of concrete materials. In the present work, Gf was measured by means of compact tension test (Fig.2). For the test, the block specimen (100 x 100 x 100 mm) was notched with 50 mm deep and 0.15 curvature radius, cemented with a steel end plate on the notched end, and affixed by a clip gauge with knife edges in order to monitor the crack mouth opening displacement. The loading speed was 0.05 mm/min by an Instron testing machine. The method and apparatus used were described in the ISRM (1988).

3 RESULTS AND DISCUSSION

3.1 Influence of quartz particle size on the tobermorite formation

(A) Reaction products and rate

Fig. 3 shows X-ray powder patterns for block-samples (A, B, C) autoclaved for various times. All products gave 1.1 nm tobermorite by X-ray examination and plate crystals by electron microscopy with a trace of C-S-H for all block samples even 1 h treatment, and further increase their crystallinity with increase of curing time; higher for the block samples used coarser quartz. The results show

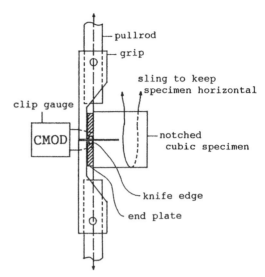

Fig. 2   Schematic representation of vertical loading CT test.

no distinctive effect of the quartz particle size on the tobermorite formation, contrasting with those reported by Mitsuda, Sasaki and Ishida (1992-b). They prepared the different quartz particles by sedimentation method; 3-7 μm, 8-12 μm and 18-22 μm. The results showed that all the quartz samples (A, B, C) contain smaller quartz particles, for example even the coarse sample (C) contains about 10 % quartz particles less than 4 μm diameter, which is almost the same as those of the fine quartz particles (3-7 μm) used in the previous work. For both works, the raw mixtures contain enough quartz to make a tobermorite, therefore smaller quartz particles contribute to the reaction in early stage. In the CaO - quartz - H_2O system, the reaction is controlled by the quartz dissolution and give a decrease in radius of the quartz at 0.85 μm/h at 180 °C (Chan et al., 1978). This could extend to the ALC processing, using cement - lime - quartz materials. After 2 h autoclaving at 180 °C, the quartz particles less than 4 μm diameter could be disappeared by the reaction with lime or cement materials to produce calcium silicate hydrates.

For the Al-bearing materials of the quartz sample (A), kaolinite was quickly reacted with lime, but sericite was unreacted, remaining in the products by X-ray. The results agree to those previously reported by Sakiyama and Mitsuda (1977).

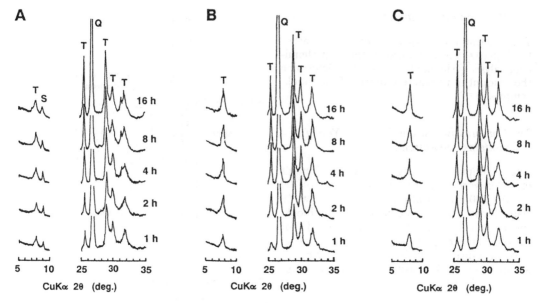

Fig. 3 X-ray powder patterns of the block-samples (A, B, C) cured for various times. T = tobermorite. Q = unreacted quartz. S = sericite.

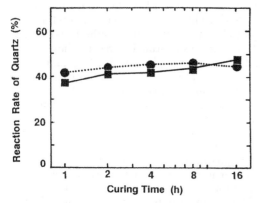

Fig. 4   Reaction rate showing reacted quartz/ total quartz for the block-samples A(●) and C (■) cured for various times.

Fig. 4 shows the reaction rate of quartz for the block-samples A (fine) and C (coarse). The results showed that after 1 h for both samples the reaction was retarded and almost stopped. This agrees to the X-ray and ATEM results. However, the reaction rate results show much higher reactivity for the coarse quartz (C) than for the fine quartz (A). This could be due to the retarding solubility of silica by kaolinite in the fine quartz sample (A).

(B) *Changing chemical composition of tobermorite during autoclaving*

Fig. 5 shows the ATEM results by plotting the major element ratios against the curing time for the block-samples (A) and (C). In the figure, the Ca/(Al + Si) and Al/(Al + Si) ratios are based on the tobermorite structure and on the Al substitution rate for Si, respectively. For the change of the Ca/(Al + Si) ratios, all the tobermorite products gave 0.92 mean ratio for both samples after 1 h autoclaving, and further decrease to their ratios to 0.88 - 0.83 for the fine quartz samples and to 0.83 - 0.78 for the coarse quartz sample both after 8-16 h, respectively. This ATEM evidence agrees to the X-ray results showing the crystallinity difference of tobermorites. Regarding the Al substitution rate for Si in tobermorite, it is obvious that the fine quartz block-samples (A) gave higher ratios than the coarse ones (C) for every curing time. This is due to the reactive Al source of kaolinite in the fine quartz block-samples (A).

The minor elements in the tobermorites were in the ranges of 0.01 - 0.03 of Mg, Fe, K, and S based on Si. Under the electron microscope, all samples included trace of round-grained particles to be hydrogarnets, indicating Ca, Al, and Si as the main elements, together with Fe, Mg, S, and K.

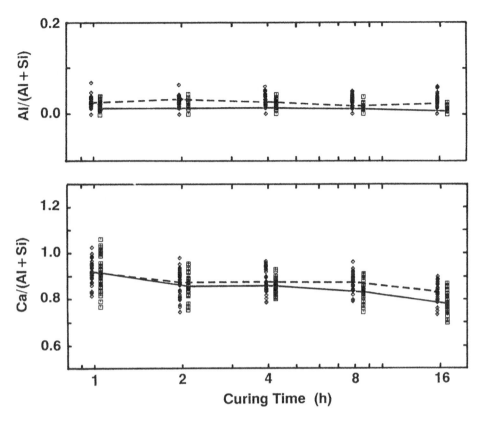

Fig. 5 Variations in the chemical composition of tobermorite crystals for the block-samples A(◇) and C(□) cured for various time. Top and bottom, showing Al-substitution and major element ratios, respectively.

3.2 *Influence of particle size on the mechanical properties*

(A) *Compressive strength and Young's modulus*

Fig. 6 shows the compressive strength and Young's modulus results with the bulk density ones. All block-samples were made to be 500 kg/m^3 bulk density. However, some block-samples showed the deviation: smaller values (468-474 kg/m^3) for the block-sample (A) cured for various times: larger ones (508-525) for (B) and (C) after 16 h. In general, compressive strength is mainly affected by the bulk density and tobermorite content of the sample. The measurement results showed that the block-samples used A, B, and C quartz particles, even after 1 h treatment of autoclaving, reached up the maximum values (4.8-5.8 MPa) and gradually decreased their values with increase of curing time except of the coarse

quartz block-sample (C) after 16 h, although the block-sample (ALC) gave 2 MPa after 1 h, which quickly increased up their values with increase of curing time. For the block-sample (ALC), the reaction product initially gives Ca-rich C-S-H, which further reacts with silica and then crystallize to 1.1 nm tobermorite after 4 h (Mitsuda, Sasaki et al., 1992-a). These results agree to those examined by X-ray, ATEM, the reaction rate of quartz.

It is obvious that the relations between the Young's modulus and compressive strength results have a linearity with some difference by the silica source. This difference as shown in Fig. 7 could be caused by the occurrence of quartz source; crystallinity, crystal size and texture.

(B) *Fracture toughness*

Fig. 8 gives the fracture energy (Gf), compar-

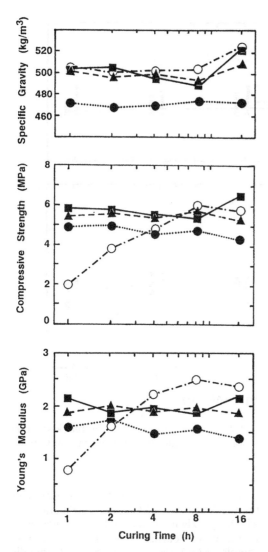

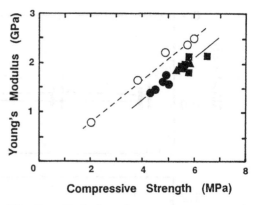

Fig. 7 Relation between compressive strength and Young's modulus for the block-samples ALC(○), A(●), B(▲), and C(■).

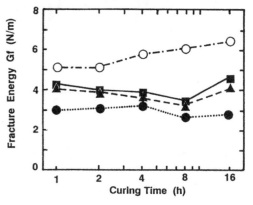

Fig. 8 Fracture energy (Gf) values for the block-samples ALC(○), A(●), B(▲), and C(■), cured for various times.

Fig. 6 Bulk density (top), compressive strength (middle) and Young's modulus (bottom) as a function of curing time. Showing the block-samples ALC(○), A(●), B(▲), and C(■).

ing with those of the block-samples (ALC) cured for various times. The Gf value is defined to the total energy of main and micro cracks formation, and an effective parameter to evaluate the toughness of concrete materials. The results showed that the block-samples (A, B, C) used from fine to coarse quartz give almost the same trend and similarity, contrasting with the block-samples (ALC) results. The block-samples (ALC) showed higher Gf

values than those of the block-samples (A, B, C), although all of tobermorite content, compressive strength and Young's modulus values after 1 - 2 h were smaller for the ALC samples than for the block-samples (A, B, C). For the Gf values, all the samples cured for various times using quartz of different particle sizes or of different source gave no affected results, comparing to other mechanical and chemical properties. The results suggest that coarser the quartz gives higher the Gf values. This is similar to the results reported by Teramura et al. (1988). They examined fracture energy for two ALC samples, both of which were almost identical for the compressive and bending strengths, pore size distributions and also unreacted quartz contents.

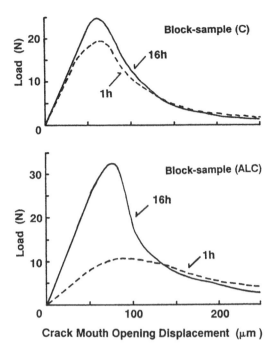

Fig. 9 Load - displacement curves for the block- samples C (top) and ALC (bottom) cured for 1 h and 16 h.

However, the difference between them was observed for remained quartz particle sizes; one (average diameter, 80 μm) is larger than the other (10-20 μm) by optical microscopy. This causes the toughness difference between them; higher the fracture energy, larger the quartz unreacted.

In the present work, the load vs. crack mouth opening displacement curves were affected by the reacted products: weak and broad for C-S-H, strong and sharp for tobermorite (Fig.9). For examples, the block-samples (C) after 1 h to 16 h show similar curves for tobermorite product, but the block-samples (ALC) after 1 h to 16 h give different results affected by C-S-H and tobermorite products, respectively.

4 CONCLUSIONS

The present work used two kinds of quartz source; one for the ALC Factory product, and the other with different particle sizes for auto-claved xonotlite product.

(1) For the tobermorite formation, the block-samples (ALC) using the coarsest quartz in the present work gave highly crystalline tobermorite after 4 h, contrasting with that after 1 h for the block-samples (A, B, C) using different quartz source with three different particle sizes smaller than the quartz sample (ALC). No effective particle size difference was observed for the block-samples (A, B, C), because all quartz samples contain smaller particle enough for tobermorite formation in the early stage of the reaction.

(2) For the compressive strength and Young's modulus, the values are increased with tobermorite content, expecting to be the same bulk density. The relations between them, show a linearity for the same quartz source.

(3) For the fracture toughness, Gf value gave no affected results of the quartz particle size, contrasting with those on other mechanical and chemical properties. The results suggest that coarser the quartz block-sample gives higher the Gf values.

ACKNOWLEDGMENTS

We thank Dr. H. Ishida, INAX Corporation, for his helpful discussion and comments.

REFERENCES

Chan, C.F. and T.Mitsuda 1978. Formation of 11Å tobermorite from mixtures of lime and colloidal silica with quartz. *Cem. Concr. Res.* 8: 135-138.

Ishii, T., C.F.Chan and T.Mitsuda 1978. Reaction kinetics of calcium oxide-quartz-water suspension system under hydrothermal condition. *Sem. Gijyutu Nempo* 32: 75-78 (in Japanese).

ISRM 1988. Suggested methods for determining the fracture toughness of rock. F.Ouchterlony working group coordinator. *Int. J. Rock Mechs. Min. Sci. and Geomech. Abstr.* 25: 71-96.

Mitsuda, T., K.Sasaki and H.Ishida 1992-a. Phase evolution during autoclaving process of aerated concrete. *J. Am. Ceram. Soc.* 75: 1858-1863.

Mitsuda, T., K.Sasaki and H.Ishida 1992-b. Influence of particle size of quartz on the tobermorite formation. *RILEM Int. Sympo. on Autoclaved Aerated Concrete, Zürich* : in press.

Mitsuda, T., T. Kiribayashi, K.Sasaki and H.Ishida 1992. Influence of hydrothermal processing on the properties of autoclaved aerated concrete. *RILEM Int. Sympo. on Autoclaved Aerated Concrete, Zürich* : in press.

Sakiyama,M. and T. Mitsuda 1977. Hydrothermal reaction between C-S-H and kaolinite for the formation of tobermorite at 180 °C. *Cem. Concr. Res.* 7: 681-686.

Teramura, S., K.Tsukiyama and H.Takahashi 1988. Evaluation of fracture toughness of autoclaved lightweight concrete by means of acoustic emission technique. *J. Acoustic Emission* 7: 1-8.

Wittmann, F.H. and I. Gheorghita 1984. Fracture toughness of autoclaved aerated concrete. *Cem. Concr. Res.* 14: 369-374.

Advances in Autoclaved Aerated Concrete, Wittmann (ed.) © *1992 Taylor & Francis. ISBN 90 5410 086 9*

Following up the setting of cellular concrete by acoustic test

C. Boutin & L. Arnaud

Ecole Nationale des Travaux Publics de l'Etat, Laboratoire Géomatériaux, E. P. du CNRS 16, Vaulx en Velin, France

ABSTRACT: In order to follow up the setting of the crude paste of cellular concrete, a non destructive vibratory test at low frequency (100-800 Hz) is developed. By analyzing the celerity of the waves, we clearly characterize the evolution of the material from the liquid to the solid phase. This efficient experimental device allows to study other formulations of crude cellular concrete and can be adjusted to the study of other compressible, heterogeneous, evolutionary materials.

1 INTRODUCTION

This study presents a new experimental method which will follow and characterize the evolutions of the crude paste of cellular concrete during the four hours of its setting before the autoclaving. The material changes from a liquid (at the moment when it is casted) to a bubbly fluid (during the first hour) and then to a porous solid (after the four hours). During this period a solid structure is being built; cracks or fissures can appear industrially. Consequently, it is interesting to follow the mechanical behaviour of the paste during these four hours.

The particular characteristics of the crude paste make its rheological study difficult: it is heterogeneous (porosity about 50 %, bubbles of 1 mm in diameter), it evolves from fluid to solid and it is very sensitive to the thermodynamical conditions. For these reasons, the conventional mechanical tests are not suitable to the study of such a paste.

Therefore, we have adjusted a non-destructive test based on plane-wave propagation at low frequency (100-800 Hz).

In the first part, we will present the specificities of the material and the difficulties in applying the conventional tests. The second part will detail the experimental device operational in the laboratory. In the third part, we will present the experimental results concerning the variations of the waves celerity during the setting. A physical interpretation of the phenomenons and the sensitiveness of the measurements to a change of the formulation are also presented.

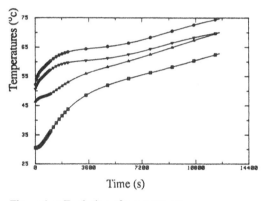

Time (s)

Figure 1-a: Evolution of temperatures.

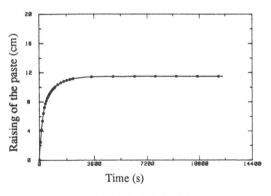

Time (s)

Figure 1-b: Evolution of the height of the cake.

2 WHICH EXPERIMENTAL METHOD CAN BE USED TO CHARACTERIZE THE PASTE ?

2.1 *The cellular paste in the laboratory*

In order to realize this study, the first difficulty was to obtain the same paste of cellular concrete as the one manufactured by SIPOREX.

We have used the same formulation, the same preparation and the same raw materials (sand mud, cement, quicklime and aluminium powder). Then, we have adjusted, in the laboratory, the manufacture process.

We have used an adiabatic container because it was the best way to model, in the laboratory, the thermodynamical conditions of the setting of the paste. The evolutions of the height and of the temperature in the container and at different points in the paste are presented. The values are similar to those measured in a SIPOREX factory (fig.1-a and 1-b).

Some samples taken in our container and autoclaved were tested. They confirmed that our material and the factory product have the same characteristics according to the conventional tests: density of 450 kg/m^3, compressive strength on a cube (ridge 10 cm) of 50 da N/cm^2 (Arnaud, 1989).

During the first hour after the casting, the aluminum powder reacts in the basic medium releasing hydrogen. The bubbles created have a diameter of 1 mm . They confer to the paste its final porosity of 50 % (cf fig.1-b). In the same time and during the following three hours, the reactions of crystallization (lime and cement) occurs and the material changes from an heterogeneous fluid to a porous solid. The kinetic of those chemical reactions are very sensitive to the thermodynamical conditions.

In conclusion, during its manufacture, the paste of cellular concrete is an <u>heterogeneous</u>, <u>evolutionary</u> material, <u>sensitive to the thermodynamical conditions</u>.

2.2 *Conventional tests*

The conventional tests are not easy to carry out and not adjusted to analyze such a material.

On the one hand, those used for the study of fluids cannot be adjusted to the study of the solid phase; and inversely, those used for the solids are unsuitable when the material is fluid.

On the other hand, those tests are realized with large strains and, consequently, lead to the collapse of the microstructure. So, they prevent following the evolution of the paste properties with time.

Because of the sensitiveness to the thermodynamical conditions, it is also very difficult to take apart, during the setting, representative samples in order to carry out those tests.

Ultrasonic tests: $\begin{cases} \text{small wavelength} \\ \quad \lambda \approx 0,1 \text{ mm} \end{cases}$

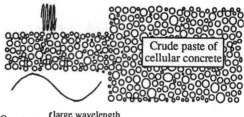

Our tests: $\begin{cases} \text{large wavelength} \\ \quad \lambda > 3\text{cm} \end{cases}$

Figure 2: Small or large wavelength.

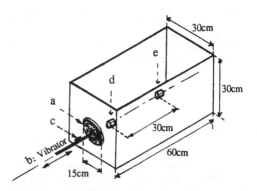

Figure 3: The experimental device.

According to non-destructive ultrasonic tests, the wavelengths (λ) considered are small compared to the size of the heterogeneities. For example, an usual frequency of 100 kHz leads in the paste to a wavelength of order of 0,2 mm because the celerity of the wave at the beginning of the setting is of order of 20 m/s. The average diameter of the bubbles is 1mm which is great compared to 0,2 mm.

Consequently, this test only gives a local measure (fig.2). And moreover, the waves scattered on the bubbles are strongly attenuated.

So, ultrasonic tests cannot be used easily.

In conclusion, the conventional tests as well as the ultrasonic one are not efficient to the study of the crude paste of cellular concrete.

3 EXPERIMENTAL DEVICE

3.1 *Principle and description (fig.3)*

We will suggest here a non-destructive vibratory test. Transient waves trains at low frequency (100-800 Hz) are propagated in the material inside a container (60 ∗ 30 ∗ 40 cm^3).

A plate of 15 cm in diameter (a) linked to a vibrator

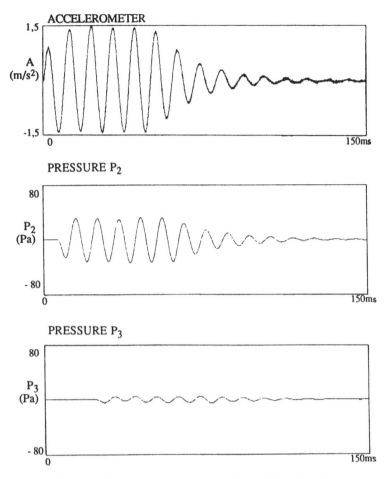

Figure 4: Signals emitted and received at t = 900 s (15 mn)

(b) generates compression plane waves in the paste. An accelerometer (c), fixed on the plate, records the signal of excitation. The detection of the waves is achieved by two piezoelectric transducers (d) and (e), plunged in the paste at respectively 10 cm and 40 cm from the plate.

A signal analyzer (HP 3567 A) allows, on the one hand, to define the signal generated by the plate movement. On the other hand, it records and stores in real time the signals detected by the accelerometer and the two pressure transducers. The measurements are made every 10 minutes during the four hours of the setting.

3.2 Characteristics of the test

Let us examine the specific characteristics of this experiment.

The waveform of the excitation signal is composed by several sine oscillations of constant amplitude, then the amplitude decreases exponentially to zero (fig.4-a presents an example).

The excitation signal has a finite duration which allows:

1. to clearly determine the times of arrival of the wave and so, to calculate the celerity in the medium;

2. to prevent, as far as possible, that the signals emitted and reflected are superposing each other.

The excitation sine wave is generated at one particular frequency ($100 \leq f \leq 800$ Hz) in order to avoid the celerity dispersion. The frequency spectrum obtained is contained in a band, whose width is very small, centered around the excitation frequency f. This frequency remains constant during one particular experiment.

Between two signals emitted, we wait till no pressure may be detected.

The imposed level of acceleration varies between

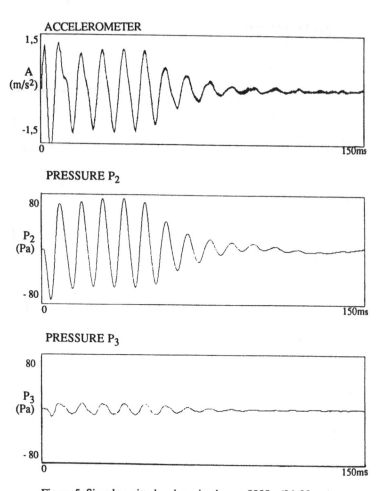

Figure 5: Signals emitted and received at t = 9000 s (2 h 30 mn)

0,5 and 5 m/s². Consequently, the maximum displacement, at a point in the paste, is about:

$$\frac{A}{\omega^2} = \frac{5}{(2\pi 100)^2} = 10 \, \mu m.$$

The recorded level of pressure varies between 10 and 200 Pa according to the rigidity of the paste.

So, we can consider that the test is non-destructive, the microstructure is not broken.

For the excitation frequencies considered, the wavelength λ is great compared to the size of the bubbles. So, this vibratory test gives us a bulk measurement of the material; the diffraction of the waves is negligible.

At the beginning of the setting, we will see that the celerity of the wave is very small (c $\approx$ 20 m/s). For the frequency used (case least favourable, f = 800 Hz), λ is of order of 4 cm which is great compared to the average diameter of the bubbles (1mm).

As the paste is setting, it becomes more rigid, the celerity increases and so does the wavelength.

The fact that the wave is plane is more convenient for the analysis of the signals received. The parameter linked to the directivity of the emitted wave field is the ratio between the size of the vibratory plate (D) and the wavelength (λ). For small values of λ/D (λ/D < 3), the pattern of the directivity function presents one principal lobe and eventually, small side lobes (Hueter & Bolt, 1955).

A plate of 15 cm in diameter is the best compromise between correct directivity and dimensions of the container for a celerity less than 200 m/s.

For higher celerities, the wave field is more complex because the plate radiates in every directions.

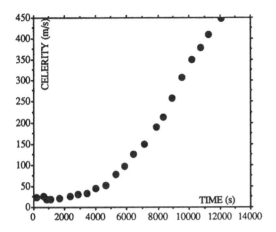

Figure 6: Evolutions of the celerity in time.

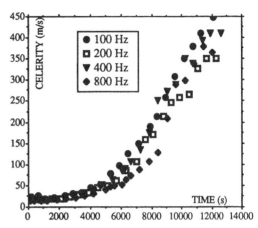

Figure 8: Evolution of the celerity for the different frequencies.

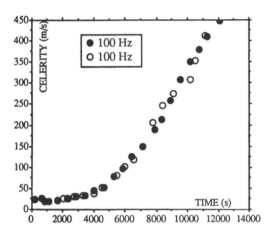

Figure 7: Comparison between two "cakes" tested at the same frequency.

4 RESULTS AND PHYSICAL INTERPRETATION

4.1 *Measurements*

Five different "cakes" were manufactured. During each experiment, the excitation frequency was constant: two "cakes" were tested at f= 100 Hz, one at f= 200 Hz, one another at f= 400 Hz and the last one at f= 800 Hz.

Figure 4 presents the signal emitted (4-a: accelerometer) and the signals received by the pressure transducers (4-b: pressure P_2 at d= 10 cm and 4-c: pressure P_3 at d= 40 cm) 15 minutes (or 900s) after the casting.

The same curves are presented fig.5 for a time of 2h30mn (or 9000s) after the casting. We observed that :

1. the arrival times of the wave on the pressure transducers at t = 900s is greater than those at t = 9000s (fig.4: Δt_1 = 11,9 ms; fig.5: Δt_2 = 1,15 ms);

2. for a same level of acceleration (a = 1,3 m/s^2), the levels of P_2 and P_3 are weaker at t = 900s than at t = 9000s (fig.4: P_2 = 31,1 Pa and P_3 = 4,1 Pa; fig.5: P_2 = 65,5 Pa and P_3 = 8,14 Pa).

The celerity and the attenuation of the wave have been calculated from these measurements.

4.2 *Celerity*

The celerity is calculated from the measurement of time delay of the wave arrivals between the two transducers at the distance d for five different "cakes". In fig.6, the typical evolution of the wave celerity c in time is shown (t = 0 s corresponds to the casting).

Two phases are observed on this curve:

1. for t < 5000s (about 1h 20 mn), the material is characterized by a very low celerity c ($20 \le c \le 50$ m/s);

2. for $5000s \le t \le 14400s$ (that is to say 4 hours), the celerity increases rapidly ($50 \le c \le 450$ m/s).

We notice fig.7, that the celerity evolutions are in very good agreement as to the two experiments achieved with f= 100 Hz.

In fig. 8, the results obtained for the cakes tested at the different frequencies are presented. The celerity curve seems to be weakly dependent of the frequency f.

4.3 *Attenuation*

The attenuation can be calculated from the ratio (P_2 /P_3) between the levels of pressure. The level of presure $|P|$ is valued from the oscillations of each signal (duration T) by the relation:

$$|P| = \sqrt{\frac{1}{T} \int_0^T P(t)^2 \, dt}$$

From the evolution of P_2/P_3, we can also distinguish two phases:

1. for t < 4000s, the ratio is high;

2. for 4000s < t < 14400s, the ratio decreases and becomes quasi constant between 3 and 8 according to the frequency.

The analysis of the attenuation is complex because when the celerity reaches 200 m/s, the signals emitted and reflected are superposing each other. A more precise study of the attenuation is being developed.

4.4 Physical interpretation

During the first phase, the paste presents a very low celerity. This is characteristic of a bubbly fluid. At the time of the casting, the material is a viscous fluid (viscosity $\mu \approx 300$ mPa.s). Then, by hydratation of the aluminium powder, bubbles are created in this fluid.

This kind of material reacts with the compressibility K of the gas corrected by the bubbles concentration and the average density of the material.

Let us valuate the celerity of the wave in such a material using the properties of an elastic medium:

bubbles of water vapor: $K = 1{,}4 \; 10^5$ Pa

concentration of bubbles: 50 %

average density: $\rho \approx \dfrac{\text{casted mass}}{\text{final volume}} \approx \dfrac{40}{5{,}4 \; 10^{-2}}$

$$\approx 750 \text{ kg/m}^3$$

So we have:

$$c = \sqrt{\frac{\dfrac{1{,}4 \; 10^5}{0{,}5}}{\rho}} = 19{,}3 \text{ m/s.}$$

This calculated celerity closely agrees with the experimental results.

The second phase is characterized by an increasing of the celerity and a decreasing of the attenuation and the paste changes from a liquid to a solid state. The crystallization of lime and cement is occurring. The

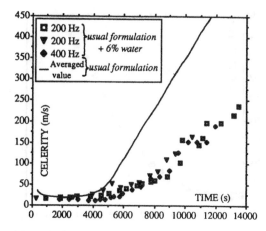

Figure 9: Comparison of the evolutions of c

crystals generated reach a size such as they begin to be connected with each other.

As a first approach, assuming that the medium is elastic, the œdometric modulus E_α can be valuated from the celerity measurements from the relation:

$$c = \sqrt{\frac{E_\alpha}{\rho}}$$

For example, four hours after the casting, the celerity and the density are:

c = 450m/s and $\rho \approx 800$ kg/m³ which gives

$E \approx 10^8$ Pa.

Let's notice that this value obtained with an elastic modeling, must be used cautiously since the measurement of the attenuation shows a significant rate of damping.

This physical interpretation of the material evolutions, based on the study of the microstructure, will be compared with a diffractometric analysis of samples taken during an experiment.

4.5 Measurement sensitiveness to a change of formulation

Three other "cakes" were manufactured with 6 % of water more than the usual formulation. The evolutions of the wave celerity are plotted fig.9. We indicate on the same graph the averaged value of the celerity obtained for the five other experiments.

On the one hand, the rapid increase of the wave celerity occurs later compared to the usual formulation (7000s or 1 h 50 mn). On the other hand, the celerity level reached four hours after the casting is smaller (250 m/s). So, the measurements are very sensitive to a change of the formulation of the crude paste.

This result leads to think that a significant change in the formulation, leading to changes of chemical reactions and consequently of the internal structure of the paste, will modify significantly the celerity evolution.

5 CONCLUSIONS

In this study, we suggest a new experimental method in order to follow and characterize the setting of cellular concrete crude paste. This non-destructive test is based on the analysis of the compression plane-wave propagation in the paste at low frequency (100- 800 Hz). By analyzing the wave celerity, we clearly determine the evolution of the material from the fluid to the solid phase (c changes from 20 to 450 m/s).

This test presents repetitive results and is sensitive to changes of the formulation. A prototype will be set up in a SIPOREX factory in order to test new formulations, new raw materials and to follow the production.

The experimental device can be adjusted to study other heterogeneous, evolutionary materials (such as foam concrete, gel,...).

REFERENCES

Arnaud, L. 1989. Etude de la pâte crue de béton cellulaire Siporex. D.E.A. E.N.T.P.E.-I.N.S.A. Lyon.
Boutin, C. & Arnaud, L. 1992. Experimental characterization of the setting of cellular concrete. (To appear in Proceeding of XIth International Congress on Rheology, Brussels, Belgium.)
Hueter, T. F. & Bolt R. H., 1955. Sonics. New York: Willey & Sons.

ACKNOWLEDGEMENTS

This research was supported by SIPOREX - Autoclaved aerated concrete producer-.

Advances in Autoclaved Aerated Concrete, Wittmann (ed.) © 1992 Taylor & Francis. ISBN 90 5410 086 9

Unit weight reduction of fly ash aerated concrete

F. Pospíšil
Porsit, Limited Leability Company, Brno, Czechoslovakia

J. Jambor
Institute of Construction and Architecture, Slovak Academy of Sciences, Bratislava, Czechoslovakia

J. Belko
Material Research Institute, Brno, Czechoslovakia

ABSTRACT: The possibilities of a significant reduction of aerated concretes unit weight – by means of modification of their components, composition and production technology – were studied. Effect of these modifications on changes in aerated concrete pore structure was investigated – with the aim to minimize their stregth drop. The results proved that under some pre-condition it is possible to reduce the unit weight of aerated concretes up to values of approximately 300 kg.m^{-3}. The strength drop of aerated concretes caused by the same unit weight reduction can be, however, partly different – in dependence on changes in their pore structure. The largest strength drop was found at aerated concretes by which the unit weight reduction has led to formation of the largest macropores. The technological measures leading to the restriction on the size of formed macropores – and at the same time promoting the increase of micropores volume share in the matrix, lead to the minimization of the aerated concretes strength drop. Effective technological measures of this typ can be the increase of fineness of used components as well as the increase of the binder amount in aerated concrete.

1 INTRODUCTION

Improvement of the thermal insulating properties of autoclaved aerated concretes requires the reduction of their unit weight, on which depends their thermal conductivity. A significant pre-condition of effective unit weight reduction – by simultanous minimization of the strength drop – is a deeper knowledge of the changes in aerated concrete pore structure, connected with the unit weight reduction, as well as the acquisition of data enabling the regulation of its development. That is why the possibilities of an improvement of pore structure and strength of fly ash aerated concrete – by means of fineness increase of used fly ash or dry mixture – were studied. Likewise the possibilities of substantial reduction of aerated concretes unit weight by means of the modification of their composition and production technology, as well as

the main properties and pore structure of aerated concretes with very low unit weight were investigated.

2 MATERIALS AND METHODS USED

The fly ash and quartz sand processed commonly in Czechoslovak plants for manufacturing of aerated concrete of the type Calsilox, were used as silicious components for this investigation. As the next components the Portland cement PC 400, lime and Al-powder PAP 1 were used. The composition of these materials is given in table 1.

The preparation of various dry mixtures and aerated concretes was realized in the main by common and constant way. The amount of mixing water was determined at every homogenized dry mixture in advance – by means of rotary viscosimeter – so that the value of the structure viscosity of all fresh mixtures was kept the same. Also the same autoclaving cycle - with isothermic dwell at $193^\circ C$ was applicated by manufacturing of all test specimens. The differences in the composition of individual mixtures are mentioned below.

Table 1. Composition of used materials.

		Fly ash	Ground quartz sand	Portland cement PC 400	Lime
SiO_2	wt. %	55.05	94.50	21.46	1.08
Al_2O_3	wt. %	28.45	2.55	6.56	0.68
Fe_2O_3	wt. %	7.81	0.56	3.02	0.24
TiO_2	wt. %	2.08	-	0.20	0.03
CaO	wt. %	2.28	1.19	61.39	91.08
MgO	wt. %	1.17	0.15	2.40	0.89
K_2O	wt. %	1.60	-	0.92	0.05
Na_2O	wt. %	0.24	-	0.21	0.03
SO_3	wt. %	0.13	-	2.67	0.15
Ignition loss at $1000^\circ C$	wt. %	1.02	0.55	0.97	5.39
Specific surface /accord. Blaine/ $m^2.kg^{-1}$		235	210	365	-

The unit weight and compressive strength were determined by using the 10 cm edge test cubes - according to Czechoslovak standard ČSN 731290. The micropore structure in the matrix of tested aerated concretes was determined by means of mercury intrusion porosimetry - using the Carlo Erba high pressure porosimeter. The macropore structure was investigated by means of automatic image analyzing apparatus Olympus CUE 4 - in connection with stereo magnifying glass Olympus SZH. The thermal conductivity was measured on the plate specimens of 2 cm thickness - by using the Shoterm apparatus.

3 RESULTS AND THEIR DISCUSSION

3.1. Effect of components fineness

Two procedures were used for the ascertainment of the effect of increased fineness of components on properties of aerated concrete. In the first case the effect of increased fineness of dry mixture and in the second case the effect of increased fineness of fly ash was studied. The dry mixture and fly ash were ground in a ball mill for the time period of 5, 10 and 15 minutes, in order that a graded fineness was achieved. The results of sieve analyses of ground dry mixture and fly ash are given in table 2.

Table 2. Results of sieve analyses of ground mixture and fly asch

	Remainder on the sieve in wt. %	
	0.063 mm	0.063-0.040mm
Dry mixture of fly ash : lime : portland cement - 70 : 15 : 15 wt. %		
Original state	35.70	13.82
Fineness grade 1	23.16	15.06
Fineness grade 2	18.76	15.78
Fineness grade 3	10.20	14.70
Fly ash		
Original fly ash	40.16	13.76
Fineness grade 1	16.14	15.14
Fineness grade 2	7.56	12.38
Fineness grade 3	2.20	7.60

Mixture of fly ash:lime:Portland cement in proportion of 70:15:15 wt. % and containing a constant dose of Al-powder 0.077 wt. % was used for this investigation. The basic properties of aerated concretes made from dry mixture and fly ash ground to various fineness grade are given in table 3.

These results show, that at the same unit weight the aerated concrete strength increases with increasing fineness of dry mixture or used fly ash - even in spite of partial increase of the water:binder ratio. The achievement of approximately 4 MPa compressive strength of aerated concrete with the unit weight of 430 $kg.m^{-3}$ shows, that the increase of fineness of used fly ash belongs to effective interventions for the strength enhancement.

The micropores size distribution in the matrix of all tested aerated concretes was very similar and the fineness increase of dry mixture or

Table 3. Basic characteristics of aerated concretes made from dry mixture and fly ash of various fineness.

| | | Dry mixture fineness grade | | |
		1	2	3
Water: dry mixture ratio		0.793	0.811	0.828
Mixture temperature in °C	$T_{init.}$	36.5	35.0	36.0
	$T_{max.}$	68.5	70.0	70.2
Unit weight kg.m^{-3}		433	425	428
Compressive strength - MPa		3.51	3.38	3.87
Total porosity %		81.5	81.8	81.7
Volume of micropores /R<7.5 μm/	mm^3.g^{-1}	559	579	582
	%	24.2	24.6	24.9

| | | Fly ash fineness grade | | |
		1	2	3
Water: dry mixture ratio		0.836	0.854	0.871
Mixture temperature in °C	$T_{init.}$	36.0	35.0	36.0
	$T_{max.}$	68.9	66.5	66.7
Unit weight kg.m^{-3}		423	425	431
Compressive strength - MPa		2.42	3.38	4.22
Total porosity %		81.9	81.8	81.9
Volume of micropores /R<7.5 μm/	mm^3.g^{-1}	596	603	607
	%	25.2	25.6	26.2

increased of 0.7 to 1.0 % of their total porosity - to detriment of macropores volume.

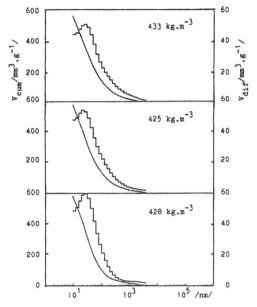

Fig. 1. Micropores size distribution curves of aerated concretes from mixture of graded fineness.

fly ash did not influence it significantly. This confirm also the curves of micropores size distribution in aerated concretes made from the dry mixtures ground to various fineness grade, which are illustrated in figure 1. Micropores with radius of 24 to 28 nm predominated in all tested aerated concretes and only minimal part of micropores with radius above 500 nm was present. On the other hand the results proved, that the total volume of micropores in aerated concrete matrix increases with increasing fineness of used components. The conversion of micropores volume in the matrix to total volume of respective aerated concrete proved, that due to higher fineness of used components the micropores volume in aerated concretes

Slight changes in the macropores structure of aerated concretes made from the component of graded fineness were found, only. The number of macropores in 1 cm^2 ranget from 214 to 234 and their average size was found within 470 to 487 um in all tested samples. The results, however, indicated, that the increase of component fineness can lead to partial improvement of homogeneity of macropores structure - due to slight size decline of the largest macropores.

The size and total volume of micropores developed in cement matrix and above all of micropores with radius below 100 nm, reflect the type and volume of binding hydration products created in the matrix (Jambor, 1973, 1976, 1990, Mindess,1984). Predominant presence of micropores with the same radius of approximately 24-28 nm in all tested aerated concretes indicates, that due to fineness increase of components does not change the type of hydration products in aerated concretes matrix. On the other hand the increase of total micropores volume - by corresponding decrease of macropores volume - indicates partial volume enhancement of hydration products - probably due to intensification of hydrothermal process. This volume enhancement of hydration products in the matrix explains the higher compressive strength of aerated concretes made from component of higher fineness (Mindess 1984; Pereira, Rice, Skalny, 1989).

3.2 Unit weight reduction chance

Unit weight reduction was achieved by means of successive changes of portland cement and Al-powder amount in the mixture, as well as by changes of water:binder ratio and fresh mixture temperature. On this way a series of fly ash aerated concretes of the unit weights from 440 to 285 $kg.m^{-3}$ was manufactured. Similarly - for comparison - the second series of aerated concretes on quartz sand basis with corresponding unit weights, was made. The composition and main properties of these aerated concretes are summarized in table 4.

Obtained results proved, that the production of aerated concretes with unit weight on the level of approximately 300 $kg.m^{-3}$ is essentialy possible also by using of current technological measures only - even though in case a real production it would be necessary to solve the problem of very slow setting of these fresh mixtures (Pospíšil 1990). Proportionately to the unit weight reduction a subtantial strength decrease at both types of aerated concretes was ascertained. The relationship between compressive strength and unit weight of tested aerated concretes, which is plotted in figure 2, however shows that the strength decrease in dependence on the unit weight reduction is not identical by both aerated concrete types, as at fly ash aerated concretes seems to be more expressive. It is probably due to about 10 wt. % lower amount of binder in fly ash aerated concretes. On the other hand fly ash aerated concretes show - at the sa-

Table 4. Composition and properties of aerated concretes with graded unit weight.

| | Aerated concretes on fly ash basis | | | | | |
	1	2	3	4	5	6
Fly ash:lime: p-cement in wt. %	69:16:15	64:16:20	59:16:25			
Al-powder wt. %	0.080	0.100	0.120	0.135	0.155	0.180
Water: dry mixture ratio	0.792	0.785	0.790	0.812	0.812	0.821
Mixture temperature $T_{init.}$ / $T_{max.}$ in °C	33.0 / 68.7	34.0 / 72.0	36.0 / 73.5	30.0 / 64.5	30.0 / 66.0	30.0 / 55.7
Unit weight kg.m⁻³	440	385	350	330	300	285
Compressive strength - MPa	3.14	2.21	1.60	1.35	1.37	0.99
Coefficient of thermal conductivity $\lambda - w.m^{-1}.K^{-1}$	0.094	0.082	0.080	0.075	0.072	0.067

| | Aerated concretes on quartz sand basis | | | | | |
	1	2	3	4	5	6
Sand:lime:p-cement wt. %	59:16:25	54:16:30	49:16:35			
Al-powder wt. %	0.090	0.110	0.130	0.142	0.160	0.185
Water: dry mixture ratio	0.700	0.700	0.700	0.719	0.719	0.714
Mixture temperature $T_{init.}$ / $T_{max.}$ in °C	29.0 / 69.5	29.0 / 70.0	30.0 / 73.0	26.0 / 64.0	25.9 / 63.5	25.5 / 54.0
Unit weight kg.m⁻³	435	370	345	320	305	286
Compressive strength - kPa	2.78	1.83	1.94	1.54	1.53	1.15
Coefficient of thermal conductivity $\lambda - w.m^{-1}.K^{-1}$	0.107	0.095	0.088	0.087	0.079	0.079

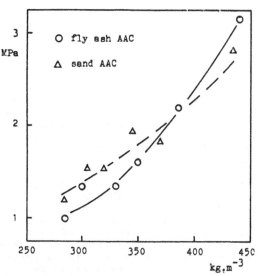

Fig. 2. Relationship between compressive strength and unit weight of tested aerated concretes.

me unit weight – more favourable
values of the coefficient of thermal conductivity.

Results of pore structure investi-
getion of tested aerated concretes
are given in table 5. Characteristic
curves of micropores size distribution in matrix of both types of tested aerated concretes – with various
unit weight – are illustrated in figures 3 and 4. Characteristic macropores size distributions of selected
aerated concretes are plotted in figures 5 and 6.

Table 5. Pore structure characteristics of aerated concretes with graded unit weight.

	Aerated concretes on fly ash basis					
	1	2	3	4	5	6
Unit weight. kg.m^{-3}	440	385	350	330	300	285
Total porosity %	81.2	83.5	85.1	85.9	87.2	87.8
Micropores R<7.5 μm						
Radius median nm	23.7	28.2	41.2	46.1	38.1	39.3
Volume mm^3.g^{-1}	589	546	600	615	628	669
Volume %	25.9	21.0	21.0	20.3	18.8	19.1
Macropores R>7.5 μm						
Total volume %	55.8	62.5	64.1	65.6	68.4	68.7
Number in cm^2	188.6	138.2	121.0	117.0	105.0	-
Average size μm	471.9	527.1	630.7	613.1	633.4	-

	Aerated concretes on quarts sand basis					
	1	2	3	4	5	6
Unit weight kg.m^{-3}	435	370	345	320	305	286
Total porosity %	81.8	84.5	85.6	86.4	87.1	87.9
Micropores R<7.5 μm						
Radius median nm	86.7	97.5	41.8	40.8	38.1	40.7
Total volume m^3.g^{-1}	554	591	547	541	575	565
Total volume %	24.1	21.9	18.9	17.3	17.5	16.1
Macropores R>7.5 μm						
Total volume %	57.7	62.6	64.1	65.6	68.4	68.7
Number in cm^2	178.6	152.0	118.7	121.3	108.0	-
Average size μm	545.2	531.8	686.1	659.2	718.2	-

Obtained results prove, that unit
weight reduction of aerated concretes is a consequence of increase of
macropores volume share in volume
unit of aerated concrete only. The
volume share of micropores, which
are as a matter of fact the co-pro-
duct of developed hydration products, diminishes in aerated concrete with its unit weight reduction
proportionate to decrease of matrix
volume. Decrease of micropores volume share in aerated concrete comes true even in the case, when the
relative micropores volume in matrix
remains constant or even increases

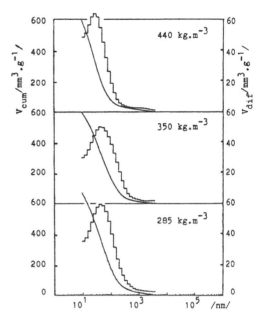

Fig. 3. Characteristic curves of micropores size distribution in fly
ash aerated concretes with graded
unit weight.

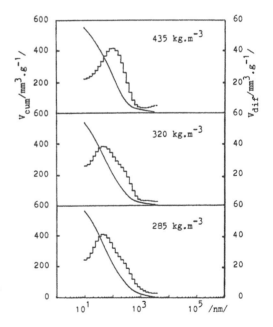

Fig. 4. Characteristic curves of micropores size distribution in quartz
sand aerated concretes with graded
unit weight.

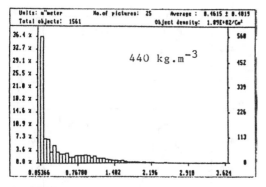

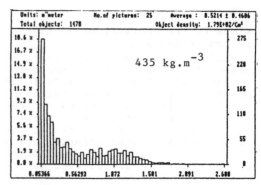

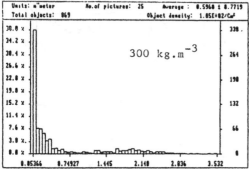

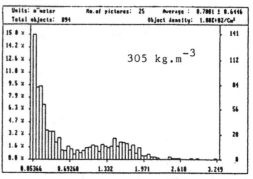

Fig. 5. Characteristic macropores size distribution in fly ash aerated concretes with unit weight of 440 and 300 kg.m^{-3}.

Fig. 6. Characteristic macropores size distribution in quartz sand aerated concretes with unit weight of 435 and 305 kg.m^{-3}.

partly. The values of micropores radius median in both types of tested aerated concretes, as well as their changes due to unit weight reduction show evident differences. While in fly ash aerated concretes with unit weight of 370-440 kg.m^{-3} the value of micropores radius median amounts to 24-28 nm, in corresponding aerated concretes on quartz sand basis this value amounts 86-98 nm. This indicates the development of partly different hydration products in tested types of aerated concretes, which corresponds also to known experience. Remarcable, however, is the ascertainment, that due to unit weight reduction the differences of micropores median value in both types of aerated concretes diminish and come together in aerated concretes with unit weight of 285 - 300 kg.m^{-3}.

Total volume of macropores in the volume unit of both types of aerated concretes increases expressively -

with the reduction of their unit weight. The increase of the total macropores volume is manifested by the size increase of individual macropores and by simultaneous decrease of their number. This is proved also by values of average size of macropores in aerated concretes of various unit weight. The values of average size of macropores, however, do not provide enough truthful image about the macropores structure. A deeper analysis of results of macropores size distribution investigation showed, that 50 to 70 % of total macropores volume - in all tested aerated concretes - is constituted by macropores with diameter up to 0.6 mm. The remaining 30 to 50 % of total macropores volume is constituted by larger macropores. The size of these largest macropores in individual tested aerated concretes differs, however, significantly. This is proved also by results given in table 6.

Table 6. Volume share and size of macropores with diameter above 0.6 mm in tested aerated concretes.

	Aerated concretes on fly ash basis				
	1	2	3	4	5
Unit weight kg.m^{-1}	440	385	350	330	300
Total volume of macropores with R>7.5 μm %	55.8	62.5	64.1	65.6	66.4
Macropores share with R>0.6 mm					
Total volume %	16.7	20.0	22.4	22.9	22.6
Average size mm	0.8	1.1	1.5	1.6	1.5
Size of the largest pores mm	1.6	2.0	2.0	2.4	2.4

	Aerated concretes on quartz sand basis				
	1	2	3	4	5
Unit weight kg.m^{-1}	435	370	345	320	305
Total volume of macropores with R>7.5 μm %	57.7	62.6	66.7	69.1	69.6
Macropores share with R>0.6 μm					
Total volume %	21.3	26.3	32.6	32.4	31.3
Average size mm	1.1	1.2	1.2	1.5	1.65
Size of the largest pores mm	1.6	1.7	1.9	2.0	2.1

These results show, that in fly ash aerated concretes - due to reducing their unit weight - the macropores with a larger average, as well as absolute size are formed - in comparison with tested aerated concretes on sand basis. It is probably the consequence of substantial lower amount of binder in fly ash aerated concretes. In every case, however, the trend of size increase of largest macropores in tested aerated concretes corresponds to the trend of their strength decrease. It is therefore possible to conclude, that more expressive strength decrease of tested fly ash aerated concretes - due to the reduction of their unit weight - is caused above all by the development of larger macropores in their pore structure. This opinion is in agreement with findings of F.H.Wittman (1979,1983) and others about the decissive influence of pores size on the possibility of their acting as centres for formation and propagation of failure cracks in sense of Griffith theory. With increasing pores size the value of critical stress decreases distinctly and for this reason the test specimen fails on the lower level of acting load. From this follows, that the way leading to the strength increase of aerated concretes with very low unit weight consists in such chan-ges of their composition and technology, which prevent the formation of too large macropores.

4 CONCLUSION

Achieved results lead to following conclusion:
Reduction of unit weight of aerated concretes on the level of approximately 300 kg.m^{-3} - by using of current technological measures only - was achieved. For real production of such aerated concretes, however, the problem of very slow setting of fresh mixture would be necessary to solve.
Compressive strengths of aerated concretes with unit weight of approximately 300 kg.m^{-3} amount to 1.0-1.5 MPa and coefficient of thermal conductivity amounts values of 0.07 to 0.08 W.m^{-1}.K^{-1}.
Unit weight reduction of aerated concretes is above all the consequence of their macropores volume increase. This is manifested first by the size increase of individual pores.
Compressive strength decrease of aerated concretes - at the same unit weight reduction can be partly different - in dependence on their composition and on changes in their pore structure. The largest strength drop was found at aerated concretes, by which unit weight reduction was connected with the formation of the largest macropores.
Technological measures leading to restriction of the size of developed macropores and at the same time promoting the increase of micropores volume share in the matrix lead to minimization of aerated concretes strength drop caused by reduction of their unit weight. The obtained results proved, that effective technological measures of this type can be the increase of fineness of used fly ash or dry mixture, as well as the increase of binder amount in aerated concrete. Both these interventions lead to the increase of hydration products volume and micropores volume share in the matrix, as well as to restriction of macropores size developed in aerated concretes with low unit weight.

REFERENCES

Jambor, J. 1973. Influence of phase
composition of hardened binder
paste on its pore structure and
strength. Pore structure and properties of materials, pp. D75-D96.
Prague: Academia.

Jambor,J. 1976. Influence of water-
cement ratio on the structure and
strength of hardened cement pastes. Hydraulic cement pastes:
their structure and properties,
pp. 175-188. Wexham Springs: Cement and Concr.Assoc.

Jambor, J. 1990. Pore structure and
strength development of cement
composites. Cement a. Concrete
Research, Vol. 20; pp.948-954.

Mindess, S. 1984. Relationship between strength and microstructure
for cement-based materials. Very
high strength cement-based materials, pp. 53-68. Pittsburgh:
Mat. Res. Soc.

Pereira, C.J., Rice, R.W., Skalny,
J.P. 1989. Pore structure and
its relationship to properties
of materials. Pore structure and
permeability of cementitious materials, pp. 3-21. Pittsburgh:
Mat. Res. Soc.

Pospíšil, F. 1990. Some aspects of
unit weight reduction of aerated
concrete, (in Czech). Stavivo, 2
pp. 48-51.

Wittman, F.H. 1979. Micromechanics
of achieving high strength and
other superior properties. Proc.
Workshop on high strength concrete. Chicago: Ed.Univ.Illinois.

Wittman, F.H. 1983. Fracture mechanics of concrete. Amsterdam: Elsevier.

2 Pore structure and properties

Advances in Autoclaved Aerated Concrete, Wittmann (ed.) © 1992 Taylor & Francis. ISBN 90 5410 086 9

Pore structure and moisture characteristics of porous inorganic building materials

S.Tada

Texte, Inc. & Nihon University, Japan

Within the framwork of classical thermodynamics, this paper deals with three topics with examples of the hardened cement paste (HCP) and autoclaved aerated concrete (AAC). Interaction of moisture with pore structure, principles of measuring techniques of pore structure, and degradation of building materials associated with the microscopic migration of moisture in the porous media.

1 INTRODUCTION

One of the major role of building is to provide men and manufacturing equipments with a working environment enabling users to attain a higher efficiency. It is therefore necessary for the interior environment to have a smaller variation than that of the natural environment. As a result, an envelope or shelter which lies in between two different environments is inevitably exposed to transport phenomena of various mass and energy; the dissipasion of free energy.

A set of phenomenological equations describing these transport phenomena have been developed, and the resulting macroscopic profiles of moisture and temperature in a building envelope seem to cover the practical range. Prediction of phenomenological coefficients as well as other moisture characteristics from the known pore structure of materials is therefore of primary importance for the next stage.

Taking hardened cement paste HCP and autoclaved aerated concrete AAC as examples, the present paper aims to cover three topics within a framework of classical thermodynamics; Interaction of moisture with pore structure, Principles of measuring techniques for pore structure, and Degradation of building materials associated with microscopic migration of moisture in the porous media.

2 THERMODYNAMICS OF MOISTURE WITHIN POROUS MEDIA

Many building materials develop pore structure inside and can be regarded as a thermodynamic system opened to the outer environment. This open system tends to move towards a new equilibrium condition in accordanace with the change in pressure P, temperature T of the system and chemical potentials of mass μ which approach the system.

Among the components, moisture seems to have a decisive effect on properties and durability of the porous material since it is not only incorporated in the structure of the material but also shows phase transition within the range of working temperature.

2.1 *Chemical potential and water vapor sorption isotherm*

Relative humidity represents an energy state of moisture held in porous materials while the concept of moisture content is simply an expression of quantity of moisture, and it consequently treat bulk water in saturated state and adsorbed water in dry state in the same manner. If we consider a HCP and an AAC in equilibrium with an equal relative humidity, they are identical energetically in freezing point and in suction pressure to exhibit no moisture transport between them though the moisture content is, of course, different each other. This can be attributed to the equivalency in chemical potentials of water μ_w adsorbed in the materials.

A reversible compression work W, for a mole of water vapor at temperature T and pressure P to reach the saturated pressure P_s is given by

$$w=-\int_P^{P_s} Vdp = RT\int_{P_s}^P \frac{dP}{P} = RT\ln(P/P_s) = \mu_g, \qquad (1)$$

where R is gas constant, V is the volume of the

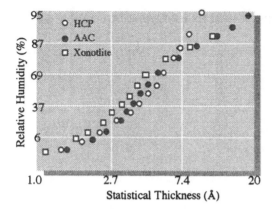

Fig.1 A modified sorption isotherm for HCP, AAC and Xonotlite

system and of course P/P_s denotes relative humidity. A chemical potential of water in liquid phase μ_w held in the material can be known from the chemical potential of the gas phase μ_g determined by relative humidity, since chemical potentials are equal without regard to their phase ($\mu_g = \mu_w$) in the thermodynamic equilibrium condition. In this case, μ_w shows the deviation from the level of the chemical potential of the isothermal pure bulk water.

Specific surface area of HCP or AAC is large enough for adsorbed water to be measured gravimetrically, and such adsorption phenomena depends upon the relative humidity of ambient air. Resulting equilibrium value gives the water vapor sorption isotherm.

We will consider a relationship between chemical potential of adsorbed water μ_w and the statistical thickness of adsorbed water film h as a thermodynamic interpretation of the sorption isotherm. A natural log of relative humidity representing a chemical potential, $\ln(P/P_s)$, and a distance from the solid surface, h, obtained by dividing an amount of adsorbate by the BET specific surface area are employed for Y and X axis respectively in Fig.1. The data for HCP is quoted from Badmann et al.(1982).

The figure illustrates qualitatively a considerable lowering of the chemical potential of adsorbed water molecule with respect to the isothermal pure bulk water as it approaches to the solid surface (h=0), and the rate of lowering $d\ln\mu/d\ln h$ is made by a characteristic manner corresponding probably to the number of adsorbed layer. The isothermal pure bulk water can be done considerable work by such adsorbed water with lower chemical potential since a water molecule with higher chemical potential is induced towards a water molecule with lower chemical potential.

2.2 Interaction of moisture with solid surface, capillary and solute

Moisture characteristics of porous inorganic building materials can be attributed to the large hydrophilic surface area, and as a result, nearly all evaporable water exist in such a manner as an adsorbed film. At a moisture content about 40 % by weight of HCP, every water molecules as well as the other mass distributed in the solution exist in a film as thick as 20 Å under the strong force field of the solid surface.

The forces are mainly composed of van der Waals interaction between water molecules and the solid surface, and an interaction between dipole of water molecules and the electrostatic field of the solid surface. As a consequence, chemical potential lowering for the water molecule depends on the distance from the solid surface.

The disjoining pressure Π, as a function of the film thickness h, has been then claimed to meet thermodynamic requirements of the uniformity of chemical potentials within an adsorbed film at equilibrium (Derjaguin and Churaev 1976). Assuming the moler volume of water v is independent to h, Π is given by

$$RT\ln(P/P_s) / v = \Pi(h). \qquad (2)$$

It should be noted that chemical potential of adsorbed water $RT\ln(P/P_s)$, implies potential work to be done by the isothermal free bulk water, and the disjoining pressure cannot be measured as a hydrostatic pressure until the adsorbed water comes in contact with a water molecule with higher chemical potential.

Although such adsorbed water is strongly bounded perpendiculary to the solid surface to exhibit loadbearing characteristics, it can move horizontally along with the solid surface. Under a long term stress, surface diffusion of adsorbed water may take place associated with the flow of the solid matrix.

Introduction of very short term load sufficiently greater than disjoining pressure can release the adsorbed water. The squeezing of pore water by triaxial compression more than 3 kbar has been conducted on the basis of this principle analogous to the ultrafiltration. Diamond and Barneyback applied 5.5 Kbar to investigate ion composition for the study of alkali silica reaction (Barneyback and Diamond

1981). Page and Vennesland traced change in chloride ion content in a silica-fume blended cement pastes by a pressure of 3.75 Kbar (Page and Vennesland 1983). Sakuta et al. studied surface tension of pore water to check the virtue of a drying shrinkage reducing chemical agent by a pressure of 6 Kbar (Sakuta et. al. 1984).

The shape of liquid-gas interface and the energy state of pore water can be associated by the Kelvin equation,

$$RT\ln(P/P_s) / v = - 2\Phi_\omega / r, \qquad (3)$$

where hydrostatic pressure, derived from the young-Laplace formula with radius of curvature of liquid-gas interface r and the surface tension Φ_ω, is set equal to the disjoining pressure. The coexistence of adsorbed film and meniscus simply means an equilibrium state of liquid-gas interface when a disjoining pressure at the surface of the film is equal to the hydrostatic pressure of the meniscus.

It follows that the meniscus does not contribute to the lowering of chemical potential of pore water but is merely an expression of liquid-gas interface under a pressure of $\Pi=-2\Phi/r$.

Powers roused attentions to very narrow gaps specific to HCP gels which prevent water vapor from free adsorption (Powers 1965). Resulting pressure disjoining the gap may be due to the "escaping tendency" of water molecule to the narrower space where chemical potential of water is lower.

The adsorbed film can be regarded as a dilute solution, where chemical potential of water can be lowered also by the presence of solute. A pressure of adsorbed water applied by the water with higher chemical potential is the osmotic pressure, and it is known that, in a bulk solution, the osmotic potential cannot be measured as a hydrostatic pressure without a semi-permeable membrane.

In some concrete accompanied by subsequently formed alkali silicate gels, considerable swelling has been occasionally observed when free water is applied from outside of the system. In these problems directly attributed to volume change, a substantial parameter is the lowering of chemical potential with respect to the isothermal pure bulk water, and the disjoining pressure comprises the osmotic pressure.

3 MEASUREMENT OF PORE STRUCTURE WITH RESPECT TO MOISTURE

Among characters representing pore structure, total porosity (volume fraction of pore), specific surface area and pore size distribution are of particular importance. An unified method to measure the pore size of building materials ranging over the order of 7th and to give a pore size distribution of whole range is expected, while at present, several methods capable of measuring a limited range must be combined.

Since moisture is always present in the microstructure of building materials at their working conditions, and affects decisively to properties of the material, it is realistic to measure pore structure by means of moisture with moisture content by volume representing cumulative pore volume and with chemical potential of water representing pore geometry.

3.1 *Water vapor sorption isotherm*

A general theory of gas adsorption without regard to particular pore geometry has been presented as the Modelless method by Brunauer and co-workers (Brunauer, Skalny and Odler 1973)(Kondo and Daimon 1974). A more general formulation with rigorous definition of thermodynamic system is, however, necessary to represent pore structural profile in conjunction with the other method such as mercury injection porosimetry.

In the measurement of water vapor sorption isotherm, amount of adsorbed water is measured in accordance with changes in relative pressure P/P_s, i.e. the ratio of saturated vapor pressure P_s at the temperature with respect to the actual vapor pressure P.

We will consider a thermodynamic system composed only of moisture in a material (denoted as l and g for liquid and gas phase respectively) under normal atmospheric pressure as illustrated schematically in Fig. 2. Under constant absolute temperature T and total pressure of the system P, changes in interfacial Gibbs free energy of adsorbed water and solid sl, water vapor and solid sg and adsorbed water and water vapor lg will take place as moisture comes into the system associated with changes in relative pressure. The total differential of Gibbs free energy of the system is given by (Dufay, Prigogine, Bellemans and Everett 1966)

$$dG=-SdT+VdP+A_{sg}d\Phi_{sg}+A_{sl}d\Phi_{sl}+A_{lg}d\Phi_{lg}+\mu_w dn, \qquad (4)$$

where S is entropy of the system, V is volume of the system, A is surface area of interface, Φ is interfacial tension and μ_w is chemical potential of water given by $RT\ln(P/P_s)$ with respect to isothermal pure bulk water. R is the gas constant and n is the number of

55

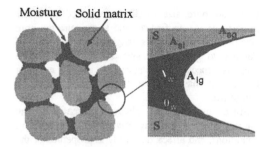

Moisture　　Solid matrix

Fig.2 Definition of the system for water vapor
adsorption

mole of water which approach the system. we have
$dn=dV/v$ with V for the volume of the newly
adsorbed water and v for the specific volume of
water.

When the system reach equilibrium condition
($dG=0$) isothermally ($dT=0$) under constant
atmospheric pressure ($dP=0$), terms representing
reversible work are only present in eq.4;

$$A_{s\,g}d\Phi_{s\,g}+A_{s\,l}d\Phi_{s\,l}+A_{lg}d\Phi_{lg}+RTln(P/P_s)\cdot dV/v = 0. \quad (5)$$

If we assume that Φ_{lg} can vary associated with
adsorption while change in $\Phi_{s\,l}$ is negligible, $d\Phi_{s\,l}$
can be set as zero. Differentiating Young-Dupre
equation $\Phi_{s\,g} = \Phi_{s\,l}+\Phi_{l\,g}\cos\theta_w$ gives $d\Phi_{s\,g}=$
$d\Phi_{lg}\cos\theta_w$, where θ_w is contact angle between
adsorbed water and solid. Then eq.5 can be written
as

$$(A_{s\,g}\cos\theta_w+A_{lg})d\Phi_{lg} + RTln(P/P_s)\cdot dV/v = 0. \quad (6)$$

Substituting $A_{s\,g}\cos\theta_w+A_{lg}$ by A, A represents
interfacial area　which undergoes actual change
associated with the progress of adsorption and
condensation of adsorbate. Integration of eq.6 with
initial condition of $V=0$ and $\Phi_{lg}=0$ at the start of
adsorption, and $\Phi_{lg}=\Phi_w$ at $V=V_w$, gives a generalized
Kelvin equation,

$$RTln(P/P_s) / v = -\Phi_w/R_H, \quad (7)$$

where Φ_w is surface tension of water and R_H is
hydraulic radius defined as V_w/A; volume of
adsorbed water divided by effective surface area.

In this derivation, any configuration or phase
geometry　between adsorbete and capillary
condensate is not assumed. At the initial stage of
adsorption below monolayer, it is proper for Φ_{lg} to

be zero. When progress of adsorption extends over
the hindered adsorbed region where free adsorption
of water molecule is difficult and over the capillary
condensation region, $A_{s\,g}$ becomes zero and dA_{lg} is
negative, and subsequently, A_{lg} will approach the
apparent surface area of the specimen. The area A
($=A_{s\,g}\cos\theta_w+A_{lg}$) has a finite value though it is
negligibly small compared to the specific surface area
inside of the specimen, and R_H will not become
infinite when P/P_s approaches to 1.

3.2 *Mercury intrusion porosimetry*

Derivation of a generalized equation for mercury
injection techniques, as Kiselev equation for gas
adsorption quated by Brunauer, has been made by
Rootare and Prenzlow (1967). Since the aim of the
equation was to calculate specific surface area,
generalized formulation for mercury injection
techniques without any pore geometry could not be
derived.

We will consider a thermodynamic system
composed only of mercury inside and outside
(superscript i and o) of a specimen as illustrated
schematically in Fig.3.

In the procedure of mercury injection to specimen,
intruded volume of mercury dV can be measured
with respect to the pressure P, by which area work
between mercury and specimen (solid phase denoted
as s) will be done. The total differential of Helmholz
free energy of the system is given by (Tada, Tanaka
and Matusnaga 1985)

$$dF = -SdT - P^o_l dV^o_l - P^i_l dV^i_l - P^i_g dV^i_g$$
$$+\Phi_{lg}dA_{lg}+\Phi_{s\,g}dA_{s\,g}+\Phi_{s\,l}dA_{s\,l}+\mu dn. \quad (8)$$

With an assumption of constant volume of mercury
in liquid phase $dV^i_l+dV^o_l=0$ can be given. Since the
pore volume and the internal surface area of the
specimen are constant, $dV^i_g+dV^i_l=0$ and $dA_{s\,g} + dA_{s\,l}$
$= 0$ can be held respectively.

With contact angle between mercury and the solid
surface θ_m , Young-Dupre equation gives
$\Phi_{sg}-\Phi_{sl}=\Phi_{lg}\cos\theta_m$. Since no mass comes into the
system, dn is equal to zero and when the system
approaches isothermally ($dT=0$) to an equilibrium
condition ($dF=0$), we have

$$(P^o_l-P^i_l+P^i_g) dV^i_l + (dA_{lg}-dA_{s\,g}\cos\theta_m)\Phi_{lg} =0, \quad (9)$$

where we can set $P^o_l=P^i_l=P^i_g=P$ and $dV^i_l=dV$, and A_{lg}
can vary from a constant value "a" ,which is nearly
equal to an apparent surface area of the specimen, to

A_{lg} associated with change in V from 0 to V_m. Integration of eq.9 with these initial condition gives

$$PV_m = -\Phi_{lg}(A_{lg} - A_{sg}\cos\theta_m) + a, \qquad (10)$$

where a is negligibly small. Substituting $R_H = V_m/A'$ with $A' = A_{lg} - A_{sg}\cos\theta_m$ in eq.10, we have

$$P = -\Phi_m / R_H, \qquad (11)$$

where Φ_m is interfacial tension between mercury and vapor of mercury.

Above formulation is generalized Rootare and Prenzlow equation and can be a modelless theory of mercury porosimetry.

During the water vapor adsorption, eq.7 is valid and R_H and V_m can be obtained by measurements of relative humidity and weight of specimen. Naturally, the finer the pore is, the sooner it is filled with adsorbed water. During the mercury injection, on the other hand, eq.11 is held and amount of mercury injected up to the pressure P can be measured, when the coarser the pore is the sooner it is filled with mercury. To combine these two techniques for the wide range pore-size analysis, we assume relationships between V_w and V_m, θ_w and θ_m at any R_H as follows;

$$V_w + V_m = V_p, \quad \theta w + \theta m = \pi, \qquad (12)$$

where V_p is the total porosity of the specimen which can be easily known by vacuum saturation method.

The left side term in eq.7 is negative disjoining pressure which is denoted as Π, and at the same R_H, the hydrostatic pressure of mercury P in eq.11 can be translated to the disjoining pressure of adsorbed water as follows,

$$\Pi = P \cdot \Phi_w / \Phi_m. \qquad (13)$$

The amount of adsorbate corresponding to V_m at the disjoining pressure Π can be calculated from $V_p - V_m$. Consequently, the wide range water vapor sorption isotherm can be obtained, and hence the chemical potential of water at a given moisture content is derived.

3.3 Dew point depression

Dew point measurement based on thermodynamical principle is suitable for determining the water activity ranging from 0.960 to 0.999. Thermocouple dew point psychrometer employed for this measurement is an instrument prevailing in agricultural science

Mercury   Solid matrix

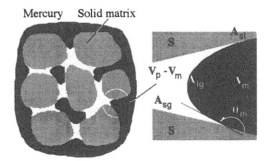

Fig.3 Definition of system for mercury injection

(Richards 1965). The measuring principle is as follows: the dew point drop ΔT is measured in such a way that one junction of a thermocouple in the probe is cooled down to the dew point temperature by Peltier effect, and the temperature difference with the other junction of the thermocouple is obtained by Seebeck effect.

For a mole of water vapor at temperature T, dew point temperature T_d and with high relative pressure P/P_s, both $\Delta P = P_s - P$ and $\Delta T = T - T_d$ are sufficiently small, then clausius-Clapeyron equation gives

$$\Delta P / \Delta T = \Delta H / TV_g = P_s \Delta H / RT^2, \qquad (14)$$

where V_g is the volume of water vapor which is larger enough than that of water, ΔH is the enthalpy change associated with the phase transition of water and the relation of ideal gas $P_s V_g = RT$ is employed.

The thermoelectromotive force E of the thermocouple wired between cooled junction and normal junction with a coefficient ε can be written as

$$E = \varepsilon \Delta T = \varepsilon [RT^2 / \Delta H] [(P_s - P) / Ps]$$
$$= \varepsilon [RT^2 / \Delta H](1 - P/P_s). \qquad (15)$$

Substituting P/P_s by $\exp(-\Delta\mu/RT)$ in eq.15 gives

$$E = \varepsilon [RT^2 / \Delta H][1 - \exp(-\Delta\mu/RT)]$$
$$= \varepsilon [RT^2 / \Delta H][\Delta\mu/RT - (\Delta\mu/RT)^2 / 2!] = \varepsilon T\Delta\mu / \Delta H. \qquad (16)$$

If the vapor pressure does not vary, the enthalpy change ΔH associated with the phase transition of water does not depend on temperature and ΔT is small, chemical potential of water vapor, as well as that of water in materials, is thus approximately obtained from the dew point depression. The measurable range is $-5 \times 10^3 \sim -10^2$ (J /Kg)

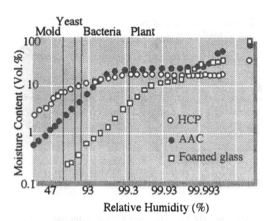

Fig.4 Moisture characteristic curve for HCP, AAC and foamed glass with respect to the viability of microorganisms

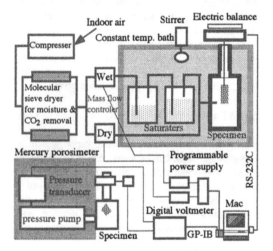

Fig.5 Schematic diagram of experimental system for the wide range of sorption isotherm.

depending upon both cooling capacity and accuracy in the measurement of microvoltage.

3.4 *The moisture characteristic curve*

A relationship between moisture content of porous materials ϕ and chemical potential of water μ_w with respect to isothermal pure bulk water has been studied mainly in soil physics in terms of moisture characteristic curve, which has wide range of application field over plant physiology, food science as well as building science (Tada and Nakano 1983).

The gradient $d\mu/d\phi$ can be directly applied for the transformation of phenomenological transport equations based on moisture content as a driving force.

Prediction of unsaturated hydraulic conductivity from the moisture characteristics curve has been developed in soil science, and in food science, knowledge of minimum water activity required for the viability of microorganisms and the corresponding moisture content is indispensable.

Examples of moisture characteristics curve for latex-modified cement paste, AAC and foamed glass in conjunction with prediction of viability of microorganisms on building wall is shown in Fig.4 (Tada 1983). Measurement of the moisture characteristic curve is based on water vapor adsorption combined with mercury injection as described in the previous chapter. The schematic diagram of the experimental apparatus is shown in Fig.5 (Tada 1990).

4 DEGRADATION OF POROUS MATERISALS BY MICROSCOPIC MIGRATION OF WATER

4.1 *Drying shrinkage*

It has been considered that pore water forms meniscus at liquid-gas interface. The pressure difference between inside and outside of the meniscus with radius of curveture r is given by Laplace equation

$$\Delta P = P_1 - P_2 = -2\Phi / r . \qquad (17)$$

The capillary tension theory of drying shrinkage of porous building materials claims that the water in the pore exists in tension and this create an attractive force between the wall of the pore. As already stated, the meniscus and adsorbed water film can coexist, and the disjoining pressure of the film is equal to the capillary tension at liquid-gas interface. Though the meniscus can no longer be stable at relative humidity below 40 %, substantial shrinkage can be observed. Experimental results, however, have often shown good agreement with the capillary tension theory at high relative humidity in combination with the disjoining pressure theory proposed primarily by Powers (1960),

$$E (\Delta L/L) = RT \ln(P/P_s) / v \quad (= -2\Phi/ r) , \qquad (18)$$

where E is a constant including elastic modulus and creep coefficient, and $\Delta L/L$ is the length change rate. The concept of disjoining pressure introduced by Powers is somewhat different from the pressure in the adsorbed film, and was prepared to explain the shrinkage at relative humidity below 40 %. He

58

considered very narrow gaps, specific to HCP, where the free adsorption is no longer possible, and the water outside of the gap tends to move towards the narrower gap in order to equalize the film thickness. As a result, the pressure disjoin the gap and becomes smaller at low relative humidity as shown in Fig.6.

On the basis of the capillary tension theory, attempts have been made to reduce the shrinkage by lowering the surface tension of pore water (Sato, Goto and Sakai 1983). However, the disjoining pressure $RT\ln(P/P_s)$ may be unchanged in a constant relative humidity, so that in eq.18, smaller Φ results merely smaller r, and a value $-2\Phi/r$ is unchanged. An apparent difference in the application of such shrinkage reducing agent is the amount of released water since the meniscus can be in equilibrium at a relative humidity with smaller r.

The concept of disjoining pressure employed by Wittmann and Setzer, as well as this study, is originated from the work of Derjaguin. It is a pressure in adsorbed water film and can explain the shrinkage at above two adsorbed molecular layers. For the drying shrinkage at adsorbed water film less than monolayer, change in surface free energy has been considered to be a main cause. The change in surface free energy of solid particle is given by Gibbs equation (Wittmann 1976),

$$\Delta F = RT \int \Gamma \, d\ln(P/P_s) , \qquad (19)$$

where Γ is the thickness of adsorbed water. The volume change of a solid particle in accordance with the change in surface free energy is given by Bangham equation,

$$\Delta L/L = \lambda \Delta F , \qquad (20)$$

where λ is a constant. In this theory, hardened cement paste particles are considered to be isolated from water and has no water molecule inside. However this drying hydrogel can actually contain water molecules in their structure, and can be applied the osmotic pressure by the adsorbed water since the water molecules inside the particle have lower chemical potential than that of the adsorbed water.

In the field of polymer physics, theory of volume change for ionic polymer hydrogel has made a remarkable progress. Following the Flory-Huggins equation, Tanaka calculated forces acting as osmotic pressure and derived an equation of state for the hydrogels (Tanaka et.al. 1980),

$$\Pi = - (NKT / v)[\phi + \ln(1-\phi) + \Delta F\phi^2 /2KT]$$
$$+ \nu KT[\phi / 2\phi_0 - (\phi / \phi_0)^{1/3}] + \nu fKT(\phi / \phi_0) , \qquad (21)$$

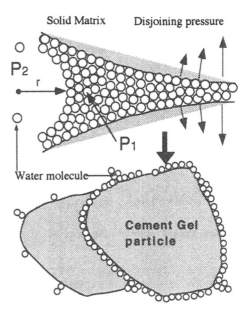

Solid Matrix   Disjoining pressure

P_2
r
P_1

Water molecule

Cement Gel particle

Fig.6 Schematic diagram for liquid-gas interface

where N is Avogadro's number, K is Boltzmann's constant, T is the temperature, v is the moler volume of the solvent, ϕ is the volume fraction of the network, ΔF is the free-energy decrease associated with the formation of a contact between polymer segments, ϕ_0 is the volume fraction of the network at the condition when the constituent polymer chains have random-walk configurations, ν is the number of constituent chains per unit volume at $\phi=\phi_0$, and f is the number of dissociated hydrogen ions per effective chain.

This equation showed excellent result in explaining second order phase transition of polymer gels associated with the change in chemical potential of solvent.

4.2 Phase transition of adsorbed water

The lowering of chemical potential of adsorbed water can be attributed to van der Waals interaction dipole-electrostatic field interaction and solute concentration. How is it ice-like even at room temperature as the study of clay-water system has often reported?

The differential scanning calorimetric analysis at temperature down to liquid Nitrogen have been made for HCP and AAC equlibrated with various relative humidity; i.e. with various adsorbed film thickness.

As well as HCP, AAC shown in Fig.7 exhibited no phase transition (at least first order) up to second

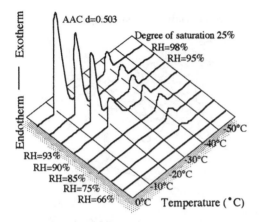

Fig.7 Phase transition of adsorbed water in AAC

molecular layer available by curing under the relative humidity of 66 %, however, third to sixth molecular layer, corresponding to relative humidity of 90%, actually froze at approximately -40 °C (Tada 1991). The measured ΔH for AAC was 7.96 (cal/g) while that of HCP (Stockhausen, Dorner, Zech and Setzer 1979)(Sellevold and Barger 1980) and diatomaceous brick (Nakamura et.al. 1985) were 7.17 and 8.18 respectively.

It has been studied that the contribution of van der Waals interaction to the lowering of chemical potential of adsorbed water of first and second molecular layer is ten times larger than that of third and fourth layer (Fujii and Nakano 1984), and the steric effect of hydrogen bond with the surface silanol group is substantial in the first two layer (Conway 1977). Thus the practically non freezable water and unfrozen water are actually present in between solid matrix of the material and ice (frozen bulk water) at a working temperature.

We will take a state of water in equilibrium with ice and vapor under a temperature of 0 °C and a pressure of 1 atm as a reference. When a small deviation of temperature take place from the reference state, and then unfrozen adsorbed water denoted "a" and ice "i" become again in equilibrium, change in chemical potential of unfrozen adsorbed water is given by

$$d\mu_a = -s_a dT + v_a dP_a,$$ (22)

where s_a and v_a are moler entropy and moler volume of the adsorbed water respectively.

The same manner may be valid for the derivation of the change in chemical potential of ice. If the moisture content is low enough for ice to be opened to atmospheric pressure, pressure of ice may be kept 1 atm ($dP_i=0$). Within a small temperature deviation

dT, an equilibrium condition $d\mu_a=d\mu_i$ gives

$$-s_i dT = -s_a dT + v_a dP_a.$$ (23)

If the moler heat of fusion of adsorbed water q has little difference from that of bulk. substitution of $s_a - s_i$ by q/T and integration between T_o and T of eq. 23 give a pressure of adsorbed water against ice,

$$\Delta P = q \Delta T / v_a T_o.$$ (24)

This pressure difference under unsaturated moisture condition can be a cause to drive the pore ice to a vacant space provided by entrained air bubbles, and at the same time, change in disjoining pressure by the presence of ice with lower chemical potential results little contraction of the solid matrix.

However at higher moisture content, even entrained air bubbles (air pores) may be partially occupied by adsorbed water, and during freezing process, ice formation takes place primarily in air pores. it follows that air pores are not in contact with atmospheric pressure, and the ice pressure P_i as well as P_a can no longer maintain a pressure of 1 atm. In this condition $P_i=P_a$, the Clausius-Clapeyron equation for adsorbed water and ice is given by

$$\Delta P = q \Delta T / (v_a-v_i) T_o.$$ (25)

This pressure difference acts as a disjoining pressure between ice and solid.

4.3 *Moisture migration during freezing*

Since moisture content plays decisive role in frost resistance of porous building materials, a

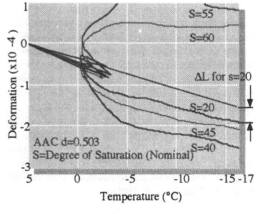

Fig.8 Change in length of AAC with different moisture content during freezing

measurement of change in length during freezing of AAC specimen with various moisture content is performed. Specimens with high degree of saturation (moisture content based on pore volume) showed considerable expansion while specimens with lower degree of saturation shrunk even more than expected thermal contraction as shown in Fig. 8.

The critical degree of saturation for frost resistance S_{cr} of AAC, defined as the minimum moisture content above which internal damage due to frost can be detectable, is determined as 44 % by pore volume according to the test method of RILEM (Fagerlund 1978). In this moisture condition, water intrusion extends over almost all micropores which exist in the solid matrix of AAC and over smaller air pores. It is very important to have an image of what pore region is the pore water extending at certain moisture content as shown in Fig.9.

The cryo-SEM photograph shows a microscopic ice segregation on the surface of air pores of AAC with moisture content 47.5 % by pore volume. Since no bulk water was present in air pores before freezing, the observed ice on the surface of air pores must be considered to come from the micropores of matrix part by the unfrozen film flow. As far as the free ice formation can take place on air pores, pressure of the system does not increase. However at moisture content above S_{cr} , smaller air pore with diameter less than 75 μm shown at the bottom of Fig.10 is filled with ice and pressure of this region may increse.

The freezing behavior of porous materials with moisture content around S_{cr} is schematically shown in Fig.11.

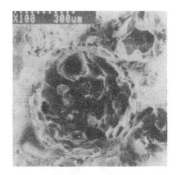

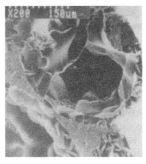

Fig.10 Microscopic ice segregation on air pores of AAC

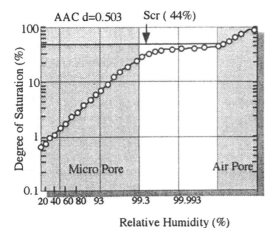

Fig.9 Pore structure and the critical degree of saturation for frost resistance of AAC

5 CONCLUSIONS

With respect to the pure bulk water, the chemical potential of water in porous materials is lowered mainly due to the surface forces of the solid. The degree of binding is thus a function of the distance from the solid surface. However in equilibrium, chemical potential of water must be equal everywhere without regard to the distance from the solid surface. The disjoining pressure, which becomes highest at solid-liquid interface and zero at liquid-gas interface, is then claimed to compensate the thermodynamical requirement of the uniformity of chemical potentials. The disjoining pressure is an osmotic pressure as it is applied to a water molecule with lower chemical potential by the other water molecule with higher chemical potential, i.e. closer to the chemical potential of isothermal pure bulk water.

It is therefore important to measure the relationship between moisture content and the chemical potential of water held in the porous material, the moisture characteristic curve, which represents both energy state of water and cumulative pore volume. With thermodynamic formulation, mercury injection technique can be interpreted in terms of water vapor adsorption, and resulting wide range isothermal sorption gives the moisture characteristic curve.

During freezing, change in the thickness of

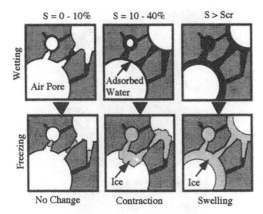

S = 0 - 10%	S = 10 - 40%	S > Scr

Wetting

Air Pore / Adsorbed Water

Freezing

Ice / Ice

No Change — Contraction — Swelling

Fig. 11 Microscopic migration of adsorbed water during freezing

adsorbed film causes microscopic water migration since the water in 3 rd to 6th molecular layer is kept unfrozen at temperature down to -35 ℃. As far as the film flow is maintained, ice formation in large pores can take place without any restriction, and the disjoining pressure directed to solid surface may decrease to result contraction of the material. If moisture content is high enough to prevent unfrozen water from migrating for ice formation, pressure of the system may increase.

Since the microscopic moisture migration in porous media is closely related to problems of durability such as drying shrinkage, alkali silica reaction and frost damage, it should be future tasks to derive generally the cause for lowering the chemical potential of water in porous system, and to establish "equation of state" for the system of porous building materials and water.

REFERENCES

Badmann, R. Stockhausen, N. & Setzer, M. J. 1982. The statistical thickness and the chemical potential of adsorbed water films. *J. Colloid Interface Sci.*82: 534-542.

Barneyback, R. S. Jr., Diamond, S. 1981. Expression and analysis of pore fluids from hardened cement pastes and mortars. *Cem. Concr. Res.*11:279-285.

Brunauer, S., Skalny, J., Odler, I. 1973. Complete pore structure analysis. *Proc. Int. Symp. Pore struct. Prop. Mater.* Modry, S., Svata, N., eds: C3-C26. Prague.Academia.

Conway, B. E. 1977. The state of water and hydrated ions at interfaces. *Adv. Colloid Interface Sci.* 8:91-212.

Derjaguin, B. V., Churaev, N. V. 1976. Polymolecular adsorption and capillary condensation in narrow slit pore. *ibid.* 54:157-175.

Dufay, R., Prigogine, I., Bellemans, A., Everett, D. H. 1966. *Surface tension and adsorption.* Wiley.

Fagerlund, G. 1978. The critical degree of saturation method assessing the freez/thaw resistance of concrete. *Mater. Constr.*10:217-229.

Fujii, K., Nakano, M. 1984. Chemical potential of water adsorbed on Bentonite. *Trans. JSIDRE.* 112: 43-53.(in Japanese)

Kondo, R., Daimon, M. 1974. Analysis of pore size distribution by ML and MP method. *Surface* 12: 377-386.

Nakamura, M., et.al. 1985. Phase transition of water in brick during cooling. *Am. Ceram. Soc. Bull.* 64:1567-1570.

Page, C. L., Vennesland, O.1983. Pore solution composition and chloride binding capacity of silica-fume cement pastes. *Mater. Constr.*16: 19-25.

Powers, T. C. 1960. Physical properties of cement paste. *Proc. 4th Int. Symp. Chem. Cem.* 2: 577-609.

Powers, T. C. 1965. Mechanism of shrinkage and reversible creep of hardened cement paste. *Proc. Int. Symp. Struct. Concr:* 319-344. London.

Richards, L. A. 1965. A thermocouple psychrometer for measuring the relative vapor pressure of water in liquid or porous materials, in Wexler, A. ed., *Humidity and Moisture*, vol.4:13-18 NY: Reinhold publ. Co.

Rootare, H. M., Prenzlow, C. F.1967. Surface area from mercury porosimeter measurements. *J. Phys. Chem.*71: 2733-2736.

Sakuta, S. et. al. 1984. A study in the application of drying shrinkage reducing chemicals for concretes. *Prep. Ann. Meeting AIJ.* 489-490. (in Japanese)

Sato, T., Goto, K., Sakai, K. 1983 . Mechanism of drying shrinkage reducing chemicals for hardened cement pastes. *Ann. Rep. Cem. Technol.* 37: 65-68 (in Japanese)

Sellevold, E-J., Barger, D. H. 1980. Low temperature calorimetry as a pore structure probe, *Proc. 7th Int, Congr. Chem. Cem.* 4: 394-399.

Stockhausen, N., Dorner, H., Zech, B., Setzer, M. J. 1979. Untersuchung von Gefriervorgangen in Zementstein mit Hilfe der DTA. *Cem. Concr. Res.* 9: 783-794.

Tada, S., Nakano, S.1983. Microstructural approach to properties of moist cellular concrete, in Wittmann, F. H., ed., *Autoclaved Aerated Concrete - Moisture and Properties*: 71-89. Amsterdam.Elsevier.

Tada, S. 1983. Moisture and microstructure characterization of porous building materials, in Wittmann, F. H., ed., *Proc. Int. Symp. Mater. Sci. Restor.* 41-45. Edition Lack+Chemie.

Tada, S., Tanaka, M., Matusnaga, Y. 1985. Measurement of pore structure of aerated concrete. *Proc. Beijing Int. Symp. Concr.* vol.3. 384-393.

Tada, S. 1990. Microstructure and moisture characteristics of autoclaved lightweight concrete. *Concr. Res. Technol.* 1: 155-164 (in Japanese)

Tada, S. 1991. Pore structure and freezing behavior of autoclaved lightweight concrete. *Concr. Res. Technol.* 2: (1)95-103.

Tanaka, T., et.al. 1980. Phase transition in ionic gels. *Phys. Rev. Lett.* 45: 1636-1639.

Wittmann, F. H. 1976. The structure of hardened cement paste - A basis for a better understanding of the materials properties. *Proc. Hydraulic cement pastes: their structure and properties.* Univ. Sheffield: 96-117.

Advances in Autoclaved Aerated Concrete, Wittmann (ed.)© 1992 Taylor & Francis. ISBN 90 5410 086 9

Application of image analysis to the estimation of AAC thermal conductivity

J. P. Laurent & C. Frendo-Rosso
Laboratoire d'Étude des Transferts en Hydrologie et Environnement (IMG), Grenoble, France

ABSTRACT : A new approach to estimate the thermal conductivity on the basis of an image of polished section of AAC samples impregnated by a black dye epoxy resin is described. First, a topological network of all the paths in the solid matrix where the heat flux can go trough is constructed by image-analysis. Then, to compute the thermal conductivity on those networks by electrical analogy, a powerful tool from the field of electrical networks control - the "Ybus matrix" - is used. On a data base of more than 30 images of AAC samples (density ranging from 0.25 to 0.6, rate of macroporosity between 0.3 and 0.8), comparisons have been made between predicted and measured values of thermal conductivity showing a very good agreement.

1 INTRODUCTION

It is well known that the porous phase of AAC can be divided in two main classes (Prim & Wittmann, 1983, Tada & Nakano, 1983) : a microporous structure in the solid matrix and a macroporous cellular structure (the air bubbles) artificially created. From the industrial point of view, to improve the thermal performances of AAC, it is easier to modify this last class, rather than this other one which depend directly on the formulation, and a simple model can be used as guide for that purpose : see Figure 1.

Therefore, it is of great practical importance to develop tools to characterise the cellular porous phase or "macroporosity". Because the size of the air-bubbles in AAC allows direct observation by optical microscopy, image analysis is a particularly well-adapted technic.

The main aim of our contribution here is to present some image analysis methods we have especially developed for the study of the relationship between structure and thermal conductivity of AAC.

2 SAMPLES PREPARATION FOR IMAGE ANALYSIS

To obtain a good contrast between the solid microporous matrix and the macropores, the

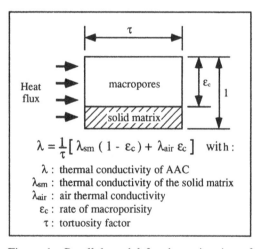

$$\lambda = \frac{1}{\tau} \left[\lambda_{sm} (1 - \varepsilon_c) + \lambda_{air} \varepsilon_c \right] \quad \text{with :}$$

λ : thermal conductivity of AAC
λ_{sm} : thermal conductivity of the solid matrix
λ_{air} : air thermal conductivity
ε_c : rate of macroporisity
τ : tortuosity factor

Figure 1 : Parallel model for the estimation of AAC thermal conductivity (Laurent, 1991).

samples have to be impregnated by an epoxy resin (Epofix™ from Struers for example) in which is incorporated a black dye. To facilitate the penetration of the resin in the material, it is useful to heat it gently to reduce its viscosity and impregnate the samples under vacuum conditions to purge them from the air they contain. During the polymerisation time, the quality of the impregnation is improved if the samples are placed in a pressure cell to force the resin to fill the smallest pores.

After the polymerisation, the samples are polished using fine grain sand-paper disks mounted on a rotating polishing machine working under wet conditions.

3 IMAGE ACQUISITION

Images are taken by means of a black and white video CCD camera mounted on an optical microscope. Then, they are transferred to a computer after digitalisation by a frame-grabber card on a matrix of 512 x 512 image points or "pixels". To cover a sufficient area, weak magnifications are used : typically, we work on 8 x 8 mm or 17 x 17 mm images.

To distinguish between pores and solid, a "threshold" has to be apply on the initial 256 grey levels images. To avoid the subjectivity of the observer, an automatic statistical method is used. At this step, "binary" black and white images like the one of the Figure 2 are ready for further analysis.

4 CHARACTERISATION OF THE MACRO-POROUS STRUCTURE

On a binary image, the rate of macroporosity can be first calculated simply just by counting the pixels of the porous phase and dividing the result by the size of the image.

Then, to determine the pore sizes distribution, a method of successive apertures by discs of increasing radius can be used (Coster & Chermant, 1989). Performing this morphological operation is analogous to sieving with a sieve of increasing mesh : at each step, i. e. for a given radius of the disc, only the pores fraction that can contain entirely that disc is left in the image and the corresponding surface can be measured. The main advantage of this method is the possibility of making a porosimetry on complex shape connected pores such as those that can be observed on a 2D image of AAC. Usually, tedious calculations are necessary to performs all these operations. To avoid that problem, while keeping the physical realism of this classical method, we have introduced sequential algorithms (instead of iterative) based on the use of quasi-Euclidian distances : the "chamfers" (Moschetto, 1991). The Figure 3 give an example of pore-size distribution obtained by this method.

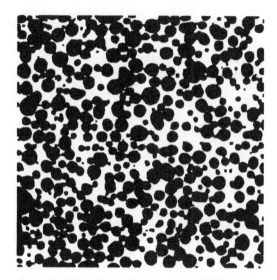

Figure 2 : Example of binary image (17 x 17 mm, pores in black) of a polished section of AAC impregnated with a black dye epoxy resin, rate of macroporoity $\varepsilon_c \approx 59\ \%$.

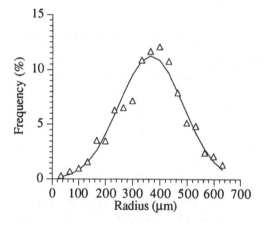

Figure 3 : Example of determination by image analysis of the pore sizes distribution (image of figure 2 above, points : measured data, solid line : corresponding Gaussian distribution).

The example of Figure 3 illustrates a fact that we have noticed on all samples of AAC we have analysed : the pore-sizes distribution in surface is very close to a Gaussian distribution. This can be used for generating numerically artificial structures of AAC to study the theoritical effect of some variations of the characteristics of the macroporous structure.

5 TOPOLOGICAL NETWORKS BUILDING UP METHOD

To estimate a transfer property like thermal conductivity on the basis of an image such as the one of Figure 2, it's theoretically possible to use this image itself as a grid for direct calculations, by a finite differences method for example, but, since artificial disconnections are present on a 2D image, the results obtained in that way are far from the measured ones.

Therefore we choose another approach which consists in the construction by image-analysis of a topological network of all the paths in the solid matrix where the heat flux can flow through including those across "necks" between connected pores. This construction is made in two steps :

- First, a skeletton of the solid matrix is obtained by computing the "median line graph" (Montanvert, 1987) of this phase : see example on Figure 4.

- Then, the missing paths - paths going beneath or above the plane of the image - are re-constructed applying criteria of minimum distance between unconnected branches (Thiel & Montanvert, 1991). This treatment is quite efficient, as can be seen on Figure 5, and it allows the construction of some kind of a "2D1/2" image.

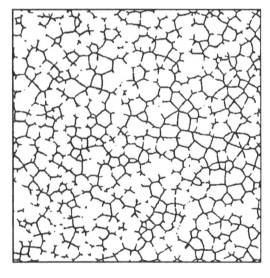

Figure 4 : Skeleton of the solid phase of the image of Figure 2.

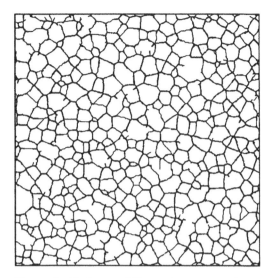

Figure 5 : Skeleton of Figure 4 reconnected.

6 CALCULATION OF THERMAL CONDUCTIVITY ON THE NETWORKS

Before any calculation on the skeleton could be undertaken, it must be described as a graph : i. e. the "nodes", intersections between branches, must be labelled and their coordinates measured and all "links" between these nodes listed. Afterwards, an electrical analogy can be applied using the formal similitude which exists between the Ohm's law and the Fourier's law :

$$U = R I \Leftrightarrow \Delta T = \frac{L}{S \lambda_{sm}} Q \qquad (1)$$

Where U is a potential difference (V), R an electrical resistance (Ω), I the intensity (A), ΔT a temperature difference (°K), L a length (m), S a surface (m²), λ_{sm} the thermal conductivity of the solid matrix (W/m°K) and Q the heat flux (W). Therefore, to estimate the apparent thermal conductivity, the thermal resistance R_{ij}, or, which is equivalent, the admittance Y_{ij} inverse of R_{ij}, of each link between nodes i et j of coordinates (x_i, y_i) and (x_j, y_j) must be calculated by :

$$Y_{ij} = \frac{\lambda_{sm} e}{\sqrt{(x_i - x_j)^2 + (y_i - y_j)^2}} \qquad (2)$$

Where e is a measure of the thickness of the

link. If the solid phase is assumed to be equally distributed on all links, this parameter can be estimated by :

$$e = \frac{N \ (1 - \varepsilon_c)}{n_l} \qquad (3)$$

Where N is the size of the image (256 or 512 pixels in our case), ε_c the rate of macroporosity and n_l the average number of links on a line.

To handle efficiently the admittances which are quite numerous as can be seen on the image of Figure 5, we have adopted the "Ybus matrix" representation that we build up using the iterative algorithm proposed by Jegatheesan (1987). This algorithm consists in two basic operations : see Figure 6 below.

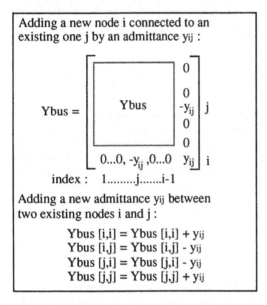

Figure 6 : The two basic operations for building up the Ybus matrix.

When these two steps (building up the graph and its representation by the Ybus matrix) are performed which in fact takes only a few minutes on a "standard" workstation, the apparent thermal conductivity λ is given by :

$$\lambda = Y_{bus}^{-1} [N,N] \qquad (4)$$

Actually, the inversion of the Ybus matrix is the most CPU time consuming operation in our approach. Nevertheless, this can be partially overcome by choosing optimised numerical technics like the "LU decomposition method" or the "Cuthill-McKee renumbering algorithm" (Frendo-Rosso, 1991).

7 VALIDATION OF OUR APPROACH

To validate our approach, we have tested it on a data base of more than thirty binary images taken on samples of AAC of densities ranging from about 0.25 to approximately 0.6 (rates of macroporosity between 0.3 and 0.8) drilled in blocks where the thermal conductivity has already been measured in our laboratory using the "line source method" (Laurent, 1989). For the calculations, we have used a value of 0.35 W/m°K for the thermal conductivity λ_{sm} of the solid matrix which has been estimated on the basis of direct measurements made on samples of autoclaved concrete realised without incorporating aluminium powder. The Figure 7 below shows a comparison between our data and the values estimated by direct calculation on topological networks. Obviously, the two sets of points on that figure are very close from each other. Therefore, we consider this justify and validate our approach.

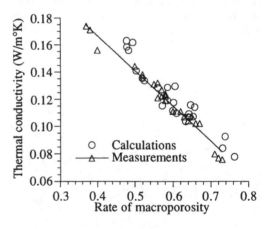

Figure 7 : Comparison between data and calculated values on topological networks.

We previously mentioned a parallel model for the estimation of thermal conductivity of AAC : Figure 1. This model considers two structural parameters : the rate of macroporosity ε_c which can be measured by image analysis (§4) and a "tortuosity factor" τ which was not measurable. With the measured values of the thermal conductivity λ, it was only possible to estimate this factor a posteriori by :

$$\tau = \frac{\lambda_{sm}\,(1\text{-}\varepsilon_c) + \lambda_{air}\,\varepsilon_c}{\lambda} \qquad (5)$$

Now, our method can be used to predict a value of τ by applying equation (5) with the value of λ calculated on the tological network. Figure 8 below shows a comparison on the same database as Figure 7 between these two ways of estimation of the tortuosity factor. Again, experimental and calculated values are in good agreement.

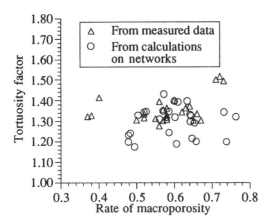

Figure 8 : Comparison between the values of the tortuosity factor that can be deduced from the measurements an those estimated on the networks.

8 CONCLUSION

We have presented a method to estimate the thermal conductivity of AAC on the basis of a topological network obtained by image-analysis technics.

This method has been tested on a database of images of existing AAC products in a wide range of density. The good agreement noted between the measured and the calculated values of thermal conductivity validates our approach.

Therefore, it is now possible to use it as a prediction tool to study the influence of some variations of the structural parameters. It should also be noticed that others transfer properties,

like permeability of electrical resistivity for example, might be handled this way.

REFERENCES

Coster, M., J. L. Chermant 1989. *Précis d'analyse d'images*. Paris: Presses du CNRS.

Frendo-Rosso, C. 1991. *Utilisation de réseaux irréguliers pour la modélisation de la relation structure/conductivité thermique du béton cellulaire autoclavé*. Grenoble: rapport de stage.

Jegatheesan, R. 1987. A new development of bus impedance building algorithm closely related to bus admittance matrix. *Electric machines and power systems* 12: 43-56.

Laurent, J. P. 1989. Evaluation des paramètres thermiques d'un milieu poreux : optimisation d'outils de mesure "in situ". *Int. J. Heat & Mass Transfer* 32: 1247-1259.

Laurent, J. P. 1991. La conductivité thermique à sec des bétons cellulaires autoclavés : un modèle conceptuel. *Matériaux et Constructions* 24: 221-226.

Montanvert, A. 1987. *Contribution au traitement de formes discrètes. Squelettes et codage par graphe de la ligne médiane*. Grenoble: Thèse.

Moschetto, C. 1991. *Caractérisation de la structure poreuse des matériaux par analyse d'images : apport des distances discrètes à la morphologie mathématique*. Grenoble: rapport de stage.

Prim, P., F. H. Wittmann 1983. Structure and water absorption of aerated concrete. *Autoclaved Aerated Concrete, Moisture and Properties*: 55-69. Amsterdam: Elsevier.

Tada, S., S. Nakano 1983. Microstructural approach to properties of moist cellular concrete. *Autoclaved Aerated Concrete, Moisture and Properties*: 71-88. Amsterdam: Elsevier.

Thiel, E., A. Montanvert 1991. Shape splitting from medial lines using the 3-4 chamfer distance. *First Int. Workshop on Visual Form*. Capri: Italy.

Advances in Autoclaved Aerated Concrete, Wittmann (ed.)© 1992 Taylor & Francis. ISBN 90 5410 086 9

Porosity and permeability of autoclaved aerated concrete

F.Jacobs & G.Mayer

Institute for Building Materials, ETH Zürich, Switzerland

Abstract: Porosity and pore size distribution were determined on different types of AAC . Due to the large range of pore sizes different techniques were used. Additionally permeability was measured by water and gas to characterize the transport properties of AAC under a pressure gradient. As can be anticipated the moisture content has a significant influence on the gas permeability. At fully saturated specimen a threshold pressure was observed. Although the porositiy differs for different types of AAC the permeability does not vary to the same extent. A simple model is presented to predict the permeability of different kinds of AAC.

1 INTRODUCTION

Four different types of autoclaved aerated concrete (AAC) were investigated: GN= normal quality, GH, GH*= high quality, GS= superior quality. GH and GH* were delivered as the same quality, but they are coming from different batches. The AAC was manufactured by using fine quartz sand, hydraulic lime, cement, water and Al-powder as an air entraining agent. Density, permeability and porosity were determined on these samples.

2 EXPERIMENTS

2.1 Sample preparation

Cores with 20 mm diameter and 20 mm length were cut for mercury intrusion porosimetry (MIP). MIP measurements were carried out at least on four identical cores . Additionally pieces of each of the four AAC types were ground and the fraction 0.25/0.5 mm were removed by sieving and used for MIP analysis.

For image analysis discs with 96 mm diameter and 30 mm thickness were cut. The pores were filled by white ZnO, the matrix was couloured by black ink.

For permeability measurements from the supplied blocks two sizes of discs were cut having a diameter of 150 and 96 mm and a height of 60 mm. At least

three discs were used for the determination of the permeability for each type of AAC. These specimens served at the same time for the determination of the density and the porosity by water saturation.

2.2 Experimental setup

Three methods were applied to determine the density and porosity:
1. vacuum saturation with water and drying at 105 °C
2. mercury intrusion porosimetry
3. image analysis

The discs with a height of 60 mm were dried at 105 °C until constant weight. Vacuum saturation was carried out by applying a pressure of 10^{-3} Torr for more than 24 hours until constant weight was reached. Deaerated water served for the water saturation.

Mercury intrusion porosimetry was carried out by using an Autopore 9220, manufactured by micromeritics, with a measuring range of 0.3 mm to 3 nm. A Hg contact angle of 130 degrees and a Hg surface tension of 485 dyn/cm were used. Corrections of the obtained data were made to neglect the thermal effects and the compressibility of the Hg and the sample holder. At 153 pressure steps intrusion was determined after an equilibrium time of 20 seconds.

Image analysis was carried out to measure the visible pores of AAC. Pores with an equivalent dia-

Table 1: Densities of AAC determined by water saturation (WS)

type	bulk density [kg/m³]	skeletal density [kg/m³]	porosity [%]
GN	390	2330	84.0
GH*	490	2320	78.9
GH	610	2420	74.8
GS	630	2440	74.2

Table 2: Densities of AAC determined by mercury intrusion (MIP)

type	bulk density [kg/m³]	skeletal density [kg/m³]	porosity [%]
GN	520	2030	74.2
GH*	570	1980	71.1
GH	650	2060	68.4
GS	680	2120	67.9

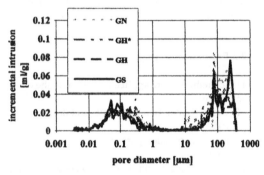

Fig. 1: Incremental pore size distribution

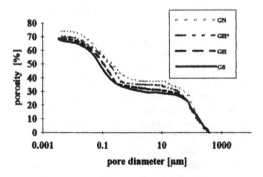

Fig 2: Cumulative pore size distribution

meter of $\Phi \geq 0.2$ mm were measured on an area of 60 mm x 60 mm. For all types of AAC, except for GS, two areas were measured. The area methods was applied to determine the porosity. The area method counts pixels of the image and distinguishs between the white and black areas.

The permeability k was determined in uniaxial flow experiments through cylindrical samples. With the applied pressure difference and the geometry of the sample, the permeability can be calculated using Darcy's law.

Gas permeability was measured on samples dried at 105 °C, on samples stored afterwards for 4 weeks at 22 °C and 60 % r.h. and finally on vacuum saturated samples. Gas permeability was determined with hydrogen gas at excess pressures of 50, 250, 500, 1000, 2000, 3000, 4000 mbar. The pressures were adjusted in an increasing order. At each pressure step we waited until a steady state flow was observed. Further details of the measuring technique are given in Jacobs & Wittmann (1992).

Water permeability was determined with aerated and deaerated tap water at excess pressures between 9 and 250 mbars. At each pressure step we waited until a steady state flow was observed. The outflow was collected in calibrated burettes.

3 RESULTS

3.1 Density and Porosity

Mean values of the density and porosities as determined by vacuum saturation with water (WS) and mercury intrusion porosimetry (MIP) on at least three identical specimens are given in table 1 and 2 respectively.

The bulk densities decrease from the quality GN to GS whereas the skeletal densities are nearly the same for all qualities. They are comparable with the skeletal densities of the C-S-H gel. Taylor (1990) found 2180 kg/m³ stored at 11 % relative humidity. Porosities determined by WS are higher than MIP porosities. Due to surface effects of the small samples the MIP bulk density is higher than WS bulk density.

In figure 1 the obtained incremental pore size distribution by mercury intrusion porosimetry is shown. Two pore maxima can be seen for all four samples, one at 30 - 300 μm and a minor one at 0.03 - 0.4 μm.

In figure 2 the cumulative pore volume shows the different porosities for the different types of AAC. In table 2 the porosity as determined by image analysis (IA) is shown. Porosity decreases from the quality GN to GS. Figure 3 shows the pore size distribution obtained by image analysis with the area method.

The results agree qualitatively with earlier measurements by Prim & Wittmann (1983) on samples with comparable densities.

3.2 Permeability

From Darcy's law it should be expected that the permeability is independent of the applied pressure, at which the flow is measured. In figure 4 the permeability is plotted as function of the mean pressure in a dry sample. As can be clearly seen, the permeability decreases with increasing pressure. This phenomena is determined by the ratio of the mean free path of the gas molecules and the mean pore diameter (Klinkenberg 1941). Here the permeability determined at the highest pressure step (4000 mbar) was taken as representative. The specimens show similar permeabilities after 105 $^{\circ}$C drying and after storage at 60 % r.h. and 20 $^{\circ}$C. The results of the gas permeability measurements are shown in table 3.

Originally water saturated samples show a lower gas permeability ($3 \cdot 10^{-15} m^2$). The gas permeability decreases with increasing water content of the samples, because the gas transport is reduced by water filled pores. Water saturated samples showed a threshold pressure, independent of the type of AAC, between 50 to 250 mbar excess pressure. In figure 5 the threshold phenomenon is shown. If the pressure is lower than the threshold pressure no gas flux can be measured. Above the threshold pressure the flux increases with increasing pressure.

The water permeability increases with the water content. This became obvious by using tap water (not deaerated) for the tests. A liberation of solved air in water takes place due to the drop of the excess pressure from 200 mbar to 0 mbar in the sample. This air blocks the water paths partially. A reduction by up to one order of magnitude compared to the measurements with deaerated water was observed.

4 MODEL OF THE POROUS STRUCTURE

Looking at the bulk densities of the different types of AAC (390 kg/m^3 - 630 kg/m^3), an influence on permeability should be expected. In the following a three-dimensional model of the porous structure of AAC is presented to show the influence of bulk density and porosity on the permeability of AAC.

Table 3: porosities determined by image analysis

type	porosity [%]
GN	34.8
GH*	32.7
GH	23.0
GS	26.4

Table 4: Water and gas permeabilities of AAC

type	bulk density [kg/m^3]	water permeability [10^{-14} m^2]	gas permeability [10^{-14} m^2]
GN	390	3.0±1.8	2.8±1.4
GH*	490	1.0±0.6	1.4±0.4
GH	610	2.0±1.5	2.4±1.6
GS	630	2.9±1.8	2.4±0.3

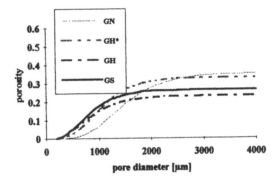

Fig 3: Porosity by Image Analysis

The porous structure of AAC can be modeled as a material consisting of two components. One component represents the spherical air voids (determined by IA , $\Phi \geq 0.2$ mm), the other one the cementitious microporous matrix. As can be seen from MIP measurements in figure 2 the porosities of the matrix ($\Phi < 20$ μm) for different types are nearly the same, too. For simplicity the air voids are modeled as cubes (fig. 6). Each of the six faces is covered by a slice of the matrix material.

If the flux through the material is uniaxial, the fluid penetrates only the part V_a of the matrix and the pores. The permeability of the whole material can be calculated approximately with a serial model:

$$\frac{V}{k} = \frac{V_p}{k_p} + \frac{V_a}{k_m} \tag{1}$$

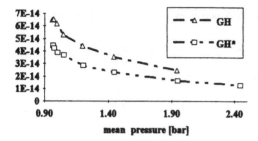

Fig. 4: Permeability as function of the mean pressure

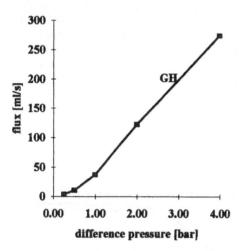

Fig. 5 Gasflux through initial water saturated sample of AAC

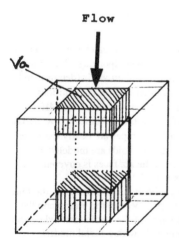

Fig. 6: Cube model of the porous structure of AAC

k : Permeability of the whole material
k_p : Permeability of the air voids
k_m : Permeability of the matrix
V : Volume of the whole material
V_p : Volume of the air voids
V_m : Volume of the matrix
V_a : Volume of the matrix which contributes to the flux

$$V_a = l_a\, l_i^2 - l_i^3 \qquad (2)$$

l_i : edge length of the inner cube
l_a: edge length of the outer cube

The permeability k_p of the air voids (> 0.2 mm) is high compared with the permeability k_m of the matrix. An estimation of the permeability ($k \approx \varepsilon\, r^2/8$)(Jacobs & Wittmann 1992) gives 10^{-9} m^2. Thus the first term on the right side of equation (1) can be neglected and we obtain:

$$k = k_m \frac{V}{V_a} \qquad (3)$$

with the volume fraction of the air voids ε_p

$$\varepsilon_p = \frac{V - V_m}{V} \qquad (4)$$

it follows:

$$k = k_m \frac{1}{(\varepsilon_p^{2/3} - \varepsilon_p)} \qquad (5)$$

From equation 5 we can conclude that the permeability is determined by the permeability of the matrix k_m and the volume fraction of the large visible pores ε_p.

5 DISCUSSION

For the interpretation of the MIP measurements the bottle neck effect (the entering diameter is smaller than the maximum diameter of a pore) must be taken into consideration. Therefore the measured porosity in the range between 0.05 to 0.2 mm is irrealistic.

Both, the water permeability of saturated samples and the gas permeability of dried samples gave nearly the same mean values for all types of AAC (table. 4). Because of the scatter of the experimental results no clear correlation between permeability and the bulk density can be observed either in gas permeability nor in water permeability measurements. This can be explained with the serial model of the porous structure of AAC. If we assume that different qualities of AAC have nearly the same permeability of the matrix, the variation of the permeability can be calculated using equation (4). The total porosity as determined by the area method (IA) was taken for the volume fraction of the air voids $\dot{\varepsilon}_p$. ε_p and the factor $1/(\varepsilon_p^{2/3} - \varepsilon_p)$ for each type of AAC are listed in table 5 .

As can be seen in table 5 no significant variation of the permeability k can be expected from the increasing fraction of air voids ε_p of the different types of AAC.

Using the mean of the water permeability the permeability of the matrix can be estimated to be approximatly $3 \cdot 10^{-15}$ m^2 . Only the degree of water saturation has a big influence on the gas permeability and the threshold pressure for all types of AAC. Water blocks the pores and limits the gas flow. Measurements of the threshold pressure indicate that, a first continuous flow path exists for a pressure of approximately 50 mbar which corresponds to a pore diameter of approximately 10 μm.

6 CONCLUSIONS

From the above results the following results can be drawn:

Water and gas permeability of AAC are approximately the same.

The artificial air pores have little influence on the permeability. A simple serial model is applied to explain this behavior

Gas permeability decreases sharply with increasing water content beyond a critical water content.

Water saturated specimen show a threshold pressure of about 50- 250 mbar.

7 REFERENCES

Jacobs, F. , Wittmann F. H. (1992): Long term Behavior of Concrete in Nuclear Waste Repositories, to be published in Nucl. Engrg. Des.
Klinkenberg, L. J. (1941): The Permeability of porous Media to Liquids and Gases.-Drilling and Production Practice, American Petroleum Institute, pp. 200-213
Prim, P., Wittmann, F.H. (1983): Structure and Water Absorption of Aerated Concrete.-

Table 5. Porosity of the air voids and the corresponding factor to calculate the permeability of AAC from the permeability of the matrix.

type	ε_p [%]	$1/(\varepsilon_p^{2/3} - \varepsilon_p)]$
GN	34.8	6.80
GH*	32.7	6.76
GH	23.0	6.86
GS	26.4	6.76

in:Autoclaved Aerated Concrete, Moisture and Properties. ed. Wittmann, F. H., Elsevier.
Taylor H. F. W.(1990) Cement Chemistry, Academic Press, London , p. 220

8 ACKNOWLEDGEMENT

We gratefully acknowledge the corporation with Mr. Iriya and Mr. Kawaguchi from Obayashi Corporation, Tokyo, Japan. They carried out the image analysis on AAC samples

Advances in Autoclaved Aerated Concrete, Wittmann (ed.) © 1992 Taylor & Francis. ISBN 90 5410 086 9

Effect of size distribution of air pores in AAC on compressive strength

G. Schober
Hebel AG, Emmering, Germany

ABSTRACT: The influence of size distribution of air pores in AAC on compressive strength was investigated. The experiments were carried out on AAC samples prepared with different aluminum powders and mixtures thereof. Within the range of variation of pore size achieved in this work the results show no significant influence of pore size distribution of the air pores on the compressive strength.

1 INTRODUCTION

There are many parameters which influence the strength of AAC. Besides recipe, quality of raw materials and autoclaving process, the pore structure of the air pores is said to have a marked effect on the compressive strength. This influence of pore size and pore size distribution and additionally the condition of the pore shells has not yet been sufficiently investigated.

In this paper we investigated the influence of pore size and pore size distribution. By using different aluminum powders and mixtures thereof we generated variations in the size distribution of the air pores. All other preparing parameters were held unchanged for all samples.

The experiments were carried out with AAC of relatively low bulk density (lower than 400 kg/m^3) to get samples with the greatest possible sensitivity to variations of the pore structure.

2 METHODS OF INVESTIGATION

2.1 Pore size distribution of air pores

The size of air pores was determined by automatic image analysis. An area of 40 x 40mm of each sample was measured. The samples were prepared by smoothing, cleaning with compressed air and illuminating with four lamps to get sufficient contrast between the air pores and the matrix material. The lower limit for detectable pore size was 0.18mm. This restriction seemed acceptable and led to increased ease of measurement.

In addition the total volume of the air pores was estimated from the amount of active aluminum metal used to generate the air pores.

2.2 Compressive strength and strength level

The compressive strength was determined using cubes with 100mm edge length and with water content between 3 and 10 Vol.-%. The cubes were cut out of larger ones of about 220mm edge length which is the usual method of producing AAC samples in the Zentrallabor of Hebel AG.

For comparing the strength of AAC samples we did not use the values of compressive strength alone. We computed a figure of merit, the strength level F, which gives a correction of the compressive strength by the bulk density:

$$F = \sigma/\varrho^2/0.22 * kg^2/Nm^4 \ (\%);$$
$$\sigma = \text{compressive strength}$$
$$\varrho = \text{bulk density}$$

The value F is the strength level of one AAC sample in terms of percent compared to the mean compressive strength of many AAC samples of a certain recipe. The value $0.22 Nm^4/kg^2$ is applicable only for the recipe used in this work. In this way, influences which have only small effects on compressive strength should be detectable.

This method is similar to the "A-Zahl" valuation first introduced by the Ytong AG, but it has been made specific to a certain recipe.

3 MATERIALS

3.1 Aluminum powders

To achieve a variation in pore size distribution of air pores different aluminum powders and mixtures thereof were used. In table 1 the relevant characteristics of three types of aluminum we used are given.

Table 1. Aluminum powder types

aluminum		particle size	
type	kind	mean (μm)	max.(μm)
Al1	paste	7.5	24
Al2	paste	48.8	200
Al3	powder	18.0	120

3.2 AAC recipe

The recipe for the AAC samples had the following composition given in mass proportions:
  sand : cement : lime : anhydrite
  16 : 10 : 5 : 2
  The bulk density was adjusted to 380 kg/m^3 by the addition of necessary quantity of aluminum powder. Table 2 shows the types or mixtures of aluminum used in each sample.

Table 2. AAC samples with different aluminum types and mixtures thereof (mass proportions of active aluminum).

AAC sample	aluminum type proportion Al1 : Al2 : Al3		
PB1	1	0	0
PB2	0	1	0
PB3	0	0	1
PB11	1	1	0
PB15	1	5	0
PB31	0	1	1
PB35	0	5	1

4 RESULTS

The following figures show the size distribution of the air pores in the AAC samples made with only one type of aluminum.

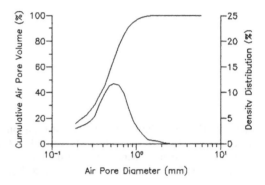

Fig.1 Cumulative volume and density distribution of air pores in PB1

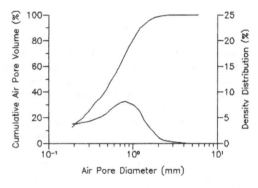

Fig.2 Cumulative volume and density distribution of air pores in PB2

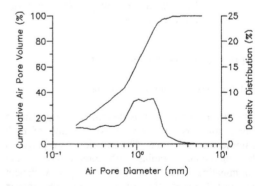

Fig.3 Cumulative volume and density distribution of air pores in PB3

The corresponding material values to the figures 1 to 3 are given in table 3.

Table 3. Material values of PB1, PB2 and PB3.

	bulk density (kg/m³)	air pore vol. me.	est.(%)	mean pore size (mm)	F (%)
PB1	380	41.7	56.7	0.59	119.0
PB2	378	47.4	56.7	0.80	98.6
PB3	354	52.6	65.1	1.08	104.8

The differences between measured and estimated pore volume is due to the lower limit for detecting pore sizes with image analysis. The graphs of cumulative volume and density distribution are corrected to the estimated air pore volume. So the curves are cut off at 0.18mm air pore diameter.

Figures 4 to 7 show the results of pore size measurement with mixtures of aluminum powders.

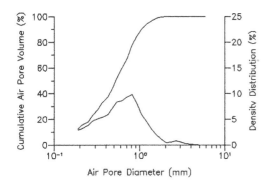

Fig.4 Cumulative volume and density distribution of air pores in PB11

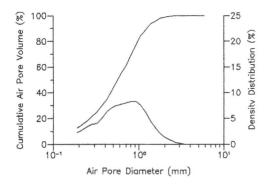

Fig.5 Cumulative volume and density distribution of air pores in PB15

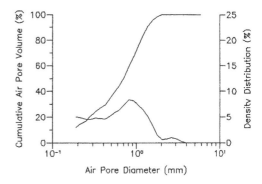

Fig.6 Cumulative volume and density distribution of air pores in PB31

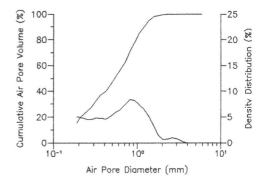

Fig.7 Cumulative volume and density distribution of air pores in PB35

In table 4 the material values of the samples PB11, PB15, PB31 and PB35 are listed.

Table 4. Material values of samples PB11, PB15, PB31 and PB35.

	bulk density (kg/m³)	air pore vol. me.	est.(%)	mean pore size (mm)	F (%)
PB11	380	46.0	56.7	0.72	106.4
PB15	380	45.0	56.7	0.76	103.6
PB31	370	51.0	61.4	0.90	102.3
PB35	378	46.3	58.4	0.76	98.9

Table 5 shows all samples arranged with respect to the variation in aluminum powder quality.

Table 5. Comparision of all AAC samples

	bulk density (kg/m3)	mean pore size (mm)	F (%)	counts
PB1	380	0.59	119.0	1700
PB11	380	0.72	106.4	1200
PB15	380	0.76	103.6	900
PB2	378	0.80	98.6	1000
PB35	378	0.76	98.9	1000
PB31	370	0.90	102.3	850
PB3	354	1.08	104.8	600

The last column (counts) gives the total number of pores detected by automatic image analysis within an area of 40 x 40mm.

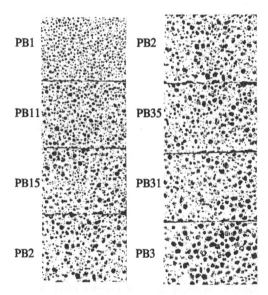

Fig.5 Copy of the pore structure of all AAC samples (actual size).

5 DISCUSSION AND CONCLUSIONS

As the results show there is only one sample that showed a marked change in compressive strength due to the change of size distribution of the air pores. All other samples do not show a significant change in strength.

The sample PB1 which is 19% higher in strength level was prepared with the aluminum type giving the smallest air pores. This seems to confirm the common opinion that smaller air pores cause higher compressive strength (Tang Luping, 1986, He Tianchun, 1988). On the other hand all the other samples are very close together in their strength levels, even though the mean pore size varies from 0.72mm to 1.08mm. Looking at the size distribution graphs and the pictures of pore structure, one recognizes that the differences are more striking than the values of mean pore size indicate. Therefore the results of this work lead to the conclusion that size and size distribution of air pores in AAC do not have a significant effect on the compressive strength.

An effect of air pore size on compressive strength as found by He Tianchun (1988) could not be confirmed. He Tianchun reports that an increase in airpore size causes a strong decrease in compressive strength.

It is probable that in the case of pore structure neither the size nor the size distribution of the air pores but the mechanical condition of pore shells is the main factor influencing strength in AAC. In this work it was assumed that because of stadardized sample preparation no differences in the mechanical condition of pore shells were produced. But there are no suitable measuring methods so this assumption was not verified. Further work on the influence of pore structure on strength of AAC will have to take this into account.

So the improvement of strength in the case of sample PB1 may be due to effects just discussed but at this point it can not be explained definitively.

ACKNOWLEDGEMENTS

The author would like to thank the Institut für WW3, Glas und Keramik, Universität Erlangen, Germany for measuring pore size distribution and Fa. Eckart, Fürth, Germany for supplying different aluminum powder types.

REFERENCES

He Tianchun, 1988. Festigkeits- und Verformungs-verhalten von Gasbeton. Dissertation Freiberg.

Tang Luping, 1986. A study of the quantitative relationship between strength and pore-size distribution of porous materials. Cement and concrete research 16: 87-96.

3 Heat and mass transfer

Advances in Autoclaved Aerated Concrete, Wittmann (ed.) © 1992 Taylor & Francis. ISBN 90 5410 086 9

The thermal performance of European autoclaved aerated concrete

W. R. Millard
Marley Building Materials Limited

ABSTRACT:
Data from major European manufacturers of AAC is used together with test results obtained on their materials in NAMAS accredited laboratories within the UK in order to determine the parameters which influence Thermal Conductivity and to quantify them.

Some history of measurement and correction methods for thermal conductivity is given along with a survey of current practices in various countries.

Conclusions are that:

1) AAC can be regarded as a generic material for thermal design purposes.
2) Tabulated values for thermal conductivity of this material can be supplied to represent the whole of Europe.
3) The thermal conductivity of AAC can be taken as a linear function of the oven dry density of the material and as a linear function of the moisture content expressed as a percentage by mass.
   Universal factors for the correction of thermal conductivity values to different moisture contents and densities are derived.
4) Equilibrium moisture contents are suggested for AAC in use based on laboratory and field test data.

1. HISTORY OF THERMAL CONDUCTIVITY MEASUREMENTS ON INSULATING MATERIALS

The first determination of the thermal conductivity of a relatively poor conductor of heat was carried out by Lees in 1898. His apparatus, which subsequently became known as Lees' disc used two relatively thin disc specimens clamped either side of an electrical heater plate and between two water cooled plates. The thermal conductivity was calculated directly from measurements of electrical energy input, specimen dimensions and specimen surface temperatures. By using thin disc specimens, the loss of heat from the specimen edges was reduced, this was important because it was assumed in calculations that all the heat from the heater plate passed axially through the specimens. Lees' apparatus was subsequently improved by Poensgren, working in Germany, who added an annular heater, known as a guard heater, around the heater plate.

The temperature of this guard heater plate could be controlled independently of the centre, allowing the temperature to be the same as the heater plate. The heat loses from the edges of the heater and the specimen could thereby be largely eliminated. In this way it was possible to carry out tests on thicker specimens and on brittle, hard materials which could not be prepared as thin discs, such as concrete. An apparatus for testing rock specimens was built by Callendar at around the same time.

Further modifications to existing technology were made by Jakob, whose name is still associated with conductivity/moisture correction factors. The current standard test method used throughout the world known as the Guarded Hot Plate, uses the same principals with the addition of more sophisticated electronics to control temperatures and voltages and for the handling of data. The details of guarded hot plate design are given in International Standards such as ISO8301, British Standard 874 and DIN 52612 which also give details of how to use the apparatus and derive results. Latest designs depend on the research which has been carried out to determine the optimum thicknesses and area of specimens as well as relative guard widths and edge insulation. For thick, low density insulation measurements, metal edge guards are clamped between the heater and guard plates to provide a linear temperature gradient edge guard which further reduces edge losses as they become more significant with such low conductivity materials.

2. THE CURRENT SITUATION

At present, the only common factor concerning thermal measurement and calculation across Europe in the use of the Guarded Hot Plate as the reference test method.

There are differences in specimen sizes, conditioning regimes, oven drying temperatures, plate temperatures, correction methods (for density and moisture content) and design calculation methods.

In most countries, there are tabulated values which can be used in the absence of test data. These values are invariably significantly higher than the true conductivity of AAC so there is a strong incentive to carry out testing.

The following examples show the differences between existing practices in some countries.

2.1 United Kingdom

National Building Regulations [1] require that three Guarded Hot Plate tests be carried out in accordance with the National Standard [2] at NAMAS [3] accredited laboratories. At least one of the tests has to be less than one year old and the mean of the three results is used as the Thermal Conductivity value.

Specimens measure 305 x 305 x 50mm and are tested after conditioning to equilibrium (constant mass) with the laboratory environment. Plate temperatures are nominally 10°C and 30°C. Following testing, the specimens are oven dried to constant mass at 105°C to determine the moisture content at the time of test.

The measured value is then corrected to 3% moisture by volume (for protected walls) using factors derived by JAKOB [4] which are used for all concrete materials.

It is estimated that the correction process amounts to an increase of 50% on the oven dry conductivity in this case to obtain a design value for a protected house wall.

2.2 Sweden

Specimens for Guarded Hot Plate testing measure 400mm x 400mm and are tested at plate temperatures of 0°C and 20°C having been oven dried at 105°C. It is necessary to seal the specimens in polyethylene film to

prevent the ingress of moisture.

The dry measured value is then corrected to a "normal moisture content" which is that at equilibrium with 30-70% R.H. This typically involves a 7% increase in the dry measured value. The "normal moisture content" value is then corrected to a design moisture content dependent on density, being 8% for an external wall protected from rain. [5]

2.3 Germany

Three Guarded Hot Plate tests to the National Standard [6] are required at different mean temperatures between 10°C and 40°C. Specimens measure 500mm square and are tested sealed in polyethylene film after oven drying at 105°C. The results are used to determine the thermal conductivity at 10°C mean and an addition is made to allow for moisture in protected house walls which depends on the conductivity value. The factor is 30% for a conductivity of 0.14W/m°K. The accepted moisture content of such walls is 4.5% by mass.[7]

2.4 Italy

Guarded Hot Plate tests are carried out on specimens measuring 500mm x 500mm in equilibrium with an atmosphere of 23°C and 50% R.H. The mean test temperature is 20°C and oven drying at 105°C after the test is used to determine the moisture content during test.

Measured data is only acceptable instead of tabulated values if it can be demonstrated as representing more than 90% of production.

National Standards [8] require an addition of 25% to the measured value to obtain the design value for protected house walls, which are assumed [8] to have a moisture content of 4% - 5% by mass.

2.5 Norway

Guarded Hot Plate tests are carried out at a mean temperature of 10°C. Typically 15% is added to the dry measured value for moisture along with other factors for ageing, uncertainty and convection according to National Standards.[9] The assumed moisture content of a protected house wall is 4% by mass.

2.6 Switzerland

There is no regular production or testing of AAC here. Products are assessed by EMPA [10] who carry out three Guarded Hot Plate tests at different temperatures. Specimens are oven dried at temperatures between 65°C and 80°C prior to testing. An addition of 12% is made to this "dry" value to give the design value for a protected house wall, which is assumed to contain 5% moisture by mass. Products are assessed individually to determine their conductivity/moisture relationship. It should be remembered that the "dry" tested specimens will contain around 2% moisture by mass relative to dry at 105 °C.

2.7 Netherlands

Specimens for Guarded Hot Plate tests measure 300mm x 300mm and can either be oven dried at 70 °C or be in equilibrium with an atmosphere of 23 °C and 50% R.H. (resulting typically in a moisture content of 3% by mass).

The mean test temperature is 10 °C. Corrections for moisture in National Standards [11] are based on investigation of specific products. Once the properties of a product are categorized, quality control testing using quick test methods is permitted unless the product is modified.

2.8 France

Guarded Hot Plate tests are carried out at a mean temperature of 10 °C on specimens sealed polyethylene after drying to constant mass at 70 °C. Tests on every declared density are carried out twice a year by an independent body [12]. A protected house wall is assumed to have a moisture content of 4% by mass.

2.9 Conclusion

Each different country is testing the same material using the same equipment in slightly different ways. It should be possible to standardize the reference test method with regard to temperature, moisture content, drying temperature and correction methods for moisture and density.

The remainder of this paper is an attempt to suggest simple ways of achieving this.

3. THERMAL CONDUCTIVITY RELATED TO DENSITY ACROSS EUROPE

3.1 Experimental method

Using AAC manufactured in Germany, France, Sweden and UK 109 determinations of Thermal Conductivity were carried out in the UK at NAMAS Accredited Laboratories over a period of several years.

3.2 Results

The results are corrected to 3% by mass for the

purpose of inter comparison and are shown on Table 1, Graph 1 shows a linear regression analysis of the data, which produces the equation $Tc = 0.000266Dd - 0.01083$ where Tc = Thermal conductivity and Dd = Dry density.

The line $TC = 0.000266Dd + 0.0003227$ drawn parallel to the regression line has been used to produce "safe" tabulated values for Harmonised European Standards. This line represents 90% of all measured values and although the values predicted by it may not always be used directly, parallel lines to it will be drawn for the correction of conductivity values for density within specific product groups.

3.3 Conclusions/Comparisons with previous work

The graph of density against λ for 100+ samples of AAC from different manufacturers across Europe all conditioned similarly and tested in the same way. Clearly there is a good relationship with a correlation coefficient of 0.952.

Manufacturers use different raw materials to produce AAC. The principal difference being that some use ground sand and some use pulverised fuel ash or a mixture of both. The correlation of these results however suggests that compared with density, the raw material influence on thermal conductivity is not significant for the purposes of this exercise.

4. THE VARIATION OF THERMAL CONDUCTIVITY WITH MOISTURE CONTENT

4.1 Experimental method

4 No. samples of AAC were prepared in the density range 370-900 kg/m³. Each pair of samples was conditioned to various moisture contents by spraying the appropriate amount of water evenly over the surfaces and storing the specimens sealed in polythene for 10 days to allow even distribution of the moisture throughout the material. In each case a measurement of λ was made at equilibrium moisture content, that is with the specimen conditioned in the laboratory environment until constant mass was established. Measurements of each sample were also made in the oven dry condition, ie. dried to constant mass at 105 °C.

Measurements of thermal conductivity were made on specimens in all the above conditions using the Guarded Hot Plate method described in BS874 and ISO 8302.

In addition a series of tests were carried out using a "probe method" [13] which can give an indication of Thermal Conductivity in a matter of minutes. The purpose of these tests was to

Table 1

GUARDED HOT PLATE RESULTS

Thermal Conductivity (W/mK)	Moisture Content (% v/v)	Dry Density (Kg/m³)	Zero Moisture Conductivity (W/mK)
0 119	1 840	463 0	0 107
0 102	1 880	397 0	0 088
0 096	1 480	392 0	0 085
0 096	1 150	389 0	0 087
0 095	1 370	397 0	0 085
0 094	1 070	395 0	0 086
0 090	1 060	374 0	0 082
0 089	0 900	374 0	0 082
0 094	1 350	385 0	0 084
0 127	3 100	468 0	0 105
0 094	1 500	375 0	0 082
0 099	1 500	388 0	0 087
0 095	1 550	375 0	0 083
0 090	1 250	375 0	0 081
0 115	1 750	453 0	0 104
0 110	1 510	433 0	0 100
0 083	0 900	376 0	0 077
0 090	1 000	374 0	0 082
0 095	1 660	403 0	0 084
0 124	1 650	501 0	0 115
0 173	2 350	685 0	0 167
0 203	2 470	784 0	0 198
0 194	2 230	791 0	0 191
0 212	2 020	830 0	0 210
0 180	1 340	741 0	0 180
0 182	1 880	700 0	0 179
0 170	1 300	707 0	0 170
0 186	3 200	713 0	0 174
0 102	1 790	388 0	0 088
0 106	2 070	389 0	0 089
0 095	1 780	400 0	0 083
0 094	2 000	411 0	0 081
0 127	2 110	517 0	0 116
0 132	2 340	544 0	0 120
0 136	2 200	565 0	0 126
0 141	1 120	583 0	0 138
0 161	1 700	680 0	0 159
0 127	0 900	432 4	0 120
0 137	0 970	471 1	0 131
0 180	1 210	582 1	0 176
0 204	1 330	681 8	0 203
0 121	1 020	473 7	0 115
0 167	1 430	633 4	0 164
0 102	1 600	398 0	0 090
0 112	2 640	401 0	0 091
0 102	2 640	401 0	0 083
0 149	2 400	567 0	0 137
0 170	2 800	668 0	0 160
0 161	2 480	620 0	0 151
0 165	2 450	634 0	0 156
0 162	2 460	639 0	0 153
0 150	2 420	646 0	0 143
0 104	2 070	392 0	0 088
0 098	1 550	389 0	0 086
0 152	2 220	638 0	0 145
0 103	1 470	410 0	0 092
0 153	2 040	638 0	0 147
0 162	2 010	658 0	0 157
0 106	1 760	385 0	0 091
0 095	1 170	379 0	0 086
0 118	2 170	405 0	0 100
0 127	3 500	397 0	0 096
0 139	1 700	538 0	0 131
0 095	1 230	380 0	0 086
0 098	1 240	390 0	0 089
0 096	1 120	394 0	0 088
0 193	2 600	720 0	0 186
0 186	3 200	713 0	0 174
0 093	0 120	343 0	0 091
0 095	1 090	343 0	0 085
0 113	0 110	418 0	0 113
0 117	1 380	418 0	0 106
0 127	0 160	463 0	0 126
0 130	1 300	463 0	0 121
0 110	0 180	463 0	0 110
0 119	2 010	493 0	0 108
0 146	0 160	651 0	0 146
0 164	2 600	636 0	0 154
0 173	3 100	590 0	0 154
0 163	2 900	654 0	0 152
0 147	1 100	630 0	0 146
0 149	1 060	624 0	0 147
0 123	1 020	594 0	0 121
0 137	1 690	513 0	0 128
0 133	1 540	578 0	0 128
0 170	2 170	596 0	0 160
0 171	2 090	589 0	0 162
0 161	1 500	700 0	0 160
0 155	2 200	644 0	0 149
0 155	1 800	651 0	0 152
0 188	0 130	771 0	0 188
0 094	1 330	381 0	0 084
0 098	1 210	385 0	0 089
0 117	1 160	435 0	0 109
0 098	1 160	394 0	0 089
0 096	1 160	381 0	0 087
0 188	1 610	635 0	0 184
0 105	1 380	436 0	0 096
0 098	1 290	388 0	0 088
0 152	1 830	614 0	0 147
0 097	0 850	383 0	0 091
0 149	1 540	597 0	0 144
0 150	1 450	597 0	0 146
0 097	1 000	386 0	0 089
0 157	2 100	610 0	0 149
0 097	1 260	389 0	0 088
0 154	2 200	605 0	0 146
0 147	2 170	659 0	0 142

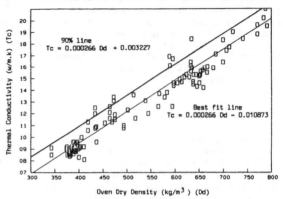

THERMAL CONDUCTIVITY RELATED TO DRY DENSITY FOR EUROPEAN AAC

90% line
Tc = 0.000266 Dd + 0.003227

Best fit line
Tc = 0.000266 Dd - 0.010873

Graph 1

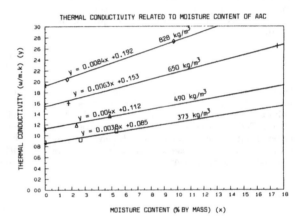

THERMAL CONDUCTIVITY RELATED TO MOISTURE CONTENT OF AAC

y = 0.0084x + 0.192   828 kg/m³
y = 0.0063x + 0.153   650 kg/m³
y = 0.004x + 0.112   490 kg/m³
y = 0.0038x + 0.085   373 kg/m³

Graph 2

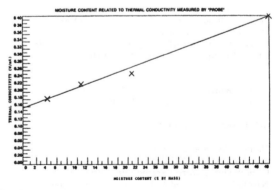

MOISTURE CONTENT RELATED TO THERMAL CONDUCTIVITY MEASURED BY 'PROBE'

Graph 3

Table 2

VARIATION OF λ WITH MOISTURE CONTENT

Oven Dry Density (kg/m³)	Moisture % by Volume	% by Mass	λ W/mk	Moisture Factor 100 $^m/_c$
373	2 0	5 4	0.108	
	1 0	2 7	0.092	4.38
	0 0	0 0	0.087	
490	2 4	4 9	0 136	
	1 2	2.4	0.119	4 12
	0 0	0 0	0 114	
650	11 4	17 5	0.265	
	1 2	1 8	0 160	3 60
	0 0	0 0	0 158	
828	8 0	9 7	0.273	
	1 4	1 7	0 204	4 52
	0 0	0 0	0 194	

Table 3

VARIATION OF λ WITH MOISTURE CONTENT PROBE TEST

Moisture Content (% by mass)	λ (W/mk)
4 0	0 18
11 5	0.22
21.0	0 25
48 5	0 40

Table 4

RESULTS OF MOISTURE CONTENT MEASUREMENTS ON THE WALLS OF HOUSES IN VARIOUS PARTS OF GERMANY

German Postal Area	Oven Dry Density (kg/m³)	Moisture Contents % by volume	% by weight
2107	540	1 49	2.75
2107	550	1 57	2 87
2741	530	1.86	3.50
2741	500	1 35	2.68
2741	500	1 86	3 71
2741	480	1 88	3 92
2741	460	1 46	3.17
2741	460	1.85	4.03
4775	498	1 40	2.81
4775	498	1 77	3 56
4775	467	1 52	3.25
6751	500	1 40	2.81
6751	500	1.29	2 59
7119	490	1 61	3 28
7119	490	1 59	3 25
7119	480	1.17	2.44
8521	500	1 13	2 25
8521	500	0.79	1 57
8521	500	1.32	2 65
8521	500	1 26	2 52
8521	480	1 16	2 36
8521	480	1 21	2 52
8899	480	1 33	2.77
Mean Results	495	1 45	2 84

determine the effect of very high moisture contents, which preclude the use of the Guarded Hot Plate apparatus because moisture migration prevents equilibrium being established. The tests were carried out on different samples of the same material conditioned to different moisture contents.

4.2 Results

The results of these tests are shown in table 2 and also on graph 2, and table 3, graph 3.

4.3 Conclusions and comparisons with previous work

Approximate "best fit" lines and equations have been shown for each density and these suggest that a linear approximation can be used for the relationship between λ and moisture content for each density.

The intercept of each line on the vertical axis is the constant in the equation for each density, having the form $y = mx + c$ where y is the λ value, m is the gradient of the line, x is the moisture content by mass and c is a constant equal to the oven dry value of λ. The percentage change in λ for a 1% change in mass can be found by calculating $^m/_c$ x 100 for each line and this value is shown in Table 2 to be close to 4% for 1% change in moisture content by mass in each case.

This finding is in broad agreement with previous work carried out by Loudon [14], Stuckes and Simpson [15] and ISO [16].

The general conclusion is that the relationship between thermal conductivity and moisture content can be approximated as a linear increase of 4% in thermal conductivity for every 1% increase in moisture content by mass.

The results of the probe tests show good agreement with the Guarded Hot Plate, which suggests firstly that the probe is a valid test method for the material and secondly that the high moisture content tests with the Guarded Hot Plate are also valid.

5. EQUILIBRIUM MOISTURE CONTENTS IN USE

5.1 Background

The thermal conductivity of AAC, as has already been established, depends upon the moisture content. In order to design buildings to achieve a certain thermal performance it is necessary to assume a moisture content for the AAC element. This moisture content will depend upon its position within the building and the climate.

5.2 Sources of Data

Digest 342 [17] produced by the Building Research Establishment in the UK suggests that the current tradition for assuming 3% moisture by volume in protected walls should be replaced by 3% by mass for low density AAC. This is clearly sensible for as the density goes down and the percentage of air in the AAC rises above 80%, the capacity to retain or absorb moisture must also fall for a given volume.

Further evidence to support the use of 3% by mass as a representative moisture content for determining design λ values is provided by measurements of moisture contents of Thermal Conductivity samples which were conditioned in controlled environmental conditions in several NAMAS laboratories throughout the UK. The results in Table 1 show a mean value of 1.6% by volume and 2.8% by mass.

Evidence was obtained from a survey carried out in several postal districts of Germany [18], where cores were removed from the walls of houses and the moisture contents determined by oven drying. The results giving a mean value of 2.84% by mass are shown in Table 4.

5.3 Conclusions

The equilibrium moisture content of walls in the UK can be taken as 3% by mass for the purposes of calculating design values for protected walls.

6. REFERENCES

1. BUILDING REGULATIONS 1985 Department of the Environment and the Welsh Office. Approved Document L 1990 Edition.
2. BRITISH STANDARD 874 Determining Thermal Insulating Properties Part 2: Tests for Thermal Conductivity and related properties: Section 2.1:1986 Guarded Hot-plate Method.
3. NAMAS - National Measurement Accreditation Service Department of Trade and Industry, U.K.
4. CIBSE - Chartered Institute of Building Services Engineers Guide A3 1980 Table A3.23.
5. BOVERKET Report - Swedish "National Board of Housing and Planning".
6. DIN German Standard 52612
7. DIN German Standard 52611.
8. UNI - Italian Standard 7357
9. NS - Norwegian Standard NS3031 and NS8046.
10. EMPA - Swiss Federal Laboratory for Materials Testing and Research.
11. NEN - Netherlands Standard 3838.
12. C.E.R.I.B.
13. Tests carried out by the Cranfield Institute of Technology U.K.
14. Loudon A.G. The Effect of Moisture on Thermal Conductivity of Autoclaved Aerated Concrete. Ed. Wittman F.H. Elsevier Scientific Publications, Amsterdam, 1983.
15. Stuckes A.D. and Simpson A. The effect of moisture on the Thermal Conductivity of Aerated Concrete. Building Services Engineers Research and Technology 6 (2) 1985 and Moisture factors and Thermal Conductivity of Concrete 7 (2): 1986.
16. ISO TR 9165:1988 Practical Thermal Properties of Building Materials and Products.
17. BRE Digest 342: Building Research Establishment March 1989 "Autoclaved Aerated Concrete".
18. Information from Ytong Germany

Advances in Autoclaved Aerated Concrete, Wittmann (ed.)© 1992 Taylor & Francis. ISBN 90 5410 086 9

Recent results on thermal conductivity and hygroscopic moisture content of AAC

E. Frey
Hebel AG, Emmering, Germany

ABSTRACT: For AAC plane block masonry the influence of the thin-bed mortar layers on the thermal resistance of the wall is negligible. A comparison of wall measurements and measurements with the guarded hot plate apparatus on small slab specimens showed corresponding thermal conductivity results. The increase of thermal conductivity with the moisture content of AAC is shown for a great number of recent wall measurements. There is nearly no influence of the apparent density of the material. A simple method for the determination of the thermal conductivity of plane block masonry in dependence on the hygroscopic moisture content of AAC is being suggested.

1 INTRODUCTION

The determination of the thermal conductivity of building materials and of uniform design values for the various groups of building materials have been the subject matter of intense work of various CEN committees for years. The present paper deals with the measurement of the thermal conductivity of AAC plane block masonry and with more recent recognitions regarding the influence of moisture on thermal conductivity. A method will be suggested hereinafter by which the thermal conductivity of plane block masonry can be determined in a simple manner in dependence on the hygroscopic moisture content.

2 COMPARISON OF WALL MEASUREMENTS AND MEASUREMENTS WITH THE GUARDED HOT PLATE APPARATUS

In addition to the fact that AAC plane block masonry can be erected easily and in short working time it has the advantage that the thin-bed mortar joints exert no influence on the thermal resistance of the wall. Owing to the high dimensional accuracy of the plane blocks, the mortar joints formed by the thin-bed mortar method have an average thickness not in excess of 1 mm. Because contrary to almost all other masonry blocks, AAC blocks have no formed voids, i.e. are massive, it can be assumed that the thermal conductivity of small slab specimens equals the thermal conductivity of the plane block masonry.

In spite of said features of plane block masonry, the German specifications for qualification still require a testing of the thermal conductivity of wall specimens in accordance with DIN 52 611 ("Determination of the thermal resistance of walls and floors"). That test is highly expensive in comparison with the testing of small slab samples with the guarded hot plate apparatus in accordance with DIN 52 612 ("Determination of the thermal conductivity with the guarded hot plate apparatus").

It has now been proved by an investigation recently conducted on request of the Association of the German Aerated Concrete Industry that the results obtained with the guarded hot plate apparatus in accordance with DIN 52 612 do not differ from those obtained from plane block walls in accordance with DIN 52 611. The results of 30 measurements on slab specimens and of 40 measurements on walls have been evaluted by the Forschungsinstitut für Wärmeschutz/ Munich and are represented in Figure 1. The German manufacturers of AAC hope that that proof will cause the requirement for a wall measurement to be cancelled an that that measurement will not be included in any European standard.

3 THE INFLUENCE OF MOISTURE ON THERMAL CONDUCTIVITY

It has been known for a long time and proved by numerous investigations that

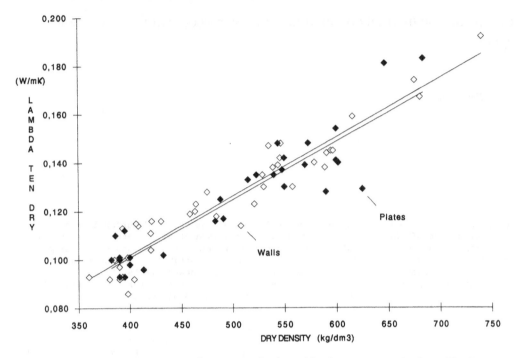

Fig. 1 Thermal conductivity ($\lambda_{10\ dry}$) of plane block masonry and of small plate specimens (with kind permission of Dr. Achtziger (1992))

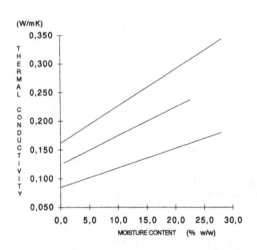

Fig. 2 Thermal conductivity as a function of moisture content of AAC in accordance with equation (1)

moist AAC has a higher thermal conductivity than dry AAC, Künzel (1971). The rise of lambda with the moisture content of the material is linear in a range which is encountered in construction practice.

For the time being in each country in which AAC is used, different methods are used to describe the influence of the hygroscopic moisture content of AAC on its thermal conductivity and to determine corresponding design values for lambda. For this reason the nationally adopted design values of the thermal conductivity for the same apparent density class differ widely in part.

In the draft European standards presently available a 4% rise of lambda per weight percent moisture content is assumed for AAC

$$\lambda_u = \lambda_{10\ dry} \cdot (1 + 0.04 \cdot u_m) \quad (1)$$

wherein u is the moisture content of the specimen in weigh percent. But that equation does not reflect the actual relationship between lambda and the moisture content of the material. It is apparent form Figure 2 that when calculated in accordance with equation (1) for specimens having a higher apparent density, where the value for $\lambda_{10\ dry}$ is higher, λ_u will rise more steeply in dependence on the moisture content than for specimens having a low apparent density. The actual rise of lambda in dependence on the moisture content is

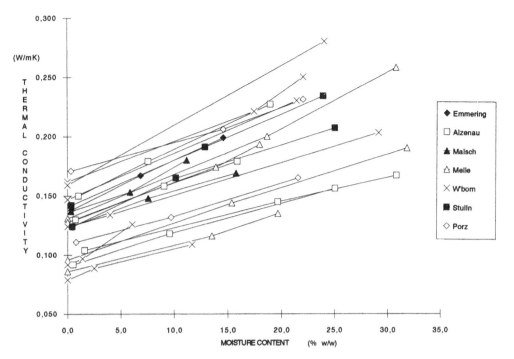

Fig. 3 Thermal conductivity as a function of moisture content from wall measurements with Hebel AAC

shown in Figure 3 for specimens having different apparent densities. That graph is based on wall measurements, officially conducted in accordance with DIN 52 611 on AAC of Hebel in 1986 to 1991. It is apparent that the wall measurements, which are held by us to be superfluous for the reasons stated above and have been made for each wall specimen with three moisture contents, afford the benefit that a large number of results on the influence of moisture is available today. By a regression analysis it can easily be checked whether the moisture-dependent rise of lambda is influenced by the apparent density. In Figure 4, the slopes of the increase of lambda with moisture are plotted against apparent density. Amounting to 0.38, the correlation coefficient r is very small.

It is directly apparent from Figure 3 that the influence of the moisture content on lambda is reflected by the following simple relationship:

$$\lambda_u = \lambda_{10\ dry} + a \cdot u_m \qquad (2)$$

rather than by equation (1).

For all specimens tested, the mean value of the slope a amounts to 0.0035 W/mK% and the 90% fractile value is about 0.0050.

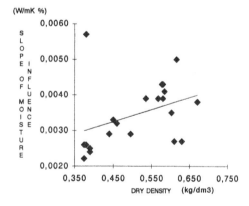

Fig. 4 Slope of moisture influence on thermal conductivity versus dry density of Hebel AAC

4 HYGROSCOPIC MOISTURE CONTENT OF AAC

A uniform value of the hygroscopic moisture content of AAC has not finally been determined in the draft European standards. Great differences may exist between the products of different manufacturers. A value of 3.5 volume percent had been adopted in Germany for a long time but is much to high, particularly for AAC having a low apparent density. This fact was already mentioned by Hums (1982). From 1987 the value of 4.5 mass percent was used in qualifications for Ytong and Hebel and the compliance with that value must be proved in a test under test conditions of 23°C and a relative humidity of 80%.

The mean value obtained in official tests of the hygroscopic moisture content of Hebel AAC since 1987 is about 3.3 mass percent and the 90% fractile value is about 4.0 mass percent. For this reason, 4.0 mass percent could be regarded as a well-established hygroscopic moisture content for AAC of Hebel. If conditions of 23°C/50% r.h. are agreed upon for European standards, even a value of 2.5 or 3.0 mass percent should be desired. The humidity in the interior of dried-out houses amounts in most cases to only 50% or even less.

But it will be shown in the next section of this paper that uniform and statistically significant values for the hygroscopic moisture content need not be established.

5 SURVEILLANCE TESTS OF THERMAL CONDUCTIVITY

The European standards will contain a table of the lambda values for all building materials, like the table in Part 4 of DIN 4108 ("Thermal insulation in buildings; characteristic values relating to thermal insulation an protection against moisture"). Being statistically significant said values are far on the safe side for the user and a manufacturer complying with said tabulated values will not be required to furnish additional proof.

But more favorable design values will be available for all building materials for which additional tests are conducted and which are to be surveilled.

A very simple method can then be adopted for AAC plane block masonry:

1. Measurement of $\lambda_{10\ dry}$ with the guarded hot plate apparatus.
2. Measurement of the hygroscopic moisture content u_m under defined climatic conditions, e.g. 23°C/50% r.h.
3. Calculation of λ_u according to equation (2)

For the products of each manufacturer, the slope a of equation (2) can be determined in dependence on results which are available regarding the influence of moisture,

e.g., as the 90% fractile value, provided that a sufficient number of measured values is available. The hygroscopic moisture content will then be represented by that value which has been determined by a measurement of the specimen used also to measure lambda.

That method affords the advantage that the specific properties of the products of each manufacturer regarding the hygroscopic moisture content and the rise of lambda in dependence on the moisture content of the material can be taken into account. The proposed method would eliminate the need for an agreement on uniform and generally accepted European factors reflecting the influence of moisture on the thermal conductivity. As a result, specific properties of the products of different manufacturers could be taken into account in spite of the use of uniform testing methods.

REFERENCES

Achtziger, J. 1992. Determination of the thermal conductivity of plane block masonry; comparison of wall measurements and small slab specimens with the guarded hot plate. Report Nr. E1-3/92 from Forschungsinstitut für Wärmeschutz e.V. München
Künzel, H. 1971. Gasbeton: Wärme- und Feuchtigkeitsverhalten (heat and moisture performance of AAC). Bauverlag Wiesbaden, 1971.
Hums, D. 1982. Relation between humidity and heat conductivity in aerated concrete. Autoclaved Aerated Concrete, moisture and properties, edited by F.H. Wittmann, 1983, Elsevier, Amsterdam.

Advances in Autoclaved Aerated Concrete, Wittmann (ed.) © 1992 Taylor & Francis. ISBN 90 5410 086 9

An experimental study on thermal transmission properties of aerated concrete composite panels

C. Liu & J. Wang

Suzhou Concrete and Cement Products Research Institute, People's Republic of China

ABSTRACT: The measurements methods and their results of steady state thermal transmission properties of aerated concrete blocks and composite wall panels thereby are chiefly presented in this paper. The authors introduce the sub-non-steady state measurement apparatus and its improved details as well as the results of experimental research on insulation properties of panels of hollow clay bricks with exterior insulation form and interior insulation form.

1 INTRODUCTION

It has been shown that in China, about one third of heat is lost through exterior walls in residential buildings in winter time. Technicians have done a considerable amount of research work in the field of energy-conserving wall systems to reduce the energy-consumption of residential buildings. On one hand, they did a lot of experimental research on various properties of those panels in laboratories; on the other hand, they constructed some trial buildings, and various types of energy-saving wall panels have been developed after practical measurement and use.

The traditional wall system of residential buildings in China is based on clay brick walls. This wall system has some advantages such as cheap and fine architectural functions. It has been widely used over a long time. However, those wall systems destroyed a lot of fertile farmland since they consumed a large amount of clay. Furthermore, owing to the unsufficient attention devoted to energy-saving projects in the past, some architectural wall systems are too thin, so that their thermal insulation properties are poor. In order to overcome these shortcomings, two schemes are put forward: one is to build composite walls using clay brick layers of variable strength; another is to build composite walls with improved insulation properties by taking the original clay brick wall as a base and combining it properly. For this purpose, we carried out some experimental studies on thermal transmission properties of composite wall systems made of clay bricks and aerated concrete in laboratory. The research project included two aspects: the first is the steady state thermal transmission property of the composite wall system; the second is to compare different kinds of wall combinations, that is, by employing an exterior insulation layer or an interior insulation layer (e.g. the aerated concrete is placed on the interior surface of the clay brick wall), then the sub-non-steady state method is used for study.

2 THE WALL SYSTEM

The wall specimen was made up of hollow clay bricks and aerated concrete blocks as shown in Fig. 1. A ferrocement mortar layer was placed on the exterior surface of the aerated concrete, a cement mortar layer was placed on the exterior surface of the clay bricks.

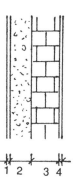

Fig 1 The composite wall system

1 Ferrocement of 2 cm in thickness;
2 Aerated concrete block, 10 cm, bulk density 500 kg/ m^3;
3 Hollow clay brick 24 cm, with 20 holes, void content 22,8 %, bulk density 1390 kg/m^3;
4 Cement mortar, 2 cm.

3 STEADY STATE TEST

We employed the standard test method, i.e. the protective hot box method which is widely used all over the world and was officially approved by ISO to measure the steady state thermal transmission properties of composite walls (In China, this method has also been officially approved as a national standard, and the standard will be issued officially in the near future).

The testing apparatus mainly consists of two sections; one section includes hot box, cold box, specimen and its frame, another section includes controlling and measuring instruments. The use of this method is based on the infinite panel thermal transmission principle. The specimen being measured is placed between hot box and ice box (see Fig 2).

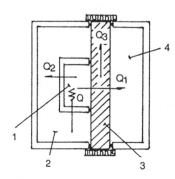

Fig. 2 The set-up of measuring system

1 hot box
2 and 4 cold box
3 specimen

The input power of the measurement box is Q; the heat quantity Q_1 penetrating through the specimen can be calculated by amending radiation heat Q_2 and Q_3. Then the data of all steady state thermal transmission properties of the wall system can be calculated on the basis of the medium temperature in all boxes and the surface temperature on both sides of the specimen.
The measurement results are shown in Table 1.

In our study, we also tested and measured the individual specimen of aerated concrete (without plaster on both sides). The results of these measurements are shown in Table 2.

4 THE NON-STEADY STATE TEST

Generally, architecture enclosure structures refer to plate walls with limited thickness. The enclosure structure made of only one material is a one-layer plate wall and the composite enclosure structure is a multi-layer plate wall. In summer, both sides of those walls can be affected by cyclic heat. In order to solve this problem, this comprehensive process is normally broken down into three individual processes which are then superimposed. All three processes are stable processes. The temperature of the interior surface in the room is constant. The cyclic thermal transmission process under the effect of outdoor thermal harmonic waves and the temperature of exterior surfaces remain constant. The cyclic thermal transmission process under the effect of indoor thermal harmonic waves is shown in Fig. 3.

Table 1

Speci-men No.	Temperature ($^\circ$C)				Power after amended Q(W)	Thermal resistance(m^2K/W)			
	hot air	hot side	cold side	cold air		t.r.* of mater-ial	r.h.t. of inter. surfa.	r.h.t.** of exter. surfa.	Total t.r.
1	25.23	22.87	−6.28	−8.04	23.20	1.29	0.11	0.08	1.48
2	24.99	22.55	−6.15	−8.01	23.04	1.28	0.11	0.08	1.47
3	24.89	22.41	−6.28	−8.19	23.04	1.28	0.11	0.09	1.48
Aver.	25.04	22.61	−6.24	−8.08	23.09	1.28	0.11	0.08	1.47

* t.r. refers to thermal resistance
** r.h.t. refers to resistance to heat transfer

Table 2

Speci-men No.	Temperature ($^\circ$C)				Power after amended Q(W)	Thermal resistance (m^2K/W)			
	hot air	hot side	cold side	cold air		t.r. * of mater-ial	r.h.t. of inter. surfa.	r.h.t.** of exter. surfa.	total t.r.
1	28.5	25.2	−1.4	−7.0	41.87	0.66	0.08	0.14	0.88
2	28.7	25.4	−1.4	−7.0	42.11	0.66	0.08	0.14	0.88
3	28.8	25.5	−1.2	−7.0	42.22	0.65	0.08	0.14	0.87
Aver.	28.7	25.4	−1.3	−7.0	42.07	0.66	0.08	0.14	0.88

* t.r. refers to thermal resistance
** r.h.t. refers to resistance to heat transfer

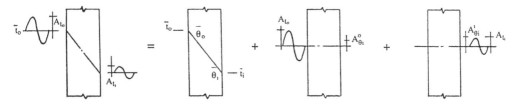

Fig. 3 The break-down of the thermal transmission process under the effects of two-way thermal harmonic waves

In order to study the non-steady state thermal transmission properties of aerated concrete and hollow clay bricks after they are combined with exterior insulation composite, we set up a set of testing apparatus. Since it is rather difficult to design an apparatus for the specimen with both sides exposed to the effects of thermal harmonic waves, the apparatus was designed in such a way that one side of the specimen experiences the effects of thermal harmonic waves and that the temperature on the other side of it will remain essentially unchanged.

As this process is not a complete non-steady state thermal transmission process, we refer to it as a sub-non-steady state thermal transmission process. The apparatus was made by taking the protective hot box as a basis, then modifying and improving it by keeping the original specimen frame and cold box and modifying the hot box properly.We replaced heating wire by infra-red lights for heating. The apparatus is shown in Fig. 4.

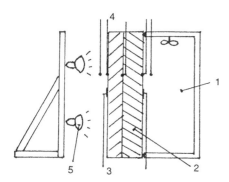

Fig. 4 Sub-non-steady state testing apparatus

1 constant temperature room;
2 specimen;
3 heat flow meter;
4 thermal couple;
5 infra-red heaters.

Heating and cooling devices are installed inside the constant temperature room, and the constant temperature is controlled by PID regulator. The infra-red heating device consists of several infra-red lights. The periodical temperature changes are controlled by a programmable setting device and PID regulator, etc.

We tested and measured a composite wall specimen with exterior insulation and a composite wall specimen with interior insulation, respectively. The temperature changes and the calculation results are shown in Fig. 5, Fig. 6 and Table 3, respectively.

The temperature amplitude on the interior surface, the total damping number of temperature wave and the delay time are three common thermal indexes. For the convenience of comparison, we quote the relevant practically measured data in situ in ref. (1) as follows (see Tab. 4):

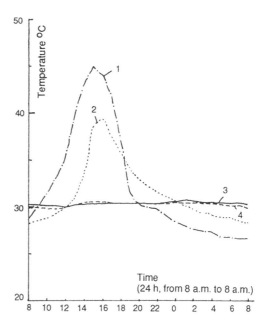

Fig. 5 The temperature curve diagram of a composite panel with exterior insulation

1 solar/air temperature;
2 exterior surface temperature;
3 interior surface temperature;
4 indoor temperature.

Table 3

		Exter. insulation	Inter. insulation
Maximum temperature (^{o}C)	exter.surface T_w	39.5	40.2
	inter.surface T_n	30.7	27.1
	temperature difference T	8.8	13.1
	indoor temper. t_n	30.5	26.8
Average temper. between day and night	inter. surface $T_{n.p}$	30.4	26.8
	indoor temperature $t_{n.p}$	30.2	26.4
Temper. amplitude of inter. surface	$T_{n.max} - T_{n.p}$	0.3	0.3
Total damping number of tempera. wave	$\dfrac{t_{z.max}-t_{z.p}}{t_{n.max}-t_{n.p}}$	43.3	37.7
Delay time	$T_{n.max}$ a.p.* $-- t_{z.max}$ a.p.	11	11
Solar radiation intensity ($km/m^2.hr$)	Max J_{max}	360	345
	Aver. J_p	195	189
Solar/air temperature (^{o}C)	Max $t_{z.max}$	45	39.5
	Aver. $t_{z.p}$	32	28.2
Indoor temperature (^{o}C)	Max $t_{w.max}$	35.3	30.1
	Aver. $t_{w.p}$	29.7	25.6

* a.p. refers to appearance time .

5 COMPREHENSIVE ANALYSIS OF TEST RESULTS

1. For the same composite wall system, no matter whether exterior or interior insulation is employed, both thermal resistance values under steady state thermal transmission are essentially identical. Therefore, in this steady state test, we adopted one kind of thermal resistance value for the composite wall system with aerated concrete and hollow clay brick.

As specified in "Thermodynamic Design Rules for Civil Building" in China (JGJ - 24-86), the total thermal resistance values of enclosure structures should not be

Table 4

	brick wall 370mm	brick wall 240mm	*	**
Interior surface temper. amplitude A_{tn} (^{o}C)	1.0	1.9	1.7	0.8
Total damping number of temper. wave ν_0	23.7	12.8	12.6	39.1
Delay time ε_0	11	10	10	10

* refers to composite exterior wall panel with fly ash and aerated concrete.
It is a big ribbed composite panel with fly ash aerated concrete as core (130 mm in thickness), and grade 200 fine stone concrete as cover; its total thickness is 200 mm.
** refers to insulation roof with aerated concrete. It consists of round hole concrete panels with 120 mm in thickness, slag ash with 70 mm in thickness, mortar with 30 mm in thickness, and aerated concrete with 100 mm in thickness; the total thickness is 350 mm.

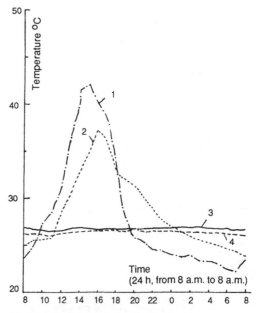

Fig. 6 The temperature curve diagram of a composite panel with interior insulation

1 solar/air temperature;
2 exterior surface temperature;
3 interior surface temperature;
4 indoor temperature.

less than the minimum total thermal resistance value determined by the following formula:

$$R_{0\ min} = \frac{(t_i - t_e)\,n}{\Delta t}\,R_i$$

where : $R_{0\ min}$ = minimum total thermal resistance of enclosure structure, m^2 k/w;

t_i : indoor temperature in winter. For ordinary residential buildings t_i is 18° C;

t_e : outdoor temperature of enclosure structure in winter. The heat inertia index for this composite wall is bigger than 6, as specified in this formula, and taking the Beijing area as an example, -9°C was chosen for t_e (in this steady state test, cold side temperature is designated and controlled according to this value);

n: the modified coefficient of temperature difference, in this test, n= 1,0;

(Δt): allowable temperature difference between indoor atmosphere temperature and interior surface temperature of enclosure structure.

As specified in this formula, the tolerable temperature difference was fixed at 6° ;

R_i: resistance of interior surface to heat transfer of enclosure structure; in this test 0,11 m^2 k/w.

On calculation, $R_{0\ min}$ = 0.50 m^2 k/w was obtained. The total thermal resistance value of the composite wall system with aerated concrete and hollow clay brick specially designed for this test was 1,47 m^2 k/w. This value is sufficient to meet the insulation requirements for enclosure structures in the Beijing area. Compared with the insulation capabilities of the 370 mm clay brick wall traditionnally employed in China (in the Beijing area, the thickness of common clay brick walls is 370 mm) (R $_{370\ clay\ brick}$ = 0,64 m^2 k/w),the insulation properties of composite walls are more than twice improved. Furthermore, composite walls reduce clay consumption by 50 %; instead, some industrial waste such as fly ash is used.

2. The temperature amplitude and delay time on the interior surface are identical for both exterior and interior insulation forms. The temperature amplitude on the interior surface in this composite wall system is 0,3°C, which is very much less than the 2,5°C temperature amplitude required for interior surfaces of common buildings by "Thermodynamic Design Rules for Civil Building" in China. Also, the total damping number of temperature waves is bigger than 37. From the preceeding analysis we can conclude that the composite wall system with aerated concrete and hollow clay brick, no matter whether exterior or interior insulation is employed, has better insulation properties. There is no big difference in insulation values between exterior and interior insulation forms. A comparison of Table 3 and Table 4, resp. of the composite wall system and the clay brick wall of 370 mm of thickness shows that both the temperature amplitude of interior surfaces and the total damping number of temperature waves of composite walls are much better than those measured in brick walls of 370 mm in thickness, except for the delay time, which is identical in both cases.

3. By summing up the points stated above, when exterior insulation or interior insulation is employed for composite wall systems with aerated concrete and hollow clay brick, the insulation properties expressed in values are essentially identical in both cases. However, when we further analyze the interior section of the composite wall structure for signs of condensation according to the relevant thermodynamic parameters (provided by the "Thermodynamic Design Rules for Civil Building" in China) and the thermal resistance values in practice during the winter heating period in the Beijing area, we find that the interior section of composite wall systems with exterior insulation shows some condensation, while that with interior insulation does not (see Fig. 7 and Fig 8). If the interior section of enclosure structures shows signs of condensation, the insulation capability of the wall system will be impaired. Therefore, when the composite wall system is used for new buildings and the aerated concrete block is used as insulation material for the renovation of old buildings in cold areas, we would recommend the use of the interior insulation form, so as to take maximum advantage of the insulation properties of the composite wall system.

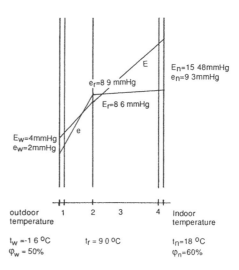

Fig. 7 Distribution of vapour pressure in a composite wall system

e: the distribution of vapour pressure within the wall;
E: the distribution of maximum vapour pressure within the composite panel;
The materials indicated by 1, 2, 3, 4 in this figure are the same as those shown in Fig. 1.

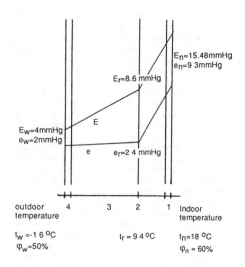

Fig. 8 e: the distribution of vapour pressure within the composite panel;
E: the distribution of maximum vapour pressure within the composite panel;
The materials indicated by 1, 2, 3, 4 in this figure are the same as those shown in Fig. 1.

REFERENCES

Jul. 1981. The summer time insulation measurement of Type 79 framed lightweight panels residential building in Shijiazhuang. Beijing Building Research Institute.
Architectural Physics Laboratory of Xian Institute of Metallurgy and Construction Engineering, 1977 Architectural thermodynamic design.
Thermodynamic Design Rules for Civil Building (JGJ 24 - 86), 1986.

Advances in Autoclaved Aerated Concrete, Wittmann (ed.)© 1992 Taylor & Francis. ISBN 90 5410 086 9

The effect of moisture on the thermal conductivity of AAC

K.F. Lippe
YTONG AG, R+D-Centre, Schrobenhausen, Germany

ABSTRACT: The effect of moisture on the thermal conductivity of AAC is frequently compensated by means of proportional addition. However, the effect of moisture is a very intricate thing, where the internal surface of AAC holds the decisive effect. Therefore it is not appropriate to use a proportional addition for the accurate determination of the thermal conductivity in the equilibrium state, but to measure the thermal conductivity in this equilibrium state if possible.

1 THE THERMAL CONDUCTIVITY OF AAC

The thermal conductivity λ of AAC in a building consists of

$\lambda_{Material}$,

$\lambda_{Radiation}$,

λ_{Water},

$\lambda_{Conductivity}$ in the water vapour/air mixture,

$\lambda_{Diffusion}$.

In the course of a research project [1] promoted by the BMFT (Federal Ministry for Research and Technology) the individual factors effecting the thermal conductivity of AAC have been examined. Within the scope of this publication it is only intended to consider the effect of the equilibrium moisture content, established in buildings, on the thermal conductivity of AAC.

2 THE EFFECT OF MOISTURE ON THE THERMAL CONDUCTIVITY OF AAC

Figure 1 shows the familiar linear dependence of the thermal conductivity on the moisture content of AAC for the relevant range. The equilibrium moisture content established in AAC-buildings depends on the internal surface of AAC (figure 2). The crucial factor for the internal surface is the portion and the crystallinity of the calcium silicate hydrates and not the macro porosity. It is thus appropriate to give the equilibrium moisture content not in vol.% but in wt.% because then it becomes independent of the bulk density (figure 3). This has repeatedly been pointed out in the literature [2]. In order to deal with the present topic, namely the effect of moisture on the thermal conductivity of AAC, it is necessary to consider λ_{cond} and λ_{diff} in detail. The following applies for the case that the temperature of AAC is < 59 °C. According to Krischer [3] λ_{diff} is decisive if the moisture content is above the hygroscopic range. Within the hygroscopic

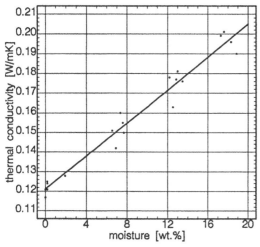

Fig. 1 Dependence of thermal conductivity on moisture content of AAC

range the water vapour pressure inside the pores is lower than the saturated vapour pressure. It is here that the transition to the dry state conductivity λ_{cond} takes place. This means, however, that the equilibrium moisture content

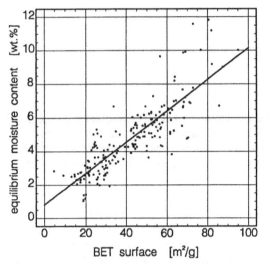

Fig. 2 Equilibrium moisture content of AAC

(for $\psi_{hygr.max}$) must effect the dependence of the thermal conductivity on the moisture content (moisture content slope α). This is outlined in figure 4. Samples with a high equilibrium moisture content (correspondingly high $\psi_{hygr.max}$) consequently must have a low increase of α.

In moist AAC there are two kinds of pores, the ones with dry walls (λ_{cond}) and the others with moist walls (λ_{cond} and λ_{diff}), that is to say, the lower the moisture content the smaller are the pores containing water. Therefore the pore size distribution should also be of importance, since in small pores the wall area to volume ratio is large. Identical pore volume results for small pores in a larger pore wall surface than for big pores. For small pores, the pore volume in the pores with moist walls is, at identical moisture content, higher than in big pores which should cause a lower increase of α. There are for instance measurements by Th. L. Madsen about the effect of moisture content on λ (soft fibre board ρ 260 kg/m³). As shown in figure 5, there is a linear dependence between λ and the relative humidity of the air [4].

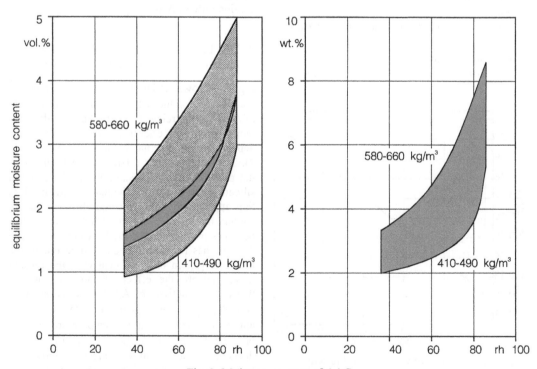

Fig. 3 Moisture content of AAC

Table 1 Effect of the measuring period on λ (moisture content ≈ 5 wt.%)

sample	O_{sp} [m²/g]	slope of the moisture content α [vol.%]	equilibrium moisture content 63 % rh , 70 °C [wt.%]	measuring period [h]	λ [W/mK]	moisture transport [wt.%]
1	76.4	0.0029	10.03	3	0.107	-
				24	0.106	-0.032
2	30.0	0.0118	3.03	3	0.123	-
				24	0.116	-0.101

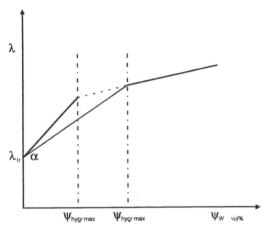

Fig. 4 Effect of the equilibrium moisture content (according to Krischer)

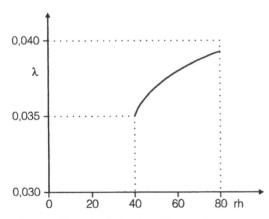

Fig. 5 Effect of relative humidity on λ

A low increase in α should therefore be reached through a high portion of CSH-phases, by fine crystalline CSH-phases and a high number of micropores.

From a long-term measurement of the thermal conductivity of AAC with strongly differing α for 24 hours, it turned out that the thermal conductivity of the sample with a low value of α and a high internal surface, according to BET, remained almost constant and the moisture content of the material decreased slightly (table 1). For the sample with a high value of α and a smaller BET-surface, however, the result was that λ decreased by 7/1000 within the 24 hours and at the same time the moisture transport was three to four times higher. The result of the experiment was that samples with a large internal surface showed a lower moisture transport than samples with a small internal surface, which agrees with the above mentioned theoretical reflections. Measurements have been performed by means of a heat flux meter according to German standard DIN 52616.

The evaluation of the measurements, carried out within the scope of the BMFT project, where AAC's with very different internal surfaces were deliberately produced, showed in agreement with the theoretical findings that the equilibrium moisture content increases with an increasing surface as per BET (figure 2) and the slope α decreases (figure 6).

3 DETERMINATION OF THE THERMAL CONDUCTIVITY IN A MOIST STATE

It can be derived from the above explanations that the simple consideration of the moisture content on the thermal conductivity of AAC by means of proportional addition cannot be very appropriate.

The moisture content has to be accounted for by means of an additional parameter which is mainly dependent on the internal surface of

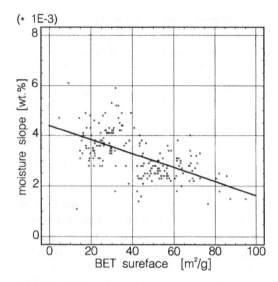

(* 1E-3)

Fig. 6 Moisture slope

AAC. On the basis of the data found by the BMFT project, the following formula can be derived.

$$\lambda = \lambda_{dry} + \rho * (0.00583 + 0.0006725 * O_{sp})$$

If, as usual practice, a constant value is taken for the equilibrium moisture content, the following formula may be deduced from the above data.

$$\lambda = \lambda_{dry} + \alpha * \rho * u_m, \quad \alpha = 0.0072$$
$$\rho \quad [kg/dm^3]$$
$$u_m \quad [wt.\%]$$

The physically most appropriate procedure so far has been suggested by Dr. H. Künzel namely to measure the thermal conductivity in the equilibrium state. While testing this suggestion, it turned out that by choosing the right measuring procedure and favourable parameters it was possible to reach values which can be compared to the previous measuring procedures. Important factors are homogeneous moisture content distribution, temperature differences ≤ 10 °C and measuring periods below 3 hours.

4 CONCLUSION

The thermal conductivity of AAC in the state of application is determined by the quantity and the behaviour of the equilibrium moisture content. Therefore the internal surface as per BET is the most important factor. The stated formulas serve to calculate the effect of the moisture content. Apparently, the most appropriate method is to measure the thermal conductivity at equilibrium moisture content.

REFERENCES

[1] Lippe, K. F. 1985. Entwicklung hochporöser CSH-Werkstoffe mit minimaler Wärmeleitfähigkeit. BMFT 03E-8052-A
[2] Künzel, H. 1986. Bestimmt der volumen- oder massebezogene Feuchtegehalt die Wärmeleitfähigkeit von Baustoffen? Bauphysik 8: 33-39.
[3] Krischer, O. and Kast, W. 1978. Die wissenschaftlichen Grundlagen der Trocknungstechnik. Berlin.
[4] Madsen, Th. L. 1969. Modified guarded hotplate apparatus for measuring thermal conductivity of moist materials. The Thermal Insulation Laboratory, Lyngby. Mitteilung No. 19.

Advances in Autoclaved Aerated Concrete, Wittmann (ed.) © 1992 Taylor & Francis. ISBN 90 5410 086 9

Application of autoclaved aerated concrete for high-temperature insulation

E. Schlegel
Bergakademie Freiberg, Germany

J. Volec
YTONG AG, Schrobenhausen, Germany

ABSTRACT: Heat-resistant AAC was produced and used in kilns and dryers as thermal insulating material at temperatures to 900 °C in three factories during the last ten years. The east-german standard TGL 33 523/02 contains the properties and demands on heat-insulation AAC. The bulk density of AAC is 300 to 600 kg/m³, the thermal conductivity at 20 °C is between 0,10 and 0,15 W/m·K, at 900 °C below 0,30 W/m·K.

1 INTRODUCTION

Several firms in the world are producing heat-insulating materials on the basis of calcium silicates for use in technical high-temperature plants like furnaces, dryers and other high temperature constructions. Usually the calcium silicate heat-insulating products are microporous high porosity materials, produced in a special press technology and mostly contained a little portion of anorganic refractory or/and organic fibres.

In East Germany such excellent heat-insulating materials was not disponsible in former times. Therefore we made a research work on high-temperature properties of AAC and on improvement of these products in last 15 years with the intension to substitute the classical diadomite bricks and the well known calcium silicate materials. An amount of about 100 000 m³ heat-resistent AAC-material was produced in 3 factories during the last 10 years and was called SILTONTHERM. Therefore in East Germany a lot of knowledge exists in production, use and properties of such macroporous calcium silicate high-temperature insulating material. The east-german standard TGL 33 523/02 contains the properties and demands on heat-resisting AAC. The first and most important field of using was the heat insulation in big damp generators of power stations.

There are several reasons of this development:
- no disposal of high quality thermal insulating materials in East Germany and high need of these products,
- fundamental suitability of AAC for high-temperature use,
- low costs of production of heat-resistant AAC.

2 PHASE STRUCTURE OF AAC

AAC consists of calcium silicate hydrates (mostly tobermorite and C_2S-hydrate) and quartz with a little amount of calcite and anhydrite (see table 1).

Table 1. Qualitative X-ray analysis after 12 h heating

temperature	AAC with bulk density of		
°C	300	500	600 kg/m³
20	Q,T,C,H	Q,T,H	Q,T,C,H
500	Q,T,C	Q,T	Q,T,C
700	Q	Q	Q
800	Q,W	Q,W	Q,W
900	Q,W,S	Q,W,S	Q,W
1000	Q,W,G,S	Q,W,S	Q,W,S

Q - Quartz, T - Tobermorite, C- Calcite, H - C_2S-Hydrate, W - β-Wollastonite, G - Gehlenite, S - Dicalciumsilicate

The high temperature properties are mostly determined by the mean components CSH-phases and quartz.

Well crystallized tobermorite or tobermorite-like phases lose the crystal water continuously in the temperature range of 100 to 600 °C. This dewatering of about 12 mass-% gives a little shrinkage, because the tobermorite structure is keeped on principle, and we have a x-ray amorphous CS-phase about at 700 °C. This amorphous CS crystalizes spontaneously at 800 °C becomes wollastonite. Thereby the density increases and we have a theoretical linear shrinkage of 8,9 %, when clean tobermorite goes to wollastonite. Therefore the limit of temperature use of pure tobermorite heat-insulating materials is 600 to 700 °C.

Quartz has a modification change at 573 °C. The decrease of density (increase of specific volume) has no essential consequences to technical properties of AAC, because the volume growth goes in the weak tobermorite binding phase. At 1000 °C quartz changes in crystobalite; this process has a very little speed and is not important for the subject of this field. Table 1 shows the crystalline phases in 3 AAC materials.

If we have a look at the phase diagramm CaO/SiO$_2$/Al$_2$O$_3$, we can see, that heat resistant AAC lies in the fields of crystobalite or wollastonite. We can expected that the first melt exists at temperature of 1165 °C, probably at somewhat deeper because of impurities in AAC. So we think that the melt at about 1100 °C gives a weak material and a high shrinkage in consequence of liquid phase sintering. So we expect that the AAC probably can used in temperature range up to 900/1000 °C.

3 PROPERTIES OF AAC AT HIGH TEMPERATURES

The technically most important property is the temperature limit of using. In table 2 we can see in several investigations that the limit of use is at temperature of 900 °C.

Table 2. Properties of AAC at high temperatures

property	measured value
shrinkage 1 %	705 - 745 °C
2 %	905 - 955 °C
begin of creep under compression	900 - 950 °C
refractoriness under load	970 - 1060 °C
hot bending strength taller than 0,05 N/mm^2	920 - 995 °C
cold compression strength after heating taller than 0,1 N/mm^2	900 - 1100 °C
thermal shock resistance, 900 °C, quenching in air	more than 10 changes
ignition loss at 1000 °C	12 - 16 %
capillary crack formation	400 °C
crack growth	900 °C
visible softening	1200 °C

In table 3 the other important basic high temperature properties of AAC are shown.

4 APPLICATION OF AAC IN HIGH TEMPERATURE INSULATION

AAC are used in several industries with high temperature factories like furnaces, kilns or dryers: power stations, metallurgy, engine-building, ceramic, glass, cement or enamel plants, chemical industry agriculture, for fuels and pipes

Table 3. Properties of AAC

property	measured value		
bulk density kg/m^3	300	500	600
linear shrinkage %			
at 600 °C	0,72	0,64	0,43
900 °C	1,96	1,86	1,92
1000 °C	2,10	2,31	2,39
compression strength N/mm^2	1,0	3,7	5,7
thermal conductivity W/m·K at 20 °C	0,098	0,135	0,150
300 °C	0,123	0,157	0,166
600 °C	0,194	0,201	0,211
900 °C	0,281	0,284	0,285
specific heat kJ/kg·K between 25 and 900 °C	1,022	1,038	1,039

technique and for laboratory kilns.

AAC with bulk density greater than 400 kg/m^3 are qualified for continuous high-temperature factories and for application, wich needs hot strength. Materials with bulk density lower than 400 kg/m^3 give a high thermal insulation and have a low heat storage; such products are qualified for discontinuous factories and for complete light building of kilns.

Today it is necessary to check the economical sense or senselessness and the possible saleability of high temperature resistant AAC.

Advances in Autoclaved Aerated Concrete, Wittmann (ed.)© 1992 Taylor & Francis. ISBN 90 5410 086 9

Experimental determination of AAC moisture transport coefficients under temperature gradients

J.-F. Daïan
Laboratoire d'Étude des Transferts en Hydrologie et Environnement (IMG), Grenoble, France

J.A. Bellini da Cunha
Laboratoire d'Étude des Transferts en Hydrologie et Environnement (IMG), Grenoble, France & Universidade Federal de Santa Catarina, Florianopolis, Brazil

ABSTRACT : Six experiments of drying of an AAC sample under various positive or negative temperature gradients were performed. The temperature and water content dependant coefficients of transport are deduced from the measurements of water content and temperature profiles. The results are compared to those of independant apparent thermal conductivity measurements. The vapor diffusion mechanisms are discussed. A multiscale pore-network model is tentatively applied to the case of AAC.

INTRODUCTION

This paper aims at four purposes:
i) To report the results of an experimental determination of the coefficients of moisture transport under temperature gradient in AAC.
ii) To compare this data to the one obtained independantly from apparent conductivity measurements.
iii) To analyse, starting from the results, the mechanisms of vapour diffusion, particularly the influence of Knudsen effect and the role of condensed water.
iv) To test a multiscale pore-network model for moisture transport in porous media developped elsewhere.

1 NON ISOTHERMAL MOISTURE TRANSPORT

Before we report the experimental process, it is necessary to breafly recall the governing equations of non-isothermal moisture transport, since the experiments aim to determine the coefficients involved.

The global moisture flux J (kg/m².s) is supposed to be a linear function of the gradients of the volumic moisture content θ and of the temperature T :

$$J/\rho_l = - D_\theta \nabla\theta - D_T \nabla T - K_l k \qquad (1)$$

ρ_l (kg/m³) is the density of liquid water. The permeability K_l (m/s) accounts for gravity (k being the vertical unit vector), and the two transport coefficients D_θ (m²/s) and D_T (m²/s.K) are temperature and moisture content dependent.

In this global flux, the particular component corresponding to vapour diffusion under thermal gradient plays an essential role in the simultaneous heat transfer. The corresponding coefficient, D_{Tv}, can be written, owing to some approximations, as :

$$D_{Tv} = \frac{\rho_{vs}}{\rho_l} \alpha \phi \zeta D_v \qquad (2)$$

where D_v (m²/s) is the coefficient of vapour diffusion in the medium, ρ_{vs} (kg/m³) is the density of saturated water vapour, α (K⁻¹) is its coefficient of variation with temperature (about 0.54 K⁻¹ at the temperature of the present experiments), ϕ is the relative humidity of air in the medium. The factor ζ accounts for a more or less important difference between the bulk temperature gradient and the temperature gradient in the gaseous phase, due to differential heat conduction (Philip and De Vries, 1957).

In transient heat transfer processes under negligible gradients of the water content, which are suposed to occur in the various measurement methods of the thermal conductivity, the coefficient D_{Tv} is involved in the apparent conductivity of the medium :

$$\lambda_{app} = \lambda + L\, \rho_l\, D_{Tv} = \lambda + L\, \rho_{vs}\, \alpha\ \phi\, \zeta\, D_v \quad (3)$$

where λ (J/m.s.K) is the Fourier conductivity and L (J/kg) is the enthalpy of vaporisation.

One of the main problems with the coefficient D_{Tv} is the value of the vapour diffusivity D_v. Its potential reference value is determined by the free diffusivity of vapour in air ($D_a \approx$ 2.6 10^{-5} m²/s in the present conditions) and the porosity ε of the medium :

$$D_{v0} = D_a\, \varepsilon$$

Besides the ordinary tortuosity factor, Knudsen effects may result in a significantly smaller value in the case where the pores smaller than the mean free path of water molecules (≈ 1.5 10^{-7} m) control the vapour paths. Another problem is the role of the capillary water in vapour diffusion. If it is considered to close the pores, D_v will be a decreasing function of the water content. On the countrary, one can consider that a condensation-evaporation process confers to the liquid phase the role of a short-circuit in vapour paths. In this case, D_v may be, in a given range, an increasing function of the water content. This effect may be particularly important when coupled with Knudsen effect. When water content increases, the highly resistant narrow pores are progressively transformed in highly conductive pores by the presence of capillary liquid.

2 THE MATERIAL STUDIED

The experiments reported below were performed with AAC Siporex of dry density 385 kg/m³. By means of moisture saturation experiments and dry density measurements performed with expanded and non-expanded material, and comparision with image analysis data (Laurent, 1992), the intrinsic porosity of the matrix can be estimated to be $\varepsilon_p = 0.60$, and the artificial porosity is $\varepsilon_c = 0.60$. The total porosity of the medium is :

$$\varepsilon = \varepsilon_c + (1 - \varepsilon_c)\, \varepsilon_p = 0.84$$

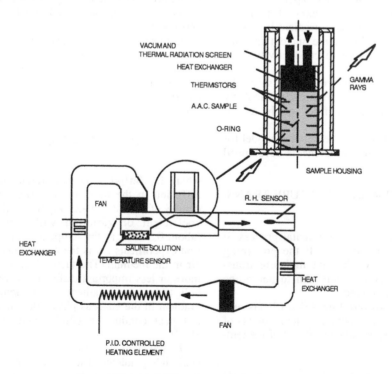

Figure 1. Experimental device.

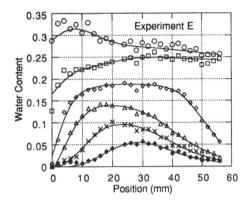

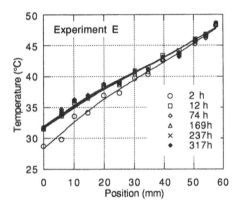

Figure 2. Water content and temperature profiles (Experiment E).

The moisture transport experiments performed are limited to values of the water content less than the water content corresponding to the saturation of the matrix, i.e. $(1-\varepsilon_c)\,\varepsilon_p = 0.24$. This moisture content is in agreement with the one obtained by means of spontaneous imbibition. Larger saturations of the material can be reached by means of thermal condensation, and the total saturation is obtained only by means of imbibition under vacuum. It appears that the artificial pores are not spontaneously filled with liquid. The water content 0.24 can be therefore considered as the maximum saturation which may occur in the normal conditions in buildings.

3 EXPERIMENTAL DEVICE

The purpose is to perform one-dimensional moisture transport under one-dimensional tem-

perature profile in a cylindrical sample of the material (40 mm in diameter, about 55 mm length). The upper and lateral surface of the sample are impervious. Fig. 1 shows the experimental device and particularly the thermal lateral insulation. The evolution of the moisture content profile along the sample is determined during the experiment by means of a mobile gamma-ray absorption device (2 mm beam thickness). The temperature profile is monitored by a set of thermistances implanted in the sample.

The sample is previously wetted by circulating wet and hot air in the loop and imposing a low temperature at the upper side of the sample. The measurements performed during the condensation phase will not be examined here. During the drying experiment itself, dry air (about 0.14 relative humidity) circulates in the loop. The temperature of air can be either larger, or smaller than the temperature imposed at the upper side of the sample. Table 1 indicates the conditions of the six experiments performed.

Table 1. The experiments performed

Exper.	Upper Temp. (°C)	Air Temp. (°C)	Average Gradient (°C/m)	Duration (hours)
A	40	40	0	300
B	32	20	140	745
C	45	20	333	540
D	22	35	-138	470
E	47	30	268	350
F	30	50	-250	500

4 EXPERIMENTAL RESULTS

Figures 2 and 3 show two typical examples of the evolutions of the water content and temperature profiles obtained. One can observe particularly that the temperature depression at the ventilated side of the sample, due to evaporation, is limited to the first hours of the experiment. The later evolution of the temperature is very slow, and a quasi-stationary heat transfer regime is achieved. The local slope of the temperature profile is more or less linked to the water content at the corresponding point, due to the dependance of the thermal conductivity on the moisture content.

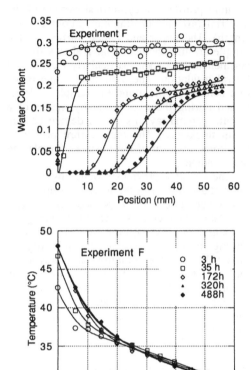

Figure 3. Water content and temperature profiles (Experiment F).

A remarquable phenomenon can be observed in the moisture content profiles in experiment E : in a large region close to the upper side of the sample, which is impervious, the negative moisture flux occurs under a negative water content gradient. This indicates that the temperature-driven flux (see Eqn 1) is strongly dominant under these circumstances.

5 CALCULATION OF THE TRANSPORT COEFFICIENTS

In Eqn (1), the two gradients can be derived from the measurements and the moisture flux can be calculated by mass balance from the moisture content profiles. A multivariate regression procedure allows one to determine the two coefficients, D_θ and D_T from a population of measurements of $\mathbf{J}$, $\nabla\theta$, ∇T obtained from the experiments. The difficulty is that the two

coefficients are supposed to be temperature and water content dependant. It is therefore necessary to select among the data those measurements which correspond to a given small range of T and θ. This is the reason why numerous experiments are required to obtain a representative population for the regression.

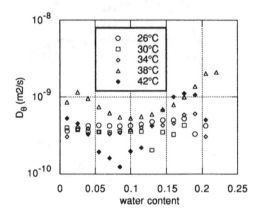

Figure 4. Moisture diffusivity D_θ.

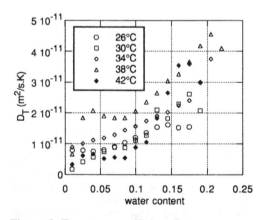

Figure 5. Transport coefficient D_T.

The results of the procedure are given in Figures 4 and 5 for the coefficients D_θ and D_T. The results for permeability are not reported, since the values obtained are statistically unsignificant. This indicates that gravity has unmeasurable effects in the present situation. For D_θ as well as for D_T, no reliable temperature dependance can be observed.

6 ANALYSIS OF VAPOUR DIFFUSION

If we assume, as it is commonly done, that the temperature driven flux is dominantly due to vapour transport, the measured coefficient $D_T = D_{Tv}$ allows one to analyse the vapour diffusion process in relation with the considerations in § 1. Eqn (2) allows to determine the product :

$$\delta_v = \phi \, \zeta \, D_v$$

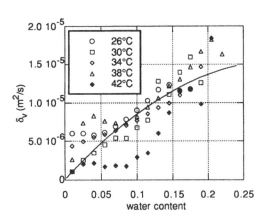

a

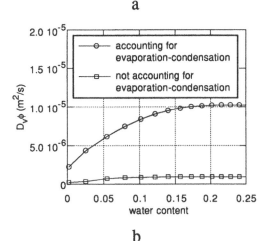

b

Figure 6. Modified vapour diffusivities. a: from experiments. b: from the pore-network model.

The results are shown in Fig. 6a. On the other hand, another estimation of this product can be derived from independant measurements of the apparent thermal conductivity per-

formed at various moisture contents and temperature on the same material (J.P. Laurent, personnal communication). According to Eqn (3), δ_v was estimated from the difference $\lambda_{app} - \lambda$, λ beeing the Fourier conductivity measured at low temperature, when vapour transport is negligible. The solid line in Fig. 6a shows smoothed results of this calculation. A satisfactory agreement with data from direct D_{Tv} measurements is obtained.

δ_v is a significantly increasing function of the water content. For all the range of saturation studied, its value is sensibly less than the estimation $D_{v0} = D_a \, \varepsilon \approx 2.2 \; 10^{-5} \; m^2/s$. The relative humidity ϕ cannot be expected to be responsible of this alone, since it is not very different from 1 for moisture contents above 0.05. The factor ζ is larger than 1 for the temperatures under consideration, since the apparent conductivity of air is smaller than the conductivity of the wet solid. We can conclude that vapour diffusion at low water content is sensibly affected by Knudsen effects. As it was discussed in § 1, increasing water content results in a reduction of this effect, which means that capillary water favourises the vapour diffusion by short-circuiting the narrowest pores.

7 TENTATIVE APPLICATION OF A MULTISCALE PORE-NETWORK MODEL

The model of moisture sorption and isothermal transport in porous media which we attempt now to apply to AAC is described with more details by Daïan (1992).

The structure of the medium is described in order to account for the presence of many orders of magnitude of the pore sizes. The pores are shared in a given number of classes, according to the size of their typical dimension. In each class, the pores are supposed to be randomly distributed, with a probability of presence depending on the class under consideration, in a 3D grid of which mesh is proportional to the typical size of pores of the class under consideration. The medium is defined by the superposition of all these grids.

The connection of the pores at a large scale is studied by means of successive renormalizations together with percolation basic principles. The pore-size distribution of the medium is defined by the number of pores of each class which are connected together or with pores of other classes. The same process is used to study the invasion of the medium by a non wetting fluid. This provides predictions of the mercury intrusion characteristic and of the desorption branch of the capillary characteristic of the medium.

The isothermal moisture transport coefficient of such a multiscale structure at a given saturation can be estimated using a modified version of Effective Medium Theory, together with renormalization to account for the presence of various scales in the medium. An hydric conductivity is assigned to each individual pore, on the basis either of Poiseuille law if the pore is filled with capillary water at the saturation under consideration, or of vapour diffusion law accounting for Knudsen effect if the pore is occupied by the gaseous phase.

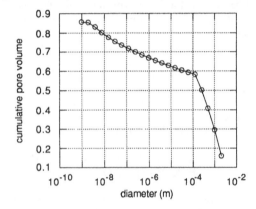

Figure 7. Hypothetical pore-size distribution used for the model.

This provides an estimation of the global coefficient of moisture transfer, fully accounting for the condensation-evaporation effect discussed in § 1. To share the respective contributions of the transport mechanisms in the two phases, the same calculation can be performed putting artificially equal to zero the hydric conductivity of the pores occupied by one of the two phases. But such a calculation supposes that there is no interaction between the two transport mechanisms, in other words, that no evaporation condensation occurs. Consequently, the sum of the two contributions obtained by this method is more or less smaller than the global coefficient of transport.

Concerning vapour diffusivity D_v, two estimations can be drawn from the model : the one which accounts for no evaporation-condensa-tion effects, and a second one which is obtained from the difference between the global transfer coefficient and the contribution of the liquid phase. The later is expected to be a better estimation of the "true" D_v, to be used for non isothermal processes.

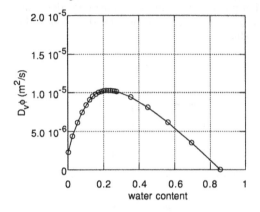

Figure 8. Modified vapor diffusivity computed from the pore-network model

The difficulty in the use of this model to describe an actual material is to chose the good pore-size distribution. For AAC, relatively objective information about the pore-size distribution of the artificial porosity can be drawn from image analysis. It is much more difficult to obtain a reliable distribution for the pores of the matrix, since it can be explored only by means of indirect methods like mercury intrusion, of which results are biased by invasion processes (Prim and Wittmann, 1983).

Among the numerous attempts which were done to define a representative pore-size distribution for the AAC under consideration, we report here the one which provided the best results from the view point of vapour diffusion. The model exhibits an extreme sensitivity to the pore-size distribution in the calculation of capillary characteristics for sorption and desorption, and of the permeability and of the diffusivity D_θ. The estimations of the vapour diffusivity are not so much influenced by the distribution. This is because the elementary hydric conductivity of the pores occupied by liquid water varies with the square of the diameter, while it is more poorly influenced by the diameter when the pores are occupied by air, even if Knudsen effect is accounted for. In any case, none of the attempted distributions provided results comparable to experimental data for coefficient D_θ, and we do not report them.

The pore-size distribution adopted is shown in Fig. 7. It accounts for the value of the total porosity (0.84) and of the artificial porosity (0.60), which pore-sizes are supposed to be within the range $1.25 \cdot 10^{-4}$ m and $2 \cdot 10^{-3}$ m. The distribution of matrix pores is relatively arbitrary, but it was chosen according to various informations available in the literature.

Figure 6b shows the estimation of the product $D_v \phi$ provided by the model, to be compared with the coefficient δ_v defined in § 5. No estimation of coefficient ζ can be drawn from the model, since it is essentially isothermal. D_v was estimated by the two methods defined above, i.e. accounting or not for the condensation-evaporation effects. It is clear from Fig. 5 that the assumption that the interaction between the two phases occurs is in good agreement with experiment, while the opposite one is not. Figure 8 shows the evolution of $D_v \phi$ provided by the model for the full range of the water content, which exhibits a shape comparable to the one obtained by Azzizi $et\ al$ (1988) by means of thermal conductivity measurements for an AAC of different density.

CONCLUSION

The experiments performed on a particular AAC provide reliable values for the two coefficients of mass transfer in the material in the range of usual water contents and of temperatures between 25 and 40 °C.

For the coefficient of mass transfer under thermal gradient, a consistent physical analysis based on the mechanisms of vapor diffusion, accounting for Knudsen transport and for the interaction with the capillary water, can be provided, owing to the comparision with the behaviour of the apparent thermal conductivity and with the results provided by a pore-network model. This analysis allows one to extrapolate the results obtained to a larger range of temperatures, having in view a building envelope model.

The coefficient of mass transfer under moisture content gradients is not so easy to analyse because of the great sensibility of the transfer characteristics of the liquid phase to the pore-size distribution of the medium. The beforehand problem to solve is to obtain reliable informations about the pore-size distribution from indirect measurements like mercury porosimetry. The coefficient measured can however be provisionally used in the envelope models, but its temperature dependance cannot be predicted.

REFERENCES

Azzizi, S., C. Moyne, & A. De Giovanni. 1988. Approche expérimentale et théorique de la conductivité thermique des milieux poreux humides. I. Expérimentation. $Int.\ J.\ Heat\ Mass\ Transfer$, 31 (11):2305-2317.

Bellini da Cunha, J.A. & J.-F. Daïan. 1991. Experimental analysis of moisture transport in consolidate porous media under temperature gradient. $Int.\ Sem.\ Heat\ and\ Mass\ Transfer\ in\ Porous\ Media$. Dubrovnik, May 20-24, 1991.

Daïan, J.-F. 1992. From pore-size distribution to moisture transport properties: particular problems for large pore-size distibutions. $8th\ Int.\ Drying\ Symp.\ IDS'92$. Montréal, Aug. 2-5 1992.

De Vries, D.A. 1958. Simultaneous transfer of heat and moisture in porous media. $Trans.\ Am.\ Geoph.\ Union$, 39:909-916.

Laurent, J.-P. & C. Frendo-Rosso. 1992. Application of image analysis to the estimation of AAC thermal conductivity. $3rd\ RILEM\ Int.\ Symp.\ on\ AAC$, Zürich, oct. 1992

Philip, J.R. & D.A. De Vries. 1957. Moisture movement in porous materials under temperature gradients. $Trans.\ Am.\ Geoph.\ Union$, 38:222-232.

Prim, P. and F.H. Wittmann 1983. Structure and water absorption of aerated concrete. In F.H. Wittmann (Ed.), $Autoclaved\ Aerated\ Concrete,\ Moisture\ and\ Properties$, p.55-69. Elsevier Scientific Publishing Company, Amsterdam.

Van der Kooi, J. 1971. $Moisture\ transport\ in\ cellular\ concrete\ roofs$. Thesis, Eindhoven.

Advances in Autoclaved Aerated Concrete, Wittmann (ed.)© 1992 Taylor & Francis. ISBN 90 5410 086 9

Determination of hydral diffusion coefficients of AAC – A combined experimental and numerical method

X. Wittmann, H. Sadouki & F. H. Wittmann
Swiss Federal Institute of Technology, Zürich, Switzerland

ABSTRACT: The time-dependent moisture field during drying of AAC can be found by solving the diffusion differential equation. Based on experimental data the diffusion coefficients of drying of three different types of AAC are numerically evaluated as function of moisture content. The moisture loss as a function of time and the moisture distribution under isothermal conditions of AAC elements are determined.

1 INTRODUCTION

Since many building materials are porous, the uptake and release of water and its transport have important consequences in building technology. Rain penetration, rising damp, water vapour condensation, drying shrinkage are just some examples. In all these cases we are concerned with a single fundamental process - the movement of water through a permeable material whose water content is non-uniform and generally below saturation. The corresponding differential equation is well known, but in general it is difficult, if not impossible, to solve it analytically for realistic boundary conditions. Also the diffusion coefficient actually may not be a constant. With the advent of computer the finite element method has become a very powerful tool for obtaining numerical solutions to problems of this kind (Wittmann et al. 1988, Roelfstra et al. 1983, Hall et al. 1982, Kasperkiewicz 1972,).

The present paper will describe the application of diffusion theory to the drying process of AAC. We assume drying to be an isothermal unsaturated flow generated by moisture gradients within a given element. Furthermore, this approach is phenomenologic and macroscopic, i.e. the theory employed does not describe the physical mechanism but examines only an average flow across sections whose dimensions are much larger than the pore size. The porous material and the water in it will be characterized by a diffusion coefficient and the local moisture content.

The aim of this study is to try to find out realistic diffusion coefficients, taking the time - dependent moisture content as the driving force. Experimental investigations on AAC include the determination of moisture change with time under isothermal conditions and the estimation of the moisture distribution in drying specimens. A computer programme based on the finite difference method combined with non-linear least square fit is developed for calculating the diffusion coefficient. Finally the results of moisture loss with time from numerical predictions and the experimental measurements are compared. It will be shown that a linear diffusion equation can not describe the drying process of AAC properly. The non-linear description of the diffusion process in drying AAC, i.e. the diffusion coefficient as a function of moisture content, gives better agreement with experimental data.

2 NUMERICAL ANALYSIS

The usually assumed mathematical law

for the moisture diffusion in porous materials is the following partial differential equation:

$$\frac{\partial U}{\partial t} = div\,[D(U)\,\overrightarrow{grad\,U}] \qquad (1)$$

where t is time, $U = U(\vec{x}, t)$ is the moisture content expressed in mass per unit volume, and $D(U)$ is the moisture diffusion coefficient. Eq. (1) is valid for isothermal conditions and homogeneous materials. In this general form Eq. (1) represents the three-dimensional moisture movement.

In the case of a drying prismatic specimen with four sealed faces, equation (1) can be simplified to describe the one-dimensional moisture movement:

$$\frac{\partial U}{\partial t} = \frac{\partial}{\partial x}\left[D(U)\frac{\partial U}{\partial x}\right] \qquad (2)$$

subject to the initial condition:

$$U(x, t = 0) = U_0$$

and boundary condition

$$U(x = 0, t) = U(x = L, t) = U_b$$

where U_0 is the moisture content in any 'elementary volume' at the beginning of the drying process; and U_b is the moisture content of the two outer layers of the drying specimen, which is in equilibrium with a given external relative humidity.

If we assume the drying of AAC to follow a linear diffusion process, i.e. the diffusion coefficient is a constant $D(U) = D$, then Eq. (2) can be written as

$$\frac{\partial U}{\partial t} = D\frac{\partial^2 U}{\partial x^2} \qquad (3)$$

In Fig. (1) direct measurements and theoretical calculations for moisture loss of drying specimen with time are shown. It can be seen that the linear equation (3) does not describe the drying of AAC realistically , because the real drying starts relatively slowly but then becomes more rapid. It follows that

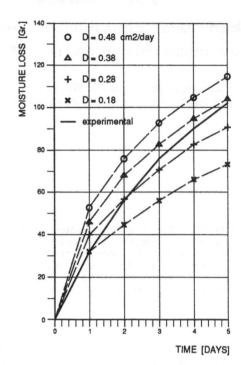

Fig. 1 Time-dependent moisture loss of AAC(GN) specimen of length 12cm

the diffusion coefficient must be assumed to be a function of the moisture content. Different forms of the dependence of D on the moisture content U have been used by different investigators for concrete (Bazant et al. 1971; Karsi et al. 1965; Pihlajavaara 1965; Mensi et al. 1988). In our case the so-called total implicit finite difference method is used (Meis et al. 1981; Marsal 1976; Wittmann et al. 1989).

In Eq. (2) the diffusion coefficient can be assumed to be the following general mathematical formular,

$$D = D(p_1, p_2, \dots, p_K) \qquad (4)$$

where $p_1, p_2, \dots, p_K$ are constants. Supposing D is differentiable until the 2nd order with respect to U and parameters $p_1, p_2, \dots, p_K$. These parameters can be obtained by fitting the calculated moisture profiles at different drying times to the experimental data, which is realized by minimizing the functional-error:

$$S(\vec{P}) = \sum_D [U_e(x, t) - U_c(x, t, \vec{P})]^2 \qquad (5)$$

114

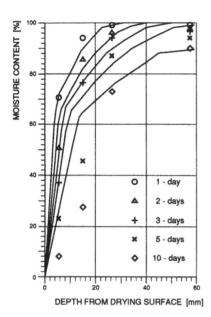

Fig. 2 Space distribution of moisture content of AAC(GN) after different drying periods

$\vec{P} = (p_1, p_2, ..., p_K)$ in Eq. (5) is a parameter vector.
D = depth from the drying surface of the specimen;
$U_e(x,t)$ = moisture profiles from experimental data at drying time t;
$U_c(x,t,\vec{P})$ = calculated moisture profiles as determined by Eq. (2) at drying time t with an initial value $\vec{P}$.

The quadratic approximation of $S(\vec{P})$ closing to point $\vec{P}_0$ can be rewritten as follows:

$$S(\vec{P}) = S(\vec{P}_0) + \sum_{i=1}^{K} (p_i \ p_{0,i}) \frac{\partial S}{\partial p_i}(\vec{P}_0)$$

$$+ \frac{1}{2} \sum_{i,j}^{K} (p_i - p_{0,i})(p_j - p_{0,j}) \frac{\partial^2 S}{\partial p_i \partial p_j}(\vec{P}_0)$$

$$(6)$$

by differentiating $S(\vec{P})$ with respect to $\vec{P}$, we obtain

$$\frac{\partial S}{\partial P_i}(\vec{P}) = \frac{\partial S}{\partial P_i}(\vec{P}_0) + \sum_{j=1}^{K} (P_j - P_{0,j}) \frac{\partial^2 S}{\partial P_i \partial P_j}(\vec{P}_0)$$

$$(7)$$

for $i = 1, 2, ..., K$. If $\vec{P}$ is a local minimum, then we can write

$$\frac{\partial S}{\partial P_i}(\vec{P}) = 0 \qquad i = 1, 2, ..., K \qquad (8)$$

Equation (7) and (8) using matrix notation lead to the following formula:

$$\vec{\nabla} S(\vec{P}_0) = -[H](\vec{P}_0) \cdot \Delta \vec{P} \qquad (9)$$

where $\vec{\nabla} S$ is the gradient of $S(\vec{P})$ with respect to $\vec{P}$; $[H]$ is the Hessian matrix,

$$[H] = \left(\frac{\partial^2 S}{\partial P_i \partial P_j} \right); \qquad \Delta \vec{P} = \vec{P} - \vec{P}_0$$

This procedure is repeated iteratively until S is less than a certain value. The l-th iteration can be calculated as

$$\vec{P}_{l-th} = \vec{P}_{(l-1)-th} - [H]^{-1}(\vec{P}_{(l-1)-th}) \cdot \vec{\nabla} S(\vec{P}_{(l-1)-th})$$

$$(10)$$

A similar technique can be used for fitting the calculated moisture loss with appropriate experimental data.

3 EXPERIMENTS

3.1 Preparation of specimens

Materials tested here are the most commonly used AAC in Europe which are usually characterized through their bulk density: GN(400 kg/m^3), GH(500 kg/m^3), GS(650 kg/m^3). In order to restrict the diffusion into a one-dimensional process, all the specimens are prepared in the way that in one dimension two opposite drying sections are kept free while in the other two dimensions the four sides are sealed with dense polymer and finally coated with copper foil. The determination of moisture distribution in the specimen as a function of time and space is based on the weight change of specimens with different lengths. The drying section of all specimens is 10×10 cm^2, and the length varies as follows:

2, 4, 6, 8, 10, 12 cm

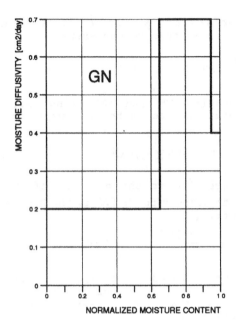

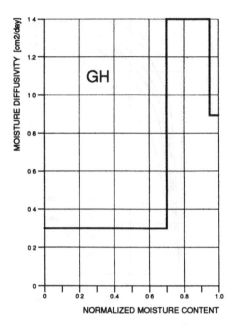

Fig. 3 Dependence of diffusion coefficient on the moisture content

Fig. 4 Dependence of diffusion coefficient on the moisture content

Before the drying process started, all the specimens were stored in water until they reached equilibrium.

3. 2 Measuring procedure

The drying conditions were fixed to be 20˚C, 60% R. H. The weight change of each specimen is measured by an electronic balance at different drying time. The moisture content $U(x_i, t)$ at drying time t and the distance from drying surface x_i is obtained from direct measurements and the following equation (Sakata 1983):

$$U(x_i, t) = \{1 - \frac{W_j - W_{j-1}}{U_0(L_j - L_{j-1})S}\} \cdot 100\% \quad (11)$$

where W_j is the moisture loss of a specimen with length L_j at drying time t; U_0 is the evaporable moisture in a unit volume; S is the area of drying surface. The evaporable moisture here means the total water content in a unit volume of AAC, which is given by the difference between the weight of specimen under water saturation and the constant weight after oven drying (105˚C).

Table 1. Moisture content of AAC under water U_0 and at 60% R. H. , 20˚C U_b in g/cm^3

	GN	GH	GS
U_0	0. 3754	0. 4806	0. 3998
U_b	0. 0142	0. 0239	0. 0206

For AAC of GN, GH, GS the U_0 and the moisture content after reaching equilibrium U_b at 20˚C and 60% R. H. are listed in Table 1.

4 RESULTS

As described above, the moisture distribution in a drying specimen at different drying time has been determined.

In Fig. 2 the moisture space distributions for GN after drying for 1, 2, 3, 5, and 10 days are shown. The solid lines in this figure are obtained by a least square fit according to the technique described above. From the fitted moisture distribution curves the moisture diffusion coefficients are obtained.

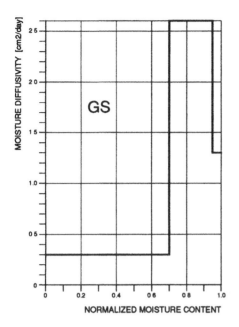

Fig. 5 Dependence of diffusion coefficient on the moisture content

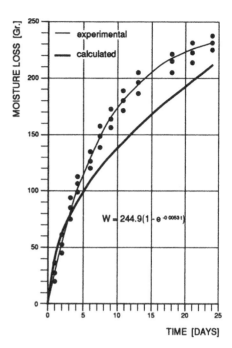

Fig. 6 Time-dependent moisture loss of AAC(GN) specimen of length 12cm

Results of the three different types of AAC under investigation are shown in Fig. 3, Fig. 4 and Fig. 5, respec - tively.

With the diffusion coefficients obtained from measurements of mois- ture space distributions, the mois- ture loss as a function of time has been calculated. A typical result is shown in Fig. 6.

Next to the numerical prediction, the results from the experimental data of prisms having a length of 12 cm is selected as an example for comparison.

The experimental data from three identical specimens can be repre- sented by the following equation:

$$W_j(t) = \alpha(1 - e^{-\beta t}) \tag{12}$$

where α and β are regression para- meters. The fit of this equation with experimental data is good. Results of specimens with a length of 4 cm are shown in Fig. 7. If only the water loss is to be determined this empirical approach gives suffi- ciently accurate predictions.

5 CONCLUSIONS

Nonlinear description of the diffu-

sion process of AAC gives better agreement with experiments than the linear one. The analysis of experi- mental results shows that the drying process of AAC may not be described as a single diffusion mechanism. The dependence of the diffusion coeffi- cient on the moisture content may be a much more complicated function than the one assumed in this paper. More investigations into the moisture movement mechanisms are necessary.

Numerically it is possible to determine AAC diffusion coefficients from moisture distribution or mois- ture loss measurements. At this moment it is not possible to formulate a continuous function in order to describe the influence of moisture content on the diffusion coefficient. For materials like AAC it seems there is no direct link between density and diffusion coefficient . Further investigations on how the pore structure of AAC, i.e. the pore size distribution and pore shape influence the diffusion process are needed.

REFERENCES

Bazant Z. P. & Najjar L. J. 1971. Drying

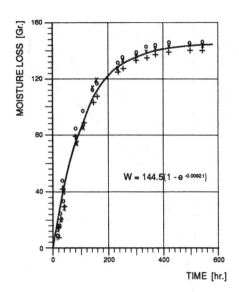

Fig. 7 Time-dependent moisture loss
of AAC(GN) specimen of length 4cm

$$W = 144.5(1 - e^{-0.00921 t})$$

concrete as a nonlinear diffusion
problem. Cement and Concrete
Research, Vol. 1, 461-473

Hall C. & Kalimeris A. N. 1982. Water
movement in porous building
materials. Building and
Environment, Vol. 17, 257-262

Kasi S. S. H. & Pihlajavaara S. E. 1965.
An approximate solution of a
quasi-linear diffusion problem.
The State Institute for Technical
Research, Helsinki. Publ. No. 153

Kasperkiewicz J. 1982. Some aspects
of water diffusion process in
concrete. Materiaux et Constructions Vol. 5, 209-214

Marsal D. 1976. Die numerische Lösung
partieller Differentialgleichungen
in Wissenschaft und Technik.
Mannheim u. a. Bibliographisches
Institut, Band XXVIII

Meis T. & Marcowitz U. 1981. Numerical
solution of partial differential
equations. Applied Mathematical
Science Vol. 32

Mensi R. , Acker P. & Attolou A. 1988.
Séchage du béton: Analyse et
modélisation. Materials and
Structures, Vol. 21, 3-12

Pihlajavaara S. E. 1965. On the main
features and methods of investigation of drying and related
phenomena in concrete. Ph. D.
Thesis, State Institute for
Technical Research, Helsinki
Pulb. No. 100,

Pihlajavaara S. E. & Vaisanen J. 1965.
Numerical solution of diffusion
equation with diffusivity concentration dependent. State Institute
for Technical Research, Helsinki
Pulb. No. 87

Roelfstra P. E. & Wittmann F. H. 1983.
Numerical analysis of drying and
shrinkage. Autoclaced Aerated
Concrete, Moisture and Properties.
edited by F. H. Wittmann, Elsevier
Scientific Publishing Company,
Amsterdam, 235-248

Sakata K. 1983. A study on moisture
diffusion in drying and shrinkage
of concrete. Cement and Concrete
Research Vol. 3, 216-224

Wittamnn F. H. , Roelfstra P. E. &
Kamp C. L. 1988. Drying of concrete-
an application of the 3L-approach.
Nuclear Engineering and Design.
Vol. 105, 185-198

Wittmann X. , Sadouki H. &
Wittmann F. H. 1989. Numerical
evaluation of drying test data.
Transactions of the 10th Interna-
tional Conference on Structural
Mechanics in Reactor Technology
(SMiRT) Vol. Q, 71-79

Advances in Autoclaved Aerated Concrete, Wittmann (ed.) © 1992 Taylor & Francis. ISBN 90 5410 086 9

Capillary suction of AAC

J. Pražák & P. Lunk

Institute for Building Materials, ETH Zürich, Switzerland

ABSTRACT: Capillary suction of different types of AAC has been studied experimentally and theoretically interpreted within the scope of the phenomenological theory. The time dependence of the total uptake of water was determined gravimetrically and the time evolution of spatial moisture distribution was visualised by means of γ-ray absorption measurements. Based on the theoretical interpretation of experimental results, it can be shown that the well known $\sqrt{t}$-law cannot fully describe water suction of AAC.

1 INTRODUCTION

The capillary water transport in AAC was investigated from a phenomenological point of view by means of a set of infiltration experiments. Vertically placed samples in form of slabs were used. The analysis of experimental results was carried out in order to answer two questions:

1. What are the features of capillary water transport specific to AAC in contrast to other building materials ?

2. What are the features of capillary water transport specific to different types of AAC?

2 MATERIALS AND EXPERIMENTS

2.1 Materials

Three types of AAC were used in the experiments. They are denoted by GN, GH and GS where N, H and S stay for normal, high and superior quality respectively. The "quality" in this context is based on the compressive strength. In order to have a comparison with typical building materials, parallel experiments were carried out with samples of burnt clay brick (CB) and normal concrete (NC).

To get the information about basic characteristics of used materials (density, porosity), samples were dried at $105°C$, then stored some for time at

Tab. 1. Basic characteristics of materials under investigation.

	bulk density $kg.m^{-3}$	porosity $\%$
GN	390	81
GH	500	79
GS	650	74
CB	1700	35
NC	2300	18

$20°C$ and 70% r.h. and finally vacuum saturated by water (Jacobs and Mayer 1992). The results of this measurements are summarised in Tab. 1.

2.2 Experiments

Vertical infiltration tests were performed. Deionised water was used. The surface in contact with water was $50 \times 150mm^2$. The samples were dried by oven-drying at $50°C$ for two days and thereafter cooled to room temperature and stored at $20°C$ and 60% r.h. to constant weight. The specimen sides parallel to the moisture transfer direction were sealed with an epoxy resin, impeding lateral evaporation and air escape. Room air temperature was conditioned to $20°C$.

Two different experimental techniques were used for observing the changes of water content

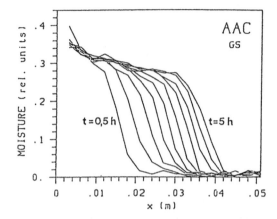

Fig. 1a. Evolution of experimentally determined moisture distribution. Time is used as a parameter and the time step is 0.5 h.

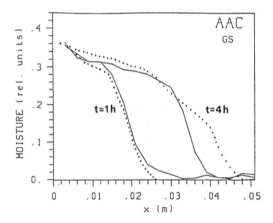

Fig. 1b. Comparison of measured and calculated moisture distribution.

in the samples during the infiltration: the computer controlled weighing and the γ-radiography. The details of the experimental set-ups are amply described in (Wittmann and Lunk,1991), but two features of special interest should be pointed out at this place:

The fully standardised procedure of weighing on a precision balance makes it possible to determine in a simple way first-order derivation of empirical suction curves. This fact is important for a more detailed analysis of experimental results.

The apparatus used for radiography of AAC has been designed for analysis of normal concrete samples. That is why it is possible to use a well focused γ-ray of sufficient intensity to get a sharp picture of the moisture distribution in a AAC sample without using the Fourier deconvolution of raw data.

3 EXPERIMENTAL RESULTS

The evolution of the water distribution in the sample during the infiltration was visualised by means of the γ-radiography. One typical set of successive moisture distribution curves in an AAC sample during the imbibition is shown in Fig. 1a.

By a simple comparison of sets of infiltration distribution curves, the essential differences which can be observed are the differences in maximal values of water content and rate of water suction for different types of AAC. The forms of the curves in all sets are similar.

The water suction into the porous materials is frequently described by means of diffusion-type nonlinear equation

$$\partial_t u = \partial_x (D(u)\partial_x u). \qquad (1)$$

The function $D(u)$ can be found in a standard way by means of Boltzmann transformation ($y = x/\sqrt{t}$) and subsequent use of Matano's algorithm (Carpenter at al. 1992). It can be parametrised in the form

$$D(u) = D_0 \exp(Cu) \qquad (2)$$

(Kießl, 1983). For the GS, the corresponding parametres have values $D_0 = 2.10^{-9} m^2.s^{-1}$, $C = 15..$ Using these parameters, a fairly good reproduction of the experimental data can be obtained for small values of time – dotted line in Fig. 1b. For bigger times, substantial divergence occurs.

Corresponding to the similarity between the forms of the infiltration curves, the values of parameters D_0 and C are similar for all investigated types of AAC.

The time evolution of total water content $m(t)$ in the sample per unit area (suction curve) was measured by precise weighing. The theoretical form of function $m(t)$ can be obtained by the integration of equation (1) over variable x for the boundary conditions corresponding to the infiltration. It leads to the well known $\sqrt{t}$-formula

$$m(t) = A\sqrt{t}. \qquad (3)$$

Typical experimental suction curves $m(t)$ for the investigated materials are displayed in Fig. 2a,b,c.

The alternative, more simple description of infiltration describes the time evolution only of the position X of the wetting front. A time dependence of the form identical to that of (3)

$$X(t) = B\sqrt{t} \qquad (4)$$

is supposed usually.

The empirical values of A's and B's for the investigated materials are summarised in Tab. 3. In order to interpret the actual moisture distribution in the scope of eq. (4), the wetting front was identified with the moisture content $u_{max}/2$.

4 INTERPRETATION OF THE EXPERIMENTS

In the scope of the description of the infiltration experiments by means of eq. (2) and (4), various types of AAC are well described by a single set of coefficients B, D_0 and C. The fact that the coefficient A in the eq. (3) is smaller for GN then for GH and GS is caused by differences in the pore structure. As can be seen from porosimetry curves of AAC (Jacobs 1992, Prim at. al 1983), GN has essentially more pore volume in the region of macropores in comparison with both GH and GS. The volume of macropores is not fully saturated during the infiltration in a gravitational field. This manifests itself by the less total water uptake for materials with a greater percentage of macropores.

The coefficients A, B, D_0 and C are all connected with the $\sqrt{t}$-behaviour of the infiltration process. A more pronounced distinction of different materials can be achieved by a comparison of the deviations from the ideal $\sqrt{t}$-behaviour.

It is common experience that by fitting the suction curves, the simple $\sqrt{t}$-dependence (3) gives good results for the medium period of the process only. Near the saturation, the imbibition takes an exponential character and deviations from the $\sqrt{t}$-behaviour can be observed at the beginning of the process as well.

The starting phase of the infiltration is worth some more attention because of the fact that building materials under influence of atmospheric conditions can be often submitted to a short-time

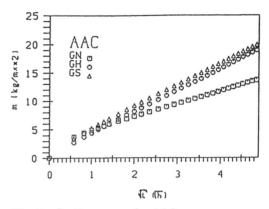

Fig. 2a. Suction curves for AAC.

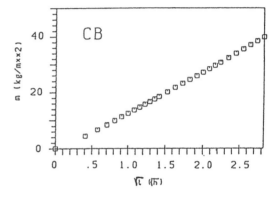

Fig. 2b. Suction curve for burnt clay brick.

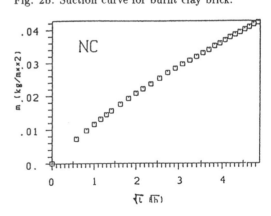

Fig. 2c. Suction curve for normal concrete.

contact with water. For a more detailed examination of the beginning of the infiltration process, the departures from the $\sqrt{t}$-behaviour can be made more visible by means of a simple transform of experimental data.

It is a specific feature of the function $\sqrt{t}$, that

121

Tab. 2. Empirical suction parameters

	A $kg.m^{-2}.s^{-1/2}$	B $m.s^{-1/2}$
GN	.038	$2.5\ 10^{-4}$
GH	.061	2.6^{-4}
GS	.066	2.8^{-4}
CB	.16	$3.^{-4}$
NC	.012	$7.^{-5}$

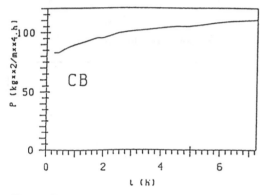

Fig. 3. Function P for burnt clay brick.

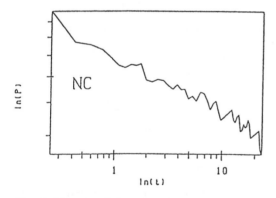

Fig. 4. Function P for normal concrete

its derivative is proportional to its reciprocal. It follows, that for materials where the time dependence of the total water uptake $m(t)$ is exactly described by equation (3), the product function $P(t)$ defined by the following relation

$$P(t) = m(t).\frac{dm(t)}{dt} \qquad (5)$$

would have a constant, time-independent value

Tab. 3. Comparison of the values of parameters A as determined experimentally with the help of eq. (3) and as calculated by tmeans of eq. (6).

	A $kg.m^{-2}.s^{-1/2}$	A_P $kg.m^2.s^{-1/2}$
GN	.038	0.041
GH	.061	0.065
GS	.066	0.069
CB	.16	0.14
NC	7.10^{-5}	$\sim 10^{-4}$

$$P(t) = A^2/2. \qquad (6)$$

On the other hand, if $P(t)$ is not constant this automatically indicates a systematic deviation from the simple $\sqrt{t}$-behaviour.

The function P differs substantially from a constant, for most porous building materials. Sometimes the divergence is limited to the beginning of the process (brick - Fig. 3). In the case of normal concrete (NC) no constant value could be observed with the measuring period.

For the normal concrete, the deviation has in many cases a systematic character, which can be expressed by a modification of the exponent of the time dependence. The more appropriate exponent of the time evolution can be found in a simple way from the function P represented in a double logaritmic scale (Fig. 4).

For the demonstrated case, the infiltration is described by the time dependence $t^{0.5-0.15}$ rather then by $\sqrt{t} = t^{0.5}$.

The P curve for the AAC has the characteristic form presented in Fig. 5.

As can be seen, the AAC is the only one of the investigated materials having a function P of approximately constant value, but after a short period (2 - 3 hours) of a rapid decrease. This behaviour may be interpreted as the suction into the whole porous space at the beginning of the process followed by the filling of a part of it in the later phase only.

5 CONCLUSIONS

Comming back to the questions formulated in

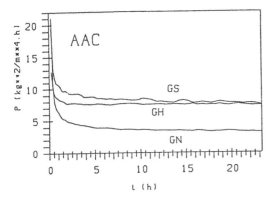

Fig. 5. Functions P for different samples of AAC

the Introduction, we can say that in the scope of the description of the infiltration process by means of equation (1), which is connected with the $\sqrt{t}$-behaviour of suction function $m(t)$, the coefficients describing the process in AAC have values similar to the corresponding coefficient of some other building materials, e.g. burnt clay brick. For the different types of AAC, only the characteristics connected with the maximal saturation water content exhibit significant difference. The shape of the distribution of water during infiltration is similar for all different types of AAC investigated in this study.

The deviations from the simple $\sqrt{t}$-behaviour, which can be described by a product function P, are more characteristic for AAC. The function P seems to have a specific shape for different building materials. The different types of AAC differ in limit values of function P.

It can be mentioned, that the use of function P and relation (6) is a convenient way of getting the parameter A for eq. (3) from experimental data. In Tab. 3, the parametres A from Tab. 2 (calculated by means of the direct fitting of the function (3) on the experimental data) are compared with those obtained by means of the function P - A_P.

REFERENCES

Wittmann F.H. und Lunk P. 1991; Beeinflussung des Feuchtigkeits- und Ionentransportes in Beton durch oberflächentechnologische Massnahmen, Schriftenreihe des Eidgenössischen Verkehrs- und Energiewirtschaftsdepartementes; Bundesamt für Strassenbau, No. 217

Kießl K. 1983; Kapillarer und Dampfförmiger Feuchtetransport in mehrschichtigen Bauteilen, Diss. Universität Essen

Prim, P. and Wittmann, F.H. 1983; Structure and water absorption of aerated concrete, in Autoclaved aerated concrete, moisture and properties, edited by F.H. Wittmann, Elsevier Company

Carpenter T. A., Davies E. S., Hall C., Hall L. D., Hoff W. D., Wilson M. A., 1992; Capillary water migration in rock: processes and material properties examined by NMR imaging, to be publisht in Materials and Structures

Jacobs F. and Mayer G. 1992; Porosity and Permeability of AAC, Proceedings of 3rd RILEM International Symposium on AAC, ETH Zürich (this issue)

Advances in Autoclaved Aerated Concrete, Wittmann (ed.) © 1992 Taylor & Francis. ISBN 90 5410 086 9

Investigation of moisture contents of autoclaved lightweight concrete walls in cold districts

T. Hasegawa
Department of Architecture, Faculty of Engineering, Hokkaido University, Sapporo, Japan

ABSTRACT

The purpose of this study is to relate moisture content of autoclaved lightweight concrete (ALC) to wall structures in cold districts. Multiple ALC wall structures of different thicknesses; with different external coatings; and with or without interior finishing materials, insulation, and vapour barriers are included in the test program. ALC wall structures are tested under three outdoor temperature (-20 °C, -10°C, -10~+5°C) and three indoor humidity (80%, 60%, 40% at +20°C).
The following conclusions are reached:
Moisture content of ALC varies significantly with the type of ALC wall structure. Moisture content in ALC wall structures is significantly higher under conditions of high indoor humidity or low outdoor temperature. Moisture accumulation in ALC wall structures is concentrated in 0°C areas within the wall cross-sections.

1 MOISTURE ACCUMULATION WITHIN ALC EXTERNAL WALLS AS A RESULT OF VAPOUR TRANSFER

Autoclaved lightweight concrete (ALC) used as part of building walls is subject to frost damage, either when moisture content in all or part of the ALC is high, so as to be susceptible to freezing and thawing; or when there is

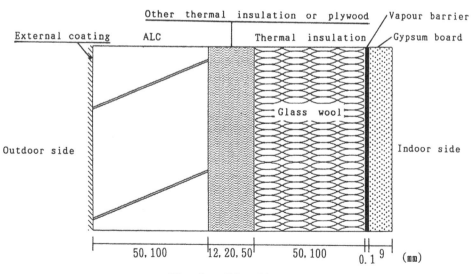

Fig. 1 ALC wall model

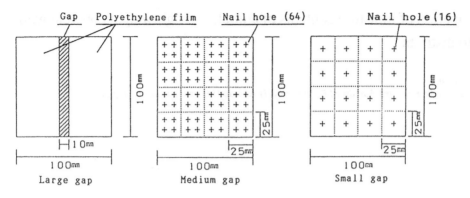

Fig. 2 Gap of vapour barrier

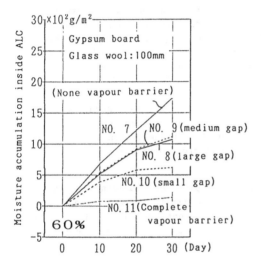

Fig. 3 Gap of vapour barrier and moisture accumulation

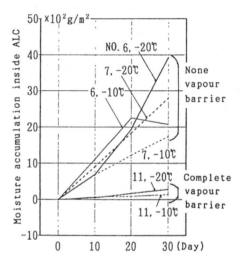

Fig. 4 Effect of vapour barrier

an area inside the ALC wall cross-section where moisture can condense and accumulate.

In winter in cold regions, when outside air is dry and its vapour pressure is low, vapour pressure inside buildings tends to be higher than the pressure outside (there are variations depending on how the indoor space is used). With vapour pressure thus different on the outside and inside of external walls, vapour continually moves from the inside, with the higher vapour pressure, to the outside, with the lower pressure, through the external wall composite materials.

If the temperature reaches the dew point

inside such a porous and permeable material as ALC, vapour in vacant spaces in the material condenses to water. Moreover, since the difference in vapour pressure remains, vapour continues to move from the high- to the low-pressure side, by which condensation water gradually accumulates to fill those vacant spaces. Accordingly, if evapo- ration is slow at the outdoor side of the ALC external wall, or if there is non-permeable wall material at the outdoor side of the permeable wall material, moisture content in the material gradually becomes higher. If the temperature goes below $0°C$ at any area within the ALC wall cross-section, the materials of the wall can be damaged or destroyed as

Table 1 Moisture accumulation inside ALC for 30 days   [×10² g/m²]

Wall NO.	EC	ALC	IP	GW	VB	GB	20℃,80% −10℃	20℃,60% −10℃	20℃,40% −10℃	20℃ 60% −20℃	20℃ 60% −10~+5℃
NO. 1		50					220.4	16.2	3.4	204.5	2.3
NO. 3	SP	"					279.9	50.6	9.6	238.3	10.2
NO. 4		10				A	35.0	0.2	−0.7	19.1	0.3
		10				B	39.4	0.8	−0.7	29.2	0.3
		10				C	37.1	3.4	−0.4	30.3	0.2
		10				D	57.3	6.3	0.6	11.5	0.1
		10				E	35.8	2.1	0.1	2.4	0.1
		50				T	204.6	12.8	−1.1	92.5	1.0
NO. 5		100					44.5	6.1	−18.8	8.0	−6.0
NO. 6		50		100			10.5	20.7	3.5	39.4	0.4
NO. 7		"		"		0	18.5	17.4	8.0	28.0	3.7
NO. 8		"		"	L	0	11.7	10.7	6.0	21.0	0.8
NO. 9		"		"	M	0	12.3	11.2	6.9	20.6	1.7
NO.10		"		"	S	0	3.7	6.2	4.2	12.8	0.6
NO.11		"		"	O	0	3.6	1.4	3.2	2.8	0.6
NO.12		"		50			57.3	53.4	16.4	84.5	4.9
NO.13		"		50		0	37.0	23.2	10.9	40.3	9.9
NO.14		"		100		0	29.7	20.4	8.9	27.2	4.0
NO.15		10		"		A	9.1	8.0	1.3	19.8	0.6
		10				B	2.0	3.2	0.9	1.3	0.2
		10				C	1.2	1.7	0.7	1.1	0.2
		10				D	0.8	1.1	0.7	0.7	0.2
		10				E	1.0	1.8	1.0	1.7	0.2
		50				T	14.1	15.8	4.6	24.6	1.4
NO.16		50		"	L	0	15.0	11.2	6.2	18.0	1.2
NO.17		"		"	M	0	14.1	10.8	7.0	15.1	2.4
NO.18		"		"	S	0	10.0	9.5	4.7	19.1	1.2
NO.19		"		"	O	0	6.5	4.3	3.1	2.0	0.7
NO.20		"		"		0	22.8	20.6	9.2	26.4	3.4
NO.21		100		"		0	45.6	17.8	11.4	24.4	6.7
NO.23		50		"	M	0	11.0	8.5	5.3	16.5	1.3
NO.24		"		"	O	0	5.2	2.5	3.3	1.1	0.7
NO.25		100		"	O	0	1.9	1.4	3.6	1.5	0.0
NO.26		50	PW	"		0	* 7.6	* 4.9	3.9	* 1.3	0.8
NO.28		"	"	"	O	0	3.0	2.5	3.7	1.7	0.7
NO.29		"	UF	"		0	* 11.3	* 2.3	3.6	* 2.2	2.7
NO.30		"	SF	"		0	* 7.3	* 6.1	8.3	* 3.1	1.1
NO.31		"	EF	50		0	* 12.0	* 4.4	6.8	* 1.5	2.1
NO.35	SP	"		"		0	40.1	25.3	13.3	34.1	4.6

(Rows NO.6–NO.19: Small mois. permeability; Rows NO.20–NO.35: Large mois. permeability)

EC: External coating, GW: Glass wool, IP: Other thermal insulation
or plywood, VB: Vapour barrier, GB: Gypsum board,
SP: Steel plate, 50,100: Thickness (mm), PW: Plywood,
UF: Rigid urethane foam, SF: Polystyrene foam, EF:Polyethylene foam,
L,M,S: Size of gap of vapour barrier,
*: Moisture accumulation is much inside wall model

the moisture in that area freezes and expands.

The purpose of this experimental study was to investigate the proper design of ACL external walls, to avoid condensation and frost damage within the ACL due to transfer of room moisture into the ALC walls. A series of tests was conducted using various kinds of ALC wall structures, under different conditions of outdoor and indoor temperature and humidity.

2 TEST METHOD

ALC wall structure models used in the tests are shown in Fig. 1. Twenty-nine varieties of wall model (see Table 1) were constructed of various materials, in different thickness and combinations. ALC segments were 100mm × 100mm of 50mm or 100mm thickness; Two segments (wall NO.4,15) were cut into 10mm slices and a layer of five 10mm slices, a thickness of 50mm in total, was used to assess the

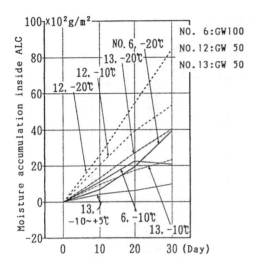

Fig. 5 Thickness of glass wool and moisture accumulation

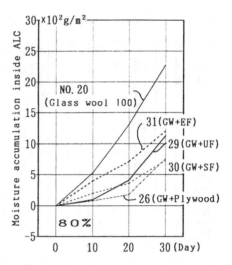

Fig. 6 Kinds of thermal insulation and moisture accumulation

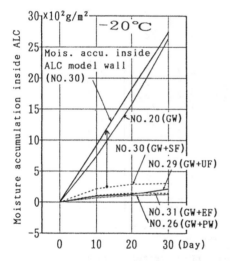

Fig. 7 Glass wool and other thermal insulation and moisture accumulation

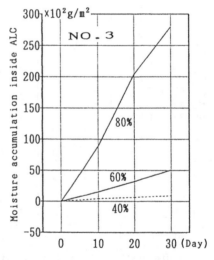

Fig. 8 External wall structure consisting only of ALC (NO.3)

distribution of accumulation. The ALC wall models were placed on external wall condition for 30 days under the following five conditions: outdoor temperature of -10°C and indoor temperature of 20°C with humidities of 80%, 60% and 40%; and indoor temperature of 20°C with 60% humidity, and outdoor temperature of -20°C or changing between -10°C and +5°C over the course of a day. Weight change of ALC and condensation within wall structures were measured.

3 TEST RESULTS AND DISCUSSION

3.1 Wall model structures/materials and condensation inside the ALC wall

① Effect of vapour barrier
(see Fig.3 and Fig.4)

• Application of a complete vapour barrier is most effective
Condensation inside the ALC wall can be prevented almost completely -- even if the indoor humidity is rather high and

Fig. 9 Crack caused by high moisture accumulation (NO.3)

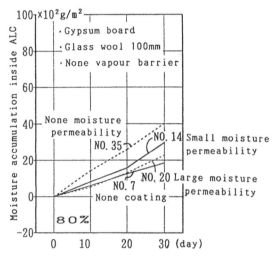

Fig.10 Moisture permeability of external coating of ALC and moisture accumulation

the outdoor air temperature is rather low -- by applying high vapour resisting material, such as polyethylene film, on the indoor side of the heat insulating material, with no gaps in the coverage.

• If the vapour barrier is insufficient (incomplete coverage, for example), the amount of condensation inside the ALC wall will vary with the degree of that insufficiency.
The amount of condensation increases when the outdoor air temperature is low and the indoor humidity is high, and when such condition is prolonged. In those circumstances, variations in condensation are primarily attributable to different degrees of incomplete coverage of the vapour barrier (see Fig.2).

If a vapour barrier is not used, there is considerable risk of condensation inside the ALC external wall.

② Vapour resisting material (not a vapour barrier) on the room side of the ALC and its effectiveness in condensation prevention (see Fig.5,6,7)
Even if a vapour barrier is not included in the wall structure, moisture accumulation within the ALC itself can be lessened by placing any material with vapour resistance on the room side of the ALC wall, such as an interior finishing or insulation material, structural plywood, etc. Some moisture accumulation may still occur, however, within the ALC wall, on the indoor side.

③ External wall structure consisting only of ALC (see Fig.8 and Fig.9)
If the external wall consists only of ALC, without an insulation layer or vapour barrier, and there are any 0°C areas inside the ALC, there is a high possibility of condensation and thus moisture accumulation, which may cause cracking with larger moisture accumulation.

④ Moisture permeability of external coating of ALC (see Fig.10)
Use of a moisture-permeable external coating on the ALC is not particularly effective in avoiding condensation if there is a 0°C area on the room side within the ALC wall, where an ice layer will form. In other conditions, however, moisture-permeable coatings are effective. On the other hand, if the external coating of ALC is highly vapour resisting, condensation can still be minimal depending on indoor and outdoor temperature and wall-structural conditions. As conditions become severe, however, the amount of condensation rapidly increases.

3.2 Indoor and outdoor temperature and humidity conditions and condensation inside a wall

① Indoor humidity conditions (see Fig.11)
If conditions support the possibility of

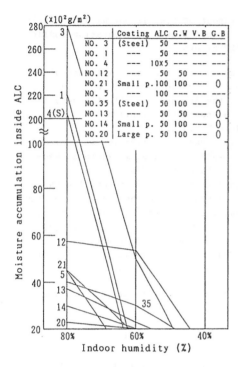

Coating	ALC	G.W	V.B	G.B
NO. 3	(Steel)	50	---	---	---
NO. 1	---	50	---	---	---
NO. 4	---	10X5	---	---	---
NO.12	---	50	50	---	---
NO.21	Small p.	100	100	---	0
NO. 5	---	100	---	---	0
NO.35	(Steel)	50	100	---	0
NO.13	---	50	50	---	0
NO.14	Small p.	50	100	---	0
NO.20	Large p.	50	100	---	0

Fig.11 Indoor humidity and moisture accumulation

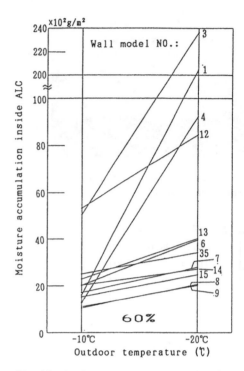

Fig.12 Outdoor temperature and moisture accumulation

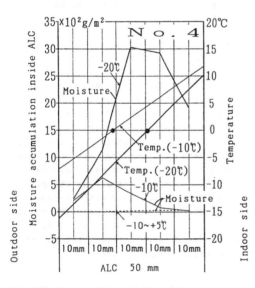

Fig.13 Temperature and moisture accumulation inside ALC (NO.4)

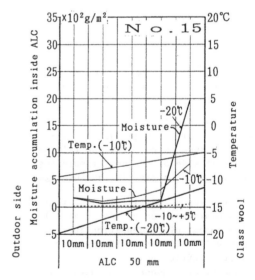

Fig.14 Temperature and moisture accumulation inside ALC (NO.15)

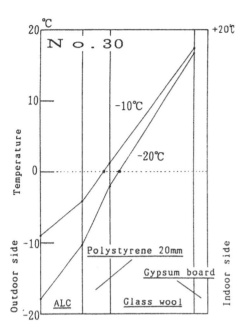

Fig.15 Outdoor temperature and temperature inside the wall model (NO.30)

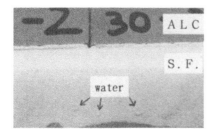

Fig.16 Outdoor temperature and place of condensation (-10℃)

condensation and moisture accumulation inside a wall, there is a clear, general tendency for the amount of such condensation to be greater as indoor humidity becomes higher. In this case, the relationship between indoor humidity and amount of condensation varies depending on the condition of ALC wall structure.

② Outdoor temperature condition
  (see Fig.12 ~ Fig.20)
If conditions support the possibility of condensation and moisture accumulation inside a wall, there is a clear, general

Fig.17 Outdoor temperature and place of condensation (-20℃)

tendency for the amount of such condensation to be greater as outdoor temperature becomes lower. In this case, the relationship between outdoor temperature and amount of condensation varies depending on the condition of ALC wall structure. Moisture accumulation in ALC wall structures is concentrated in 0°C areas within the wall cross-sections (see Fig13~). The danger of condensation inside a wall is alleviated to a considerable extent if the outdoor temperature rises periodically above 0°C (see Fig.13, 14).

4 SUMMARY

The following can be said on the basis of the test results:

• Places of condensation inside an external wall change according to the ALC wall structure, indoor/outdoor temperature and humidity conditions, and the vapour resistance of the wall materials.

• In the same building, there are places where water does and does not condense. Reasons for this include:
  • incomplete application (coverage) of vapour barriers;
  • differing indoor temperature and humidity conditions depending on how the rooms are used;
  • differing temperature and humidity conditions on the outdoor sides of the walls depending on the directions the walls face (north,

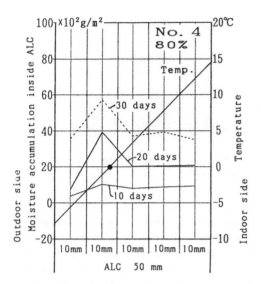

Fig.18 Temperature and change of moisture accumulation inside ALC (NO.4)

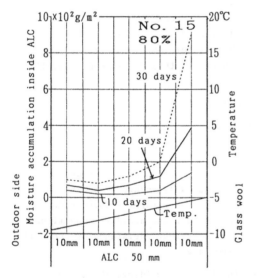

Fig.19 Temperature and change of moisture accumulation inside ALC (NO.15)

Fig.20 Frost at the surface of indoor side of ALC

region, how the building is used, and other factors.

south, etc.), and which part of the building the walls are located in (upper, lower, etc.).

• Even if the ALC external wall structure is the same, condensation and frost damage varies according to

4 Crack formation and durability

Advances in Autoclaved Aerated Concrete, Wittmann (ed.) © 1992 Taylor & Francis. ISBN 90 5410 086 9

Fracture energy experiments of AAC and its fractal analysis

W. Zhou, N. Feng & G. Yan
Tsinghua University, Beijing, People's Republic of China

ABSTRACT: Theoretical analysis and experimental results are presented to describe the micro-crack process zone and its influence on the fracture energy. Based on the matrix and defect structure of the rockmass, the microcracks ahead of the crack tip are assumed to be statistical fractal structure. Then its relation to the fracture properties, such as fracture energy G_f is deduced. Further, the splitting tests of prismatic specimen with notches of AAC (Autoclaved Aerated Concrete) are described, and the test results are analyzed and incorporated with the process zone. It is shown that this method can be used to demonstrate the fracure properties very well and may be applied to analyze the size effect of concrete fracture.

1 INTRODUCTION

In recent years, the experiments and applications of fracture mechanics concepts have been developed rapidly in concrete structure. The crack resistance properties of engineering materials are described using the critical values of fracture mechanics parameters such as the stress intensity factor K, potential energy release rate G, contour-integral J, crack tip opening displacement COD and fracture energy G_f. However most of these parameters are derived from the LEFM (Linear Elastic Fracture Mechanics) approach or PEFM (Plasto-Elastic Fracture Mechanics) approach, which are based on the assumption that there is a very small plastic yielding process zone ahead of the crack. For heterogenous materials like concrete and rock, experimental studies have indicated that a significant microcracking process zone exists ahead of the crack tip, and this fracture process zone governs the fracture properties of the materials.

For concrete and rock materials, crack properties are characterized by the formation of microcracks around a crack tip and a non-linear region where a portion of the microcracks is still interlocked. The stages of the development of the microcracks damage zone around the notch tip in rock have been given by Hoagland et al. (1973). Schmidt (1980) has given a qualitative description of the microcrack zone by the maximum normal stress criterion. On the basis of the Dugdale-Barenblatt model, Hillerborg et al. (1976) described a cohesive zone model with a constant separation surface energy (fracture energy G_f.) for concrete and rock, which is now widely used and is called FCM (Fictitious Crack Model). A blunt smeared crack-band model was used to describe the crack growth process in rock and concrete by Bazant and Cedolin (1979).

However, the characteristics of the process zone of concrete have not been extensively experimentally observed. This process zone is mainly dependent on meso-structures of the concrete and the specimen size. Moreover, the mesostructures in the concrete, such as pores, crack and inclusions, are randomly distributed, and this makes the microcracks process zone more complicated. Consequently most of the fracture property tests on concrete have been made by using a variety of specimen types and evaluation methods, and the results are in high discrepancy and size effectiveness.

The objectives of this paper are to analyze the microcracking process zone using fractal theory, and to demonstrate the calculation of process zone and the influence of it on the fracture energy. For these purposes, the fracture tests using RILEM recommended splitting method are employed.

2 FRACTAL ANALYSIS OF THE MICROCRACK PROCESS ZONE

It is essential that the development of analytic models for microcracking be guided by physical observations. Xu Shilang et al. (1991) have observed the microcrack process zone of concrete using laser speckle interferometry method and showed that this process zone grows as the load increases (as shown in Fig.1).

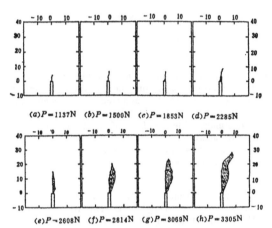

(a)P=1137N (b)P=1500N (c)P=1853N (d)P=2285N

(e)P=2608N (f)P=2814N (g)P=3069N (h)P=3305N

Fig.1. The formation and growth of microcracking process zone in cncrete under compact tension (in mm) (From Xu S., et al)

In the microlevel, many experiments (Nolen-Hoeksema et al. (1987) and H. Horii et al.(1985)) have shown that the microcracks in the concrete and rock are caused by tensile strains and grow along the direction of the minimum main principal strain direction. But due to the aggregates and matrices in the material, the microcracks actually take place in a zigzag path following aggregate boundaries. Nolen-Hoeksema R.G. et al. (1987) also observed that the microcracks in the process zone have a tree-shaped network (see Fig.2). So we can say that the fracture process zone in the concrete is a zone of microcracks initiating, propagating, proliferating and coalescending, and the microcracks are randomly distributed. To disclose the details of this zone is very difficult. In this paper the authors try to analyze this problem by using fractal theory.
Fractal theory was proposed by Manderbrot (8) in 1970's, which is used to describe the irregular and random phenomena in nature. The main concept of fractal is the fractal dimension (D), which is defined as:

$$D = \lg N/(-\lg r) \qquad (1)$$

where N represents the number of elements in fractal structure with a size larger than r. Equation (1) indicates the statistical self-similarity of fractal structures.

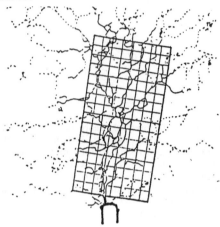

Fig.2. Microcracks notwork in Marble under tension (From Nolen-Hoeksema et al.(1987))

Based on the LEFM, a singular stress field exists around the crack tip, the stress field can be written as:

$$\sigma_{ij} = K\sqrt{2\pi r}\, f_{ij}(\theta) \qquad (2)$$

where K is the stress intensity factor. For metals there is a yielding zone at the crack tip and the stress singularity will be released. While as in concrete and rock, because of the cracks and inclusions, microcracks will initiate and grow at the crack tip, resulting in a strain-softening microcracking process zone. To analyze this process zone the following facts are considered:

1. From the micro-level observations, the microcracks are caused by the tensile strain, then using the elasticity theory, the first principal stress contours for Mode I fracture can be obtained from equation (2);

$$r = \frac{1}{2}\pi(K_I/\sigma_1)^2 \cdot \cos^2\frac{\theta}{2}(1+\sin\frac{\theta}{2})^2 \qquad (3)$$

where r and θ denote radius and angle in the polar coordinate respectively, and σ_1 is the first principal stress. It can be assumed that those contours are boundaries of microcrack process zone at the beginning of the microcrack initiation.

2. During the whole process of microcracks development, the microcracks propagate mainly along the direction perpendicular to the first principal stress. When several microcracks in this direction coalescend to form a large one, then the stress is released and some adjacent microcracks partially or even completely closed. This means that not all microcracks developed equally but those along the main crack develop first (see Fig.2).

3. Owing to the random distribution of aggregates and inclusions in concrete, the formation of microcrack network which is controlled by the mesostructure is random and stochastic.

Then it is appropriate to assume that the process zone is a tree-shaped structure and has the properties of statistical fractal. Using fractal theory, the total length of microcracks in the process zone can be written as:

$$L = AC^D \qquad (4)$$

where C is the length of the process zone, A is a constant and D is the fractal dimension of microcrack structure. D is dependent on the inner structure of concrete and can be used as a material parameter. D lies between 1 and 2.

Let W denote the dissipation energy in the formation of microcrack process zone and γ_s the surface energy of material. Using Griffith's energy theory, it can be deduced:

$$W = 2\gamma_s L = 2\gamma_s AC^D \qquad (5)$$

If G is the energy required for crack growth, then

$$G = \frac{\partial W}{\partial C} = 2\gamma_s A D C^{D-1} \qquad (6)$$

From equation (5), the fracture energy G_f can be simply related to the critical process zone length C_f by:

$$G_f = 2r_s A D C_f^{D-1} = B C_f^{D-1} \qquad (7)$$

where B can also be considered as a material parameter.

3 FRACTURE ENERGY AND SIZE EFFECT

It is clear from equation (7) that the fracture energy G_f depends on C_f. Now let us discuss this problem in detail.

In the process zone the microcracks are distributed randomly and interlocked. It is a strain-softening area, where the stresses are partially or even totally released. For the simplicity of analysis, it is assumed that the stress is completely released (see Fig. 3). Then using elasticity theory, the maximum stress at the tip of process zone in an infinite plate under Mode I fracture can be written as:

$$\sigma_{max} = \sigma(1+2\sqrt{\frac{a+c}{\rho}}) \qquad (8)$$

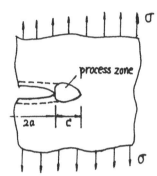

Fig.3. Crack and the stress released process zone

where ρ is radius of process zone boundary curve and σ the far-field stress.

For the specimen of width 2d, the equation (8) can be modified as:

$$\sigma_{max} = \sigma(1+2\sqrt{\frac{a+c}{\rho}}) K(\alpha) \qquad (9)$$

where $\alpha = a/d$ and $K(\alpha)$ is a function of α. Then on assuming that the process zone of length C is controlled by the crack tip stress field, the coalescence of the microcrack process zone (main crack grows) takes place only when:

$$\sigma_{max} = \sigma_t \qquad (10)$$

where σ_t is the tensile strength of concrete.

From the above analysis, it is shown that the critical process zone length C_f can not be solved analytically. But in experimental tests, the size of C_f can be measured directly.

From equation (7), the fracture energy G_f is a function of C_f. If D=1 then $G_f = B$ = constant, this is the case for very brittle homogeneous materials such as glass. When D=2, then $G_f = BC_f$, in this case the fracture energy depends strongly on C_f, and this is similar for the very ductil materials, in which the process zone grows equally in all directions. For concrete and rock, D is between 1 and 2. The smaller the fractal dimension D, the less significant the size effect is. It is generally observed that when the specimen size increases, C_f will in-crease also, and in certain level C_f will tend to be stable. From equation (7) it is shown that G_f is proportional to the (D-1)th power of C_f and tends towards stable even fast.

4. TEST AND DETERMINATION OF FRACTURE ENERGY

This paper reports an experimental study on the fracture energy and the microcrack process zone of AAC (Autoclaved Aerated Concrete), and the fracture tests were conducted using RILEM recommended splitting method.

The concrete is made of cement, lime and fly ash, its properties are shown in Table 1, the shape and size of specimen in Fig. 4.

Table 1. Properties of AAC used in the experiment

Density (kg/m3)	Compressive strength (MPa)	Splitting (MPa)	Elastic modulus (MPa)	Moisture content (%)
582	2.45	0.34	1791	21.3

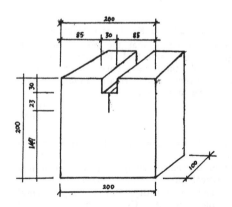

Fig.4. Specimen shape and size (in mm)

In the experiment, three specimens A-1, A-2 and A-3 are tested. Each specimen is stored in laboratory under the condition of $20 \pm 3C°$ and 65-70% RH. When the specimens reach constant weight, then they are taken out to be tested.

During the performance of tests, the splitting forces are loaded by a transmitting wedge and the roller bearing, the load rate is controlled by the crack opening displacement (COD). The vertical force Fv vs. the COD curves are shown in Fig. 5. The developement of the microcrack process zone is measured by the light speckle interferometry. The schematical shape of the process zone is shown in Fig. 6.

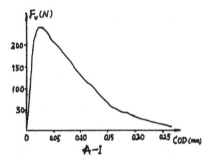

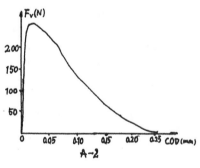

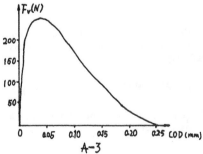

Fig.5. F_v –COD curve

The dissipation energy of the specimen splitting is evaluated according to the following formula:

$$W = \int_0^H F_v dh = \int_0^{COD_m} F_s d(co) \tag{11}$$

where H is the maximum displacement of vertical loading, COD is the crack opening displacement and Fs is the horizontal splitting force which can be determined from:

$$Fs = (Fv+Mg)/(2tg\alpha) \tag{12}$$

where M is the mass of the wedge loading part, g is the acceleration of gravity and α is the wedge angle (14°).

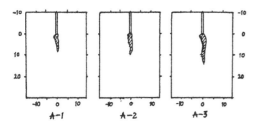

Fig.6.Schematical shape of process zone

Then the fracture energy G_f can be calculated from:

$$G_f = W/A_{lig} \tag{13}$$

where A_{lig} is the area of fracture ligament.

The measured and calculated results are shown in Table 2.

Table.2. Tests and calculated results

Speci-men	Fv,max (N)	Fs,max (N)	CODmax (mm)	G_f (N/m)	C_f (mm)
A-1	228.0	425.4	0.2627	4.53	8
A-2	240.6	449.0	0.2299	4.81	10
A-3	238.6	445.4	0.2468	5.37	12
Mean	235.7	439.9	0.246	4.90	

The fractal dimension D of process zone can be determined from the measured C_f data. From equation (7), it can be induced that:

$$\frac{G_1}{G_2} = (\frac{C_1}{C_2})^{D-1}$$

$$D = 1 + lg(G1/G2)/lg(C1/C2) \tag{14}$$

Using the data in Table 2, then the following results are obtained:

for specimens   A-1 and A-2,   D=1.27,
                 A-2 and A-3,   D=1.33,
                 A-3 and A-1,   D=1.30,

The fractal dimension of the tested concrete is D=1.30.
The other material constant B can be calculated from equation (7):

$$B = G_f/C_f^{D-1} = 9.64$$

5 CONCLUSIONS

1. Mechanical analysis and microlevel observations suggest that the initiation and growth of microcrack process zone are controlled by the mesostructure of concrete material. Owing to the random distribution of mesostructure and stress releasing of microcracks, the process zone is developing stochastically but along a dominant di-rection. Based on fractal theory, it can be described as a tree-shaped statistical fractal structure with a dimension of D in the range of 1 and 2. D can be seen as a material parameter which describes the inner structure of material.

2. The fracture energy G_f is strongly dependent on process zone and fractal dimension D by equation $G_f = BC_f$. When D = 1, this is the case for the most brittle materials such as glass, and D = 2 is the case for the most ductil materials. So fractal dimension D can also be considered as a parameter of brittleness.

3. In an actual specimen, the critical process zone size C_f is controlled by its shape and size. Consequently C_f is considered to be a basic mechanics of size effect. The larger the fractal dimension D, the more significant of size effect. The proposed model provides a good estimation of size effect but more research work is needed to disclose the details of C_f.

4. From the experimental data the fracture energy G_f of AAC is 4.90 N/m, and its fractal dimension D = 1.30.

Acknowledgements: - The presented study was supported by the Laboratory Institute of Geology, Chinese Academy of Sciences and the Scientific Fundation of China.

REFERENCES

Bazant Z. and Cedolin L., Blunt crack band propagation in finite element analysis. J. Eng.Mech.Div. ASCE, 102 (EM2), 1979.

Hillerborg A., Analysis of crack formation and crack growth by means of fracture mechanics and finite elements. Cement and Concrete Res. 6, 1976.

Hoagland R.G., Hahn G.T. and Rosenfield A.R., Influence of micro-structure on fracture propagation in rock. Rock Mech. 5, 77-106 (1973).

Horii H. and Nemat-Nasser S., Compression-induced microcrack growth in brittle solids: axial splitting and shear failure, J. Geophysical Research, Vol. 90, No. B4, 1985.

Mandebrot B.B., The fractal geometry of nature, New York, 1982.

Nolen-Hoeksema R.G. and Gordon R.B., Optical detection of crack patterns in the opening mode fracture of marble, Int. J. Rock. Mech.Min.Sci. and Gemech.Abstr.,Vol.24, 135-144, 1987.

Schmidt R.A., A microcrack model and its significance to hydraulic fracture toughness testing. Proc. 25th U.S. Symp. Rock Mech.,1980.

Xu Shilang, Fracture energy and strain field near the tip of a notch in huge concrete specimens under compact tension, J. Hydraulic Eng., No.11, 1991, (in Chinese).

Advances in Autoclaved Aerated Concrete, Wittmann (ed.) © 1992 Taylor & Francis. ISBN 90 5410 086 9

Fracture energy and strain softening of AAC

V. Slowik & F. H. Wittmann
Swiss Federal Institute of Technology, Zürich, Switzerland

ABSTRACT: Fracture energy and strain softening have proved to be useful material parameters for describing the fracture behaviour of AAC. A RILEM draft recommendation for the determination of these material parameters was outlined. According to this recommendation the fracture mechanics material parameters have been determined for different types of AAC. Additionally the influence of the loading rate on the material parameters was investigated. As in the case of normal cement concrete the rate dependence of the tensile strength can be described by a power law.

1 INTRODUCTION

It was found that cracking of AAC can be described by using concepts of non-linear fracture mechanics. The material parameters used are the fracture energy G_f and the strain softening behaviour. Brühwiler, Wang and Wittmann (1990) determined these material parameters for AAC as influenced by specimen dimensions and moisture. One main conclusion was that the fracture energy of AAC depends on specimen geometry, i.e. it increases with increasing ligament size. In order to make experimental results as determined in different laboratories comparable a RILEM draft recommendation (1992) was outlined. This recommendation is used here for the determination of the fracture mechanics material parameters for different types of AAC. For one type, additionally, the influence of the loading rate has been investigated, in order to estimate the sensitivity of the recommended test procedure to changes in the loading rate.

2 EXPERIMENTAL PROCEDURE

According to the RILEM draft recommendation wedge splitting tests have been run. Figure 1 shows the principle of this test. Wedges are pressed between roller bearings in order to split the specimens into two halves. The crack mouth opening displacement at the height of the loading points and the applied vertical load were measured. All tests have been run under crack mouth opening control. The splitting force was calculated taking the applied load and the wedge angle into consideration. For all types of AAC the rate of the crack mouth opening was the same and amounted to 60 µm/min. The experimental studies in the influence of the loading rate on the fracture mechanics material parameters were performed with specimens of type GS only. Here the crack mouth opening rate varied from 1.2 µm/min to 12000 µm/min.

Figure 2 shows the load-displacement curve for a specimen of the GS type. The specific fracture energy is equal to the area below the load-displacement curve divided by the ligament area, that is the projected area on a plane parallel to the main crack direction. The strain softening behaviour was determined by numerical simulation using the program SOFTFIT (Roelfstra and

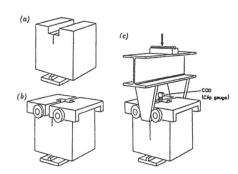

Fig. 1 Principle of wedge splitting test: (a) specimen on linear support, (b) loading devices with roller bearings, (c) traverse with wedges (Brühwiler, Wang and Wittmann 1990)

Wittmann 1986). In this program the fictitious crack model (Hillerborg, Modéer and Petersson 1976) is used for the simulation of the fracture process. The strain softening parameters were determined by fitting the calculated load-displacement curve to the experimental values. In the investigations presented here a bilinear strain softening behaviour has been assumed. The modulus of elasticity for the different AAC types was determined from the slope of the initial branch of the load-displacement curve.

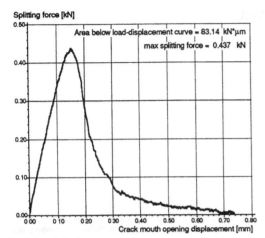

Fig.2 Load-displacement curve

Splitting force [kN]

Area below load-displacement curve = 83.14 kN*μm

max splitting force = 0.437 kN

Crack mouth opening displacement [mm]

3 PREPARATION OF SPECIMENS

All specimens had the dimensions as defined in the RILEM draft recommendation, see Figure 3. The notch width was 4 mm. The specimens were prepared by sawing and after that stored under usual climatic conditions, i.e. 65% relative humidity and 20°C, until humidity equilibrium. Table 1 contains the densities after storing for the AAC types used in this experimental investigation.

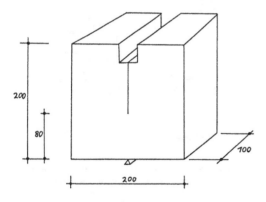

Fig.3 Specimen dimensions

Table 1. Different types of AAC investigated

designation	country of origin	density [kg/m³]
GN	Germany	480
GH	Germany	681
GS	Germany	742
Hi-Strength 7	England	823

4 EXPERIMENTAL RESULTS AND DISCUSSION

4.1 Fracture mechanics parameters of different types of AAC

Table 2 contains the experimental results for all types of AAC. The characteristic length l_{ch} was determined from the following formula:

$$l_{ch} = \frac{EG_F}{f_t^2}$$

The larger l_{ch} the less brittle and the more resistant against cracking is the material. It can be seen that the higher the density of the AAC the higher is the tensile strength and the smaller is the characteristic length, that means the more brittle behaves the material. In the case of the AAC type from England it has to be mentioned that the number of specimens was limited and that some of the specimens showed a crack pattern which was quite different from the expected one in the wedge splitting tests. Therefore the corresponding results in Table 2 are written within brackets. Figure 4 shows the strain softening diagrams for the german AAC. The calculation of the strain softening behaviour was done for one representative load displacement curve for each AAC type. These results are in good agreement with the strain softening diagrams determined by Brühwiler, Wang and Wittmann (1990) and with the tensile strengths determined in direct tension tests (Slowik 1992). The displacement values of the bilinear strain softening diagram seem to be constant for the different AAC types whereas the tensile stress values are different for the AAC types having different densities. There is only a slight difference in the behaviour of the types GS and GH. The shape of the strain softening diagrams seems to be independent on the AAC type, i.e. on the density. This can be seen in the normalised diagrams represented in Figure 5. For comparison in Figure 6 strain softening diagrams for the AAC types GS and Hi-Strength 7 as well as for normal concrete and mortar are shown.

Table 2. Fracture mechanics parameters for the AAC types

type	number of specimens	modulus of elasticity [MPa]	fracture energy [N/m]	tensile strength [MPa]	critical crack opening [mm]	characteristic length [mm]
GN	6	1800	4.96	0.42	0.037	50.6
GH	6	2200	8.56	0.66	0.042	43.2
GS	6	2500	9.12	0.74	0.042	41.6
Hi-S. 7	2	2100	(7.32)	(0.38)	(0.090)	(106.5)

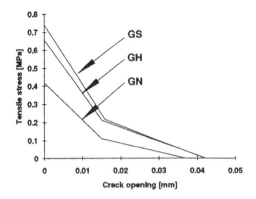

Fig.4   Strain softening diagrams for different AAC types

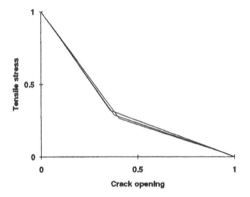

Fig.5   Normalised strain softening diagrams
        (types GS, GH, GN)

Using the experimentally determined fracture mechanics material parameters the fracture behaviour of any structural member can be predicted. Examples for the application of the experimental results presented here are given by Alvaredo and Wittmann (1992) in another contribution to these conference proceedings.

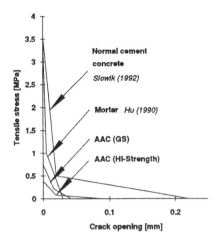

Fig.6   Strain softening diagrams for different types of concrete

4.2 Influence of crack opening rate on fracture mechanics material parameters

Table 3 contains the experimental results for the test series with varying crack opening rate. It can be seen that the tensile strength as well as the fracture energy increase with increasing crack opening rate. Figure 7 shows the fracture energy versus crack opening rate. The characteristic length decreases with increasing rate and reaches its minimum for 15 µm/min, see Figure 8. It is assumed that in the case of very slow crack opening (1.2 µm/min) the creep of the material outside the fracture process zone may not be neglected. That means the real fracture energy and tensile strength are underestimated in very slow tests. In the range of usual laboratory test rates no significant rate effects could be observed. Figure 9 contains strain softening curves for different crack opening rates. Both the stress values and the crack opening values are rate dependent. It can be seen that the curve for 1.2 µm/min is significantly different to those for higher rates. This can be explained by the creep of the material outside the fracture process zone. Mihashi and Wittmann (1980) presented a stochastic theory for fracture of concrete materials and concluded that the influence of rate of loading on the

Table 3. Fracture mechanics parameters for different crack opening rates (AAC type GS)

crack opening rate [µm/min]	number of specimens	fracture energy [N/m]	tensile strength [MPa]	critical crack opening [mm]	characteristic length [mm]
1.2	3	6.67	0.537	0.0302	58.08
15	6	9.39	0.717	0.0400	45.66
30	6	9.41			
60	4	9.12			
90	6	9.93			
120	5	10.77	0.741	0.0420	49.04
1200	5	9.84	0.719	0.0424	47.58
12000	3	12.38	0.748	0.0450	55.32

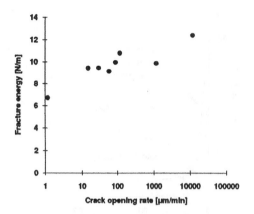

Fig.7 Fracture energy versus crack opening rate

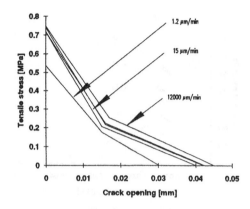

Fig.9 Strain softening diagrams for different crack opening rates

strength can be described by a power law:

$$(\sigma/\sigma_0)=(\dot{\sigma}/\dot{\sigma}_0)^{1/(\beta+1)}$$

σ_0 and $\dot{\sigma}_0$ are reference mean values of strength and of loading rate, respectively. β is a material parameter. For the tensile strength of cement mortar β amounts to approximately 20. The power law given above is used for the representation of the tensile strength for AAC, see Fig. 10, whereby the tensile strength for 120 µm/min served as reference value. The material parameter β amounts to 32.8 for this AAC type. That means the tensile strength is less rate dependent for this material as compared with cement mortar.

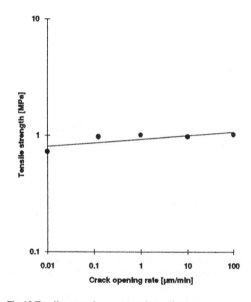

Fig.10 Tensile strength versus crack opening rate

5 CONCLUSIONS

The specific fracture energy of AAC can be determined in wedge splitting tests according to the corresponding RILEM draft recommendation. It was confirmed experimentally that the higher the density of the AAC type investigated, the higher is the fracture energy. The strain softening curves for AAC can be determined by numerical simulation using the fictitious crack model. It was found that the displacement values of these curves are less influenced by density than the stress values. The tensile strength is higher for denser AAC. The shape of the strain softening curve is nearly the same for different AAC types.

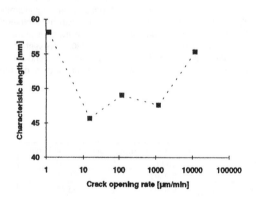

Fig.8 Characteristic length versus crack opening rate

Both the fracture energy and the tensile strength become higher with increasing crack opening rate. In the range of usual laboratory test rates no significant rate effects could be observed. As in the case of normal cement mortar and concrete, the rate dependence of the tensile strength can be described by a power law.

REFERENCES

Alvaredo, A.M., Wittmann, F.H. 1992. Influence of fracture energy on failure of AAC-elements. 3rd RILEM International Symposium on Autoclaved Aerated Concrete. ETH Zurich, Switzerland, October 14th-16th, 1992. (in this volume).

Brühwiler, E., Wang, J. and Wittmann, F.H. 1990. Fracture of AAC as influenced by specimen dimensions and moisture. Journal of Materials in Civil Engineering 2(1990)3, 136-146.

Helbling, A., Slowik, V. 1992. Determination of fracture mechanics material parameters of AAC. Internal report. ETH Zurich, Institute for Building Materials.

Hillerborg, A., Modéer, M. and Petersson, P.E. 1976. Analysis of crack formation and crack growth in concrete by means of fracture mechanics and finite elements. Cement and Concrete Research 6(1976), 773-782.

Mihashi, H., Wittmann, F.H. 1980. Stochastic approach to study the influence of rate of loading on strength of concrete. Heron 25(1980)3.

RILEM Draft Recommendation 1992. Determination of the specific fracture energy and strain softening of AAC. AAC 14.1.

Roelfstra, P.E., Wittmann, F.H. 1986. Numerical method to link strain softening with failure of concrete. Fracture toughness and fracture energy of concrete. ed. F.H. Wittmann. Elsevier, Amsterdam, 163-175.

Slowik, V. 1992. Beiträge zur experimentellen Bestimmung bruchmechanischer Materialparameter. Postdoctoral report, ETH Zurich, Institute for Building Materials.

REFERENCES

Advances in Autoclaved Aerated Concrete, Wittmann (ed.)© 1992 Taylor & Francis. ISBN 90 5410 086 9

Influence of fracture energy on failure of AAC-elements

A. M. Alvaredo & F. H. Wittmann
Swiss Federal Institute of Technology, Zürich, Switzerland

ABSTRACT: It is shown how non-linear fracture mechanics can be applied to analyse the failure of AAC. Some important phenomena observed experimentally, as the influence of structural size, rate of loading, and type of loading on the fracture of cementitious materials, are explained by making use of fracture energy based models. In particular, it is demonstrated that the difference between direct and bending tensile strength can be better understood by taking the strain-softening behaviour of AAC into consideration.

1 INTRODUCTION

Traditionally, many structural materials are classified according to their compressive strength. Within a limited range, it is possible to predict the tensile and the bending strength of these materials with the help of empirical formulae. Recently, further development of non-linear fracture mechanics has made it possible to apply this theory successfully to the description of crack formation and failure of cementitious materials. A test method to determine fracture mechanics parameters of AAC has been suggested by a RILEM Draft Recommendation (1992). Fracture mechanics parameters, certainly, will not replace the conventional material properties, such as compressive strength, in the near future. The application of fracture mechanics to AAC is still in its infancy.

Nevertheless, there are examples which clearly show the limits of a simple strength criterion. AAC-elements are often exposed to both severe hygral and thermal gradients. The risk of crack formation under the resulting eigenstresses cannot be assessed with a maximum-tensile-stress criterion. It has been shown recently how non-linear fracture mechanics can be applied to solve this problem (Alvaredo and Wittmann 1992).

This contribution is aimed at illustrating how fracture mechanics can be used as a powerful tool to predict the conditions for brittle failure of AAC-elements and to explain why the nominal bending strength of AAC is higher than the direct tensile strength. It can be anticipated that the application of fracture mechanics will help us to better understand the complex material properties of AAC.

2 SIGNIFICANCE OF FRACTURE ENERGY

In the simplest case, concrete-like materials may be considered to behave in a linear elastic way, in which proportionality between stresses and strains is assumed. A maximum-tensile-stress criterion may be then applied to predict failure. That is, as soon as the maximum tensile stress in the material reaches the tensile strength, complete disruption is assumed to take place. These assumptions concerning the behaviour of composite materials have one main advantage: their simplicity. They fail, however, to explain certain widely observed phenomena. It is well-known, for instance, that the size of the structure under study greatly affects the failure behaviour, with smaller elements attaining higher strength values and exhibiting a more ductile behaviour. Another important factor is the rate of loading: faster application of load results in a higher load bearing capacity. A further interesting issue is the difference observed in the experimentally

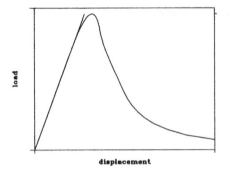

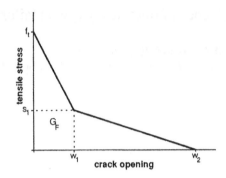

Figure 1: Schematic representation of the load-displacement diagram of cementitious materials under deformation control

Figure 2: Strain-softening behaviour of a 'fictitious crack'. The most usually employed approximation is a decreasing bilinear function

determined tensile strength values for different test types and, particularly, for direct tensile and bending tests. None of these essential phenomena can be explained if linear elasticity, in conjunction with a maximum-tensile-stress criterion, is employed to analyse the fracture of composite materials.

The main reason for the discrepancy between the predicted and the experimentally observed fracture behaviour is that cementitious materials not only depart from linearity before reaching the maximum load but, when loaded under deformation control, also exhibit a significant post-peak regime, as shown in Fig. 1. If certain conditions concerning the stiffness of the loading device are fulfilled, the load carrying capacity of the tested specimen decreases gradually with increasing applied deformation. In order to properly describe the descending branch of the load-displacement curve, energy considerations must be included in the method of analysis.

Cohesive crack models, based on non-linear fracture mechanics concepts, have proved to be outstanding tools for the description of the steady degradation experienced by concrete-like materials after peak-load. Among them, the most widely spread is the 'fictitious crack model' (FCM) developed by Hillerborg and co-workers (Hillerborg et al. 1976, Petersson 1981). In this model it is assumed that, after reaching the tensile strength, a fictitious crack or fracture process zone develops perpendicular to the direction of the maximum tensile stress. The gradual deterioration of the fracture process zone is described by a decreasing function of the tensile

stress with increasing crack opening. This post-peak behaviour is termed strain-softening and, in most practical cases, it is approximated by a bilinear function (Petersson 1981), as illustrated in Fig. 2.

An essential parameter underlying the formulation of the FCM is the fracture energy G_F. It is defined as the energy required to create a crack of unit area. Therefore, the area enclosed by the strain-softening diagram must equal G_F. A further useful parameter in fracture analysis is the characteristic length l_{ch} (Hillerborg 1983). It is defined as:

$$l_{ch} = \frac{EG_F}{f_t^2} \tag{1}$$

in which E is the modulus of elasticity. Solutions based on the FCM must be calculated numerically. The finite element formulation of the resulting system of equations is described by Petersson (1981).

3 EXPERIMENTAL DETERMINATION

The experimental determination of G_F requires to perform stable tensile tests. In particular, if direct tensile tests are carried out, all needed parameters, i.e. f_t, G_F, break-point (w_1, s_1) and critical displacement w_2 of the strain-softening diagram (see Fig. 2), can be directly obtained from the experimental records.

Due to the experimental difficulties involved in performing stable direct tensile tests, some research groups favour the use of other test types. The most usual ones are the 3-point-bending test, as described in a RILEM Draft Recommendation

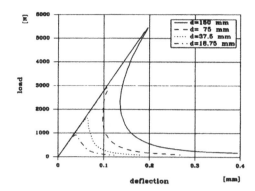

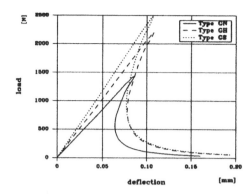

Figure 3: Load-deflection curves calculated for l/d=4 and b=250 mm. Material properties correspond to type GS in Table 1

Figure 4: Load-deflection curves calculated for l=250 mm and d=b=100 mm. Material properties correspond to the three types characterized in Table 1

(1985), and the wedge-splitting test (Brühwiler and Wittmann 1990). In these cases, only G_F can be determined directly. The parameters defining the strain-softening diagram must be numerically evaluated by a fitting technique (Roelfstra and Wittmann 1986).

A RILEM Draft Recommendation (1992) describes the obtention of fracture mechanics properties for AAC. Some characteristic values for three types of AAC obtained at a rate of loading $v = 60 \, \mu m/min$ (Slowik and Wittmann 1992) are summarized in Table 1.

4 SOME EXAMPLES

4.1 Influence of size

The calculations presented here were performed for unnotched 3-point-bending beams, their geometry being defined by span-length l, height d, and thickness b.

The load-deflection curves presented in Fig. 3 were calculated for a geometry characterized by l/d=4 and b=250 mm. The largest beam, with d=150 mm and l=600 mm, has the dimensions of the commonly used AAC-blocks. The employed material properties are those corresponding to type GS in Table 1.

It can be seen in Fig. 3 that the beams with d>37.5 mm exhibit an unstable postpeak behaviour, called snap-back. In order to achieve equilibrium in the descending branch, both the load and the displacement should be reduced. Under deformation control, sudden mate-

Table 1. Characteristic properties for three types of AAC

Type	GN	GH	GS
$\rho \; [kg/m^3]$	511	704	817
$E \; [N/mm^2]$	1800	2200	2500
$f_t \; [N/mm^2]$	0.42	0.66	0.74
$s_1 \; [N/mm^2]$	0.11	0.21	0.22
$w_1 \; [\mu m]$	15.0	15.1	15.9
$w_2 \; [\mu m]$	36.9	42.2	41.9
$G_F \; [N/m]$	5.2	9.4	10.4
$l_{ch} \; [mm]$	53.4	48.0	47.4

rial disruption occurs within the snap-back range. Larger beams, thus, fail in a brittle way.

4.2 Influence of type of AAC

In Fig. 4, the load-deflection curves calculated for the three types of AAC characterized in Table 1 are shown. The dimensions of the beams are: l=250 mm, d=b=100 mm. It can be observed that, although the f_t-values for the three types of AAC are quite different, the unstable post-peak regime is qualitatively the same for all of them. It is worth pointing out that the l_{ch}-values for the three types are nearly coincident.

In order to elucidate the influence of G_F on the failure behaviour, the calculations presented in Fig. 5 were run. The geometry of the beams is defined by l=600 mm, d=150 mm, and b=250 mm. The solid line corresponds to the material properties of type GS in Table 1. For the remaining curves, the G_F-value was increased as indicated in Fig. 5, while keeping the ratio between w_1 and w_2 constant (see Fig. 2). It can be observed that the load bearing capacity of the beams increases with G_F and that, for G_F-values greater than 35 N/m, a stable post-peak behaviour is obtained for this geometry.

4.3 Influence of rate of loading

For beams having l=600 mm, d=150 mm, and b=250 mm, calculations were performed using the material properties obtained for AAC-type GS at different rates of loading (Slowik and Wittmann 1992). It can be seen in Fig. 6 that, in good agreement with experimental findings, the load bearing capacity increases with the rate of loading.

4.4 Direct vs. bending tensile strength

A parametric study was carried out in order to assess the influence of G_F on the difference between direct and bending tensile strength. For this purpose, calculations were run for beams having l=600 mm, d=150 mm, and b=250 mm. For each of the values f_t=0.2, 0.4, 0.6, and 0.8 N/mm^2, the load-deflection curves of the beams were predicted for four values of the fracture energy: G_F=2.5, 5, 10, and 20 N/m. The numerically obtained maximum loads P_{max} were then used to compute the bending tensile strength by means of linear elasticity:

$$f_{bt} = \frac{3}{2} \frac{P_{max}\, l}{b\, d^2} \qquad (2)$$

To obtain the results presented in Fig. 7, it was further assumed that the shape of the strain-softening diagram, i.e. the ratios w_1/w_2 and s_1/f_t, remained constant. The used values of the modulus of elasticity and of the bulk density are E=2500 N/mm^2 and ρ=704 kg/m^3, respectively.

Another representation of the same results is illustrated in Fig. 8, with f_t as parameter. The difference between direct and bending tensile strength increases with G_F.

5 CONCLUSIONS

Based on the results presented above it can be concluded that:

- the same type of AAC may fail in a brittle

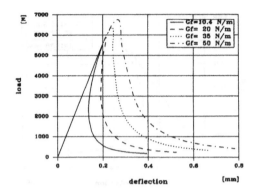

Figure 5: Load-deflection curves calculated for l=600 mm, d=150 mm, and b=250 mm, and for varying G_F-values. The solid line corresponds to the material properties of type GS in Table 1

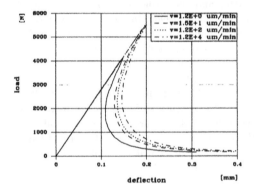

Figure 6: Load-deflection curves calculated for l=600 mm, d=150 mm, and b=250 mm, and at varying rates of loadings for AAC-type GS

or ductile way, depending on the specimen dimensions.

- for the same dimensions, the value of the fracture energy of AAC will define whether an element will fail in a brittle or in a ductile way.

- for a given tensile strength, the bending strength of a beam increases with fracture energy.

- optimisation of AAC should rather aim at increasing fracture energy than strength values.

REFERENCES

Alvaredo, A.M. & F.H.Wittmann 1992. Crack

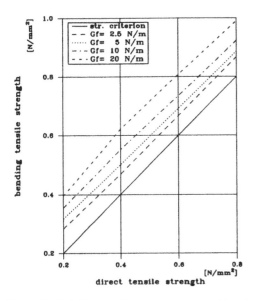

Figure 7: Relationship between direct and bending tensile strength for different values of G_F. The beam geometry is defined by l=600 mm, d=150 mm, and b=250 mm

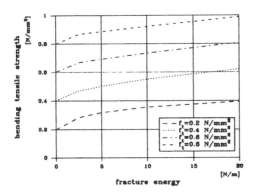

Figure 8: Influence of G_F on the bending tensile strength of AAC with different tensile strength

formation due to hygral gradients. In Z.P.Bazant (ed.), *Fracture mechanics of concrete structures*, p.60-67. Elsevier, London.

Brühwiler, E. & F.H.Wittmann 1990. The wedge splitting test, a new method of performing stable fracture mechanics tests. *Engineering Fracture Mechanics*, Vol. 35, No. 1/2/3, p.117-125.

Hillerborg, A., M.Modeer & P.E.Petersson 1976. Analysis of crack formation and crack growth in concrete by means of fracture mechanics

and finite elements. *Cement and Concrete Research*, Vol. 6, p.773-782.

Hillerborg, A. 1983. Analysis of one single crack. In F.H.Wittmann (ed.), *Fracture mechanics of concrete structures*, p.223-249. Elsevier, Amsterdam.

Petersson, P.E. 1981. Crack growth and development of fracture zones in plain concrete and similar materials. *Report TVBM-1006*, Lund, Sweden.

RILEM Draft Recommendation 1985. Determination of the fracture energy of mortar and concrete by means of three-point bend tests on notched beams. 50-FMC Committee.

RILEM Draft Recommendation 1992. Determination of the fracture energy and strain softening of AAC. 14.1-AAC Committee.

Roelfstra, P.E. & F.H.Wittmann 1986. Numerical method to link strain softening with failure of concrete. In F.H.Wittmann (ed.), *Fracture toughness and fracture energy of concrete*, p.163-175. Elsevier, Amsterdam.

Slowik, V. & F.H.Wittmann 1992. Fracture energy and strain softening of AAC. (In this volume).

Advances in Autoclaved Aerated Concrete, Wittmann (ed.) © 1992 Taylor & Francis. ISBN 90 5410 086 9

Mechanism of frost deterioration of AAC

O. Senbu & E. Kamada
Hokkaido University, Japan

ABSTRACT: Two types of frost deteriorations observed in external walls made from AAC in cold regions. One type of deteriorarion is surface scalling caused by freezing and thawing, the other is wide cracks caused by keeping the inner part of AAC at 0 ℃. It is thought that frost deterioration mechanisms of AAC are not same as those of ordinary concrete. In this paper, deteriorations of AAC caused by various test methods are analysed, and deterioration mechanisms are investigated.

1 INTRODUCTION

AAC is used for external walls in cold regions, and frost deteriorations were sometimes appeared. Two types of deterioration are observed: one is scalling of surface layers and another is wide cracks paralell to the surface. It is thought that deteriorations of AAC are not same as those of ordinary concrete.

Following investigations of the deterioration of AAC, it becomes clear that the frost deterioration of AAC is different from that of ordinary concrete. It is thought that the frost deterioration of ordinary concrete is explained by freezing of capillary water, however that of AAC is explained by freezing of water in air voids.

2 DETERIORATION OF AAC BY TRADITIONAL METHODS

2.1 Freezing and thawing test

A typical freezing and thawing method is ASTM C666. Deterioration of specimens by the method occurs in the form of scalling. Severe splitting which is observed under real conditions does not occur in the test. Figure 1 shows results of the freezing and thawing test in accordance with ASTM C666. It is clear that volume reduction by scalling becomes greater, as test conditions become more severe. It is thought that scalling occurs as surface air voids fill with water by capillary action and freezing pressure.

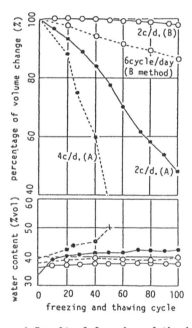

Figure 1 Result of freezig and thawing test by ASTM method (Kamada et al, 1984)

2.2 Critical degree of saturation method

A typical critical degree of saturation method is the RILEM method for concrete. A rapid method consists of freezing and thawing tests to calculate the critical degree of water content sufficient to cause damage, and water absorption tests

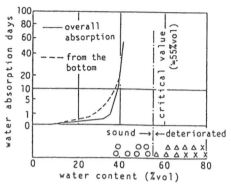

Figure 2 Test result by critical degree of saturation method (kamada et al, 1984)

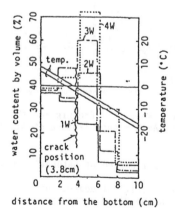

Figure 3 Test apparatus of top surface freezing test (Kamada et al, 1984)

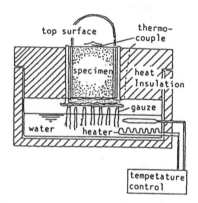

distance from the bottom (cm)

Figure 5 Changes of water content and occurance of destruction in top surface freezing test (Kamada et al, 1984)

Figure 4 Cracked specimen in top surface freezing test (Kamada et al, 1984)

to measure water absorption. In the method, frost resistance of materials is evaluated by comparing the critical degree of saturation with water content of the water absorption test. Figure 2 shows a result of critical degree of saturation method. Deterioration caused by this method is degradation of whole specimen. This type of deterioration is scarecely observed in real walls. However, critical degree of saturation is important as a measure of the quality of construction materials.

2.3 Top surface freezing test

The top surface freezing test which was proposed by Kamada et al.(1984) is a model of external wall conditions. Figure 3 shows a test apparatus of the top surface freezing test, and the test is carried out in a low constant temperature room. In the test, water is applied to the bottom surface of the specimen, the top surface is held at a constant temperature below 0 ℃, and the freezing position is fixed in the specimen.

As shown in Figure 4 and 5, evidence of deterioration during the test was a wide crack across the specimen located at the 0 ℃ level. There is little correlation between the results of the freezing and thawing test and the top surface freezing test. This means that the top surface freezing test might cause deterioration by a different mechanism than that which causes deterioration in the freezing and thawing test.

3 PROPOSAL OF FROST DETERIORATION MECHANISM OF AAC

3.1 Pore structure of AAC

Pore of AAC consist of air voids and

capillaries. The volumetric proposion of AAC (which is ordinary used in Japan) is air voids : capillaries : solid = 5 : 3 : 2. Pore size distribution of AAC and ordinaly concrete is shown in Figure 6. Roughly speaking, both AAC and ordinary concrete have a lot of pores with radiuses of 100-1000Å. A decrease of the freezing point occurs in the capillaries.

3.2 Application of cappillary theory

In the capillary theory proposed by Everett(1961), it is assumed that ice forms in air voids and capillary water is kept unfrozen when deterioration occurs. The deterioration mechanism of AAC develops as the capillary pressure differential between ice in air voids and water in capillaries increases. As shown in Figure 7 pressure (P) built up in the air void from this balance of power.

3.3 Explanation of wide cracks in the top surface freezing test

Figure 8 shows a conceptual cracked plane of a specimen in the top surface freezing test. On the plane, all the air voids are filled with ice, and pressure (P) (explained in Figure 7) builds up in all air voids. Whether or not cracks occur is determined by comparing pressure (P) per a unit area and the tensile strength of the material. Figure 9 shows the relationsips between the volumetric proportion of air voids and tensile strength and/or pressure (P) per unit area. Pressure (P) is changed by the size of the capillary connected to the air void. From the data ploted in Figure 9, capillaries greater than 200Å are thought to influence this deterioration mechanism.

3.4 Explanation of degradation of critical degre of saturation method

Although degradation is rarely observed in external walls, scalling is thought to be the degradation on infinitery small portions of the the AAC surface. By explaining this form of degradation, all the deterioration may become clear.

The self-consistent approximation method proposed by Baba(1978) was used to explain the degradation. In the method, the following items are assumed. Figure 10 shows a schematic presentation of the method.

1. One air void is small enough comparing with whole composite.

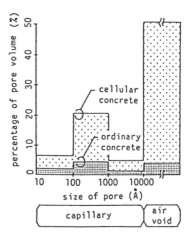

Figure 6 Distribution of pore size of AAC and ordinary concrete

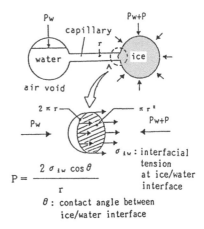

Figure 7 Pressure (P) from capillary theory

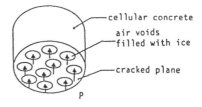

Figure 8 Model of deterioration of top surface freezing test

2. First, pressure (P) is occured at one air void located at the center of the spherical composite containing other air voids.

3. The spherical composite is thought as an elastic body, and whole pressure in the

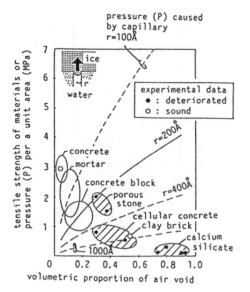

Figure 9 Relationships between volumetric proportion of air void and tensile strength of materials (or pressure (P) per unit area)

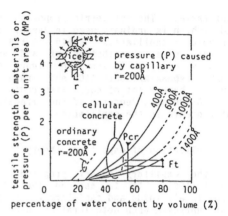

Pcr: critical water content
of cellular concrete
Ft : tensile strength
of cellular concrete

Figure 11 Relationshps between water content and tensile strength of materials or pressure occured by air voids filled with water(ice)

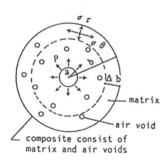

$$\sigma_m = \frac{Wp}{1-Wp} \cdot P$$

σ_m : average pressure
Wp : volumetric ratio of air voids

Figure 10 Model of degradation based on self-consistent approximation (Baba, 1978)

composite is calculated with adding all the pressure of each air void filled with ice.

Figure 11 shows the relationships between water content and the tensile strength of materials (or pressure existing in air voids filled with ice). It is assumed that deteriorations occur when average pressure in the composite becomes greater than the tensile strength. In order to cause deterioration of the composite with a critical water content of 55% by volume, capillaries 600A-1000A in size are necessary. This result reflects the actual pore structure (Figure 6) of AAC.

4 CONCLUDING REMARKS

The deterioration types which were observed in traditional test method are classified, and frost deteriorations of AAC is investigated. It is considered that frost deterioration of AAC occurs when air voids are filled with water. Pore structures of AAC are considered comparing it with that with ordinary concrete. Frost deteriorations of AAC can be explained by capillary theory.

REFERENCES

Baba,A. 1978. Drying shrinkage mechanism of building materials: BRI Research Paper NO.77. Building Reseach Institute, Ministry of Construction.
Everett,D.H. 1961. The thermodynamics of frost damage to porous solids: Trans.Faraday Soc. 57: 1541-1551
Kamada,E. Koh,Y. and Tabata,M. 1984. Frost Deterioraton of cellular concrete: Third international conference on the durability of building materials and components, Volume 3, 372-382.

Advances in Autoclaved Aerated Concrete, Wittmann (ed.) © 1992 Taylor & Francis. ISBN 90 5410 086 9

Frost resistance of increased density autoclaved aerated concrete

Y. Hama & E. Kamada
Hokkaido University, Japan

M. Tabata
Hokkaido Polytechnic College, Japan

T. Watanabe
Sumitomo Metal Mining Co., Ltd, Japan

ABSTRACT: In order to evaluate the frost resistance of various kinds of AAC, in which specific gravity, water absorption and air void size changed, a freezing and thawing test and a top surface freezing test were carried out. Further an exposure test in a cold region was also conducted out. An Increased density AAC with water repellence was more resistant to frost deterioration and a relationship was obtained between frost resistance and the material properties.

1. INTRODUCTION

In Japan, Autoclaved Aerated Concrete (AAC) is mainly used for external walls. Two types of AAC are on the market. One is a normal type, the other is a water repellent type. The specific gravity of on market AAC is defined by JIS at 0.45 to 0.55.

However, these walls are subject to frost deterioration in cold regions. Two types of deterioration are observed. One is surface layers scaling caused by freezing and thawing. The other is cracking caused by water concentration at the freezing point. It is necessary to improve the frost resistance of AAC.

In this paper, an experimental study was conducted to evaluate the frost resistance of the various kinds of AAC which had changed specific gravity, water absorption and air void size. Further, an exposure test in a cold region was carried out in order to confirm the test results on frost resistance in a laboratory.

We conclude from the experiments described above that an increased density water repellent AAC was more resistant to frost deterioration than the others.

2. EXPERIMENTAL PROCEDURE

2.1 SPECIMENS

As shown in Table 1, the experiments comprise three series. In series 1, 32 kinds of laboratory specimens were used. There were four levels of specific gravity 0.50, 0.60, 0.65, 0.70. The water absorption was obtained four levels by changing the additional rate of the water repellent. Air void size reached two levels, large and small, by using different particle-sized aluminum powder.

In series 2 and 3, four kinds of factory-line-produced specimens were used. Two of them were a normal and a water repellent type on the market. Two other kinds were increased density AAC manufactured for trial purposes.

2.2 TEST METHOD

In series 1 and 2, specific gravity, compressive strength, water absorption and pore structure as material properties were measured. And in order to evaluate the frost resistance, both a freezing and thawing test and a top surface freezing

Table 1. Type of specimens

	Series 1								Series 2, 3	
Air void size	large				small				large	small
Specific gravity	0.50	0.60	0.65	0.70	0.50	0.60	0.65	0.70	0.50	0.65
Water absorption levels	4	4	4	4	4	4	4	4	2	2

test were carried out.
In series 3, an exposure test using a model building in a cold region was carried out.

(1)TESTS OF MATERIAL PROPERTIES

Specific gravity
The size of the specimens was 10x10x10cm. The weight of the specimens was measured after drying for 2 days at 105°C and an absolutely dry specific gravity was obtained.

Compressive strength
The size of the specimens was 10x10x10cm. The compressive strength of the specimens was measured after maintaining water content level at 10%wt.

Water absorption
The size of the specimens was 10x10x10cm. The specimens were immersed in water and the weight change of the specimens was measured after 24 hours. The water content of the specimens prior to water immersion was 5%wt.

Pore structure
The pores of AAC consist of air voids and capillaries. The number of air voids and their distribution of size were measured by image analysis. The scope was 70x70mm on the specimen surface. The average value of the diameter was calculated.
The distribution of the capillary pore was measured by a porosimeter. The total pore volume and the medium diameter of the capillary in the region of 32 to 10000Å was obtained.

(2)FROST RESISTANCE TESTS

Freezing and thawing test
In order to estimate surface layer scaling, freezing and thawing test in accordance with ASTM C666 B method (freezing in air and thawing in water) was carried out. The specimen size was 10x10x20cm. In order to control the water content, the specimens were dried at the temperatures below 40°C. After immersing them in water for 48 hours, freezing and thawing cycle began. The condition of freezing was -20°C, 7hours and that of thawing was +10°C, 5hours. The repeating process of freezing and thawing was 100 cycles in series 1 and 50 cycles in series 2.
The weight of the specimens was measured in air and in water at a certain cycle of freezing and thawing, calculating the the volume change and the water content.

Top surface freezing test
The top surface freezing test was proposed by Kamada et al.(1984). It is a model of real external wall conditions. In this test, 20°C water is applied to the bottom surface of the specimen, the top surface is held constant at -20°C, and the freezing position is fixed in the specimen.
The size of the specimens was 10φx10cm. Cracking was observed, the weight of the specimens was measured and the water content was calculated. The test lasted for 80 days in series 1 and for 60 days in series 2.

(3)EXPOSURE TEST

For the exposure test in a cold region, a model building with four kinds of AAC panels which were produced in series 2 was built in Monbetsu city, in Hokkaido of Japan. Fig.1 shows a plan and a sectional plan of the model building and Fig.2 is a detail of the parts of water drip, which is subject to penetrate condensed water into the panels. The water drip corresponds to the parts cause frost deteriorations in real buildings.
Table 2 shows the type of surface coatings and water permeability for 7 days and vapor permeability for 28 days. Eight kinds of surface coating materials were used in order to estimate the influence of frost resistance.

Table 2. Surface coatings used in exposure test

Kinds	water permeability for 7days (ml)	Vapor permeability for 28days(10^2g/m^2)
A	23.4	22.3
B	0.7	4.0
C	235.3	86.2
D	37.9	60.2
E	2.2	14.6
F	1.0	7.5
G	0.2	2.5

A : Polymer emulsion coating
B : Waterproof membrane coating A
C : No coating
D : Siliceous coating
E : Polymer emulsion multi-layer coating
F : Reactive polymer solution multi-layer coating
G : Waterproof membrane coating B

The inside condition had a constant temperature and humidity (20°C and 60%RH). The exposure test began in November.1991 and deterioration was observed.

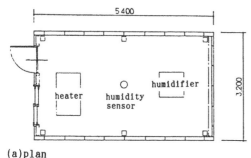

	5400

heater   humidity
          sensor   humidifier   3.200

(a)plan

water drip
(made of stainless)

(b)sectional plan

Fig.1. Model building
for exposure test

40

Fig.2. Detail of
water drip

3. RESULTS AND DISCUSSION

3.1 SERIES 1

(1)MATERIAL PROPERTIES

Table 3 shows the material properties. The absolute dry specific gravity of each specimen is 0.03 to 0.05 more heavy than the level of specific gravity. The heavier the specific gravity is, the higher the compressive strength is. The compressive strength of specific gravity level 0.5 is 4.9 to 5.7 MPa and that of increased density AAC is 7.4 to 12.8 MPa. The compressive strength of increased density AAC is about two times as high as that of on market AAC.

On the other hand, the water absorption for 24 hours is 23.6 to 57.6%wt dependent on the differences of water repellent levels. And the heavier the specific gravity is, the lower the water absorption is.

Fig.3 shows the relationship between the specific gravity and the air void size. As the effects of increased density to the pore structure, air void size lessens, the medium capillary diameter lessens and the total pore volume decreases. With an increase in the specific gravity, the pore structure forms more tightly.

Table 3. Material properties (series 1)

Sym.	Spec.	Comp. MPa	Abso. %wt	Void. mm	Pore. cc/g	Diam. Å
Air void size : Large						
5AL	0.523	5.5	57.6	0.647	0.671	370
5BL	0.529	5.3	46.7	0.670	0.618	275
5CL	0.524	4.9	44.2	0.665	0.606	289
5DL	0.528	5.0	36.2	0.655	0.608	286
6AL	0.648	8.1	56.2	0.527	0.610	309
6BL	0.635	7.6	49.6	0.509	0.542	237
6CL	0.634	7.4	36.8	0.493	0.633	282
6DL	0.643	7.7	27.1	0.495	0.596	290
65AL	0.712	10.4	51.1	0.526	0.665	301
65BL	0.707	10.1	49.4	0.519	0.576	283
65CL	0.710	9.2	36.8	0.482	0.571	281
65DL	0.687	8.9	27.5	0.499	0.597	248
7AL	0.763	12.3	52.4	0.434	0.587	250
7BL	0.751	10.0	34.9	0.459	0.594	263
7CL	0.763	11.1	30.1	0.457	0.577	276
7DL	0.742	10.3	26.9	0.431	0.597	260
Air void size : Small						
5AF	0.505	5.4	54.5	0.660	0.656	265
5BF	0.507	5.6	44.4	0.601	0.659	327
5CF	0.503	4.9	35.6	0.544	0.657	372
5DF	0.498	5.7	23.6	0.585	0.690	335
6AF	0.638	9.0	54.9	0.445	0.688	352
6BF	0.638	8.7	47.0	0.408	0.589	294
6CF	0.639	8.7	31.5	0.422	0.642	372
6DF	0.626	7.6	25.7	0.430	0.528	256
65AF	0.701	10.0	52.6	0.392	0.601	293
65BF	0.684	10.0	51.8	0.412	0.671	269
65CF	0.679	9.8	32.3	0.417	0.606	256
65DF	0.679	9.0	31.1	0.400	0.647	287
7AF	0.766	12.8	46.2	0.383	0.614	236
7BF	0.755	11.7	41.3	0.383	0.579	231
7CF	0.757	11.2	33.6	0.363	0.555	256
7DF	0.736	10.3	29.8	0.369	0.593	231

Spec. : Specific gravity
Comp. : Compressive strength
Abso. : Water absorption
Void. : Average diameter of air void
Pore. : Total pore volume
Diam. : Medium diameter of capillary

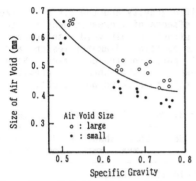

Fig.3. The Relationship between the specific gravity and the air void size

(2)FROST RESISTANCE

Table 4 shows the time until crack formation occurs on the top surface freezing test. As a result of the top surface freezing test, the lower the water absorption is, the earlier cracking occurs. No specific gravity influence was observed.

Table 4. Result of the top surface freezing test (Series 1)

Symbol	Time (days)	Symbol	Time (days)
5AL	5	5AF	20
5BL	30	5BF	50
5CL	50	5CF	70
5DL	70	5DF	no crack
6AL	20	6AF	25
6BL	25	6BF	30
6CL	60	6CF	80
6DL	50	6DF	60
65AL	25	65AF	25
65BL	50	65BF	30
65CL	50	65CF	80
65DL	50	65DF	50
7AL	40	7AF	40
7BL	60	7BF	60
7CL	60	7CF	60
7DL	50	7DF	60

Time : Time until crack formation occurs on the top surface freezing test

Fig.4 shows the volume change for 100 freezing and thawing test cycles in series 1. The volume change of the specimens with small air void size is less irrespective of the specific gravity. And the effect of the water repellent is unclear, but

the volume loss has a tendency to increase in the order of increasing water absorption. Specimens with a specific gravity level at 0.50 are inferior to the other in all combinations of water absorption levels and air void size levels.

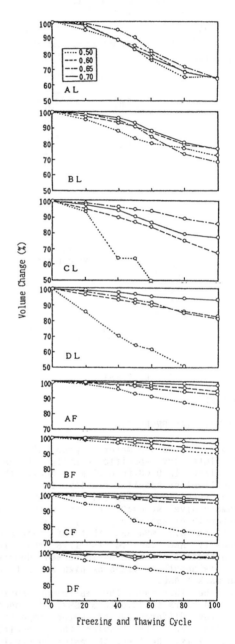

Fig.4. Volume change on freezing and thawing test (series 1)

(3) THE RELATIONSHIP BETWEEN FROST RESISTANCE AND MATERIAL PROPERTIES

Multiple regression analysis of the relationship between frost resistance and material properties was conducted. As an index of the frost resistance, the time until crack formation on the top surface freezing test and the omega-transformed value of the volume change on freezing and thawing test were used. Fig.5 shows comparisons of the measured value and the calculated value by the regression equations. The time until crack formation has a correlation with only the water absorption for 24 hours. And the volume change has a correlation with the specific gravity and the air void size.

It was confirmed that the mechanism of the two types of frost deteriorations, scaling and cracking, are different.

3.2 SERIES 2

Table 5 shows the material properties and Table 6 shows the time until crack formation. The results show the same tendency as the results in series 1. On the top surface freezing test, S0, the normal type on the market, cracking was observed, but the others which were water

Table 5. Material properties (Series 2)

Sym.	Spec.	Comp. MPa	Abso. %wt	Void. mm	Pore. cc/g	Diam. Å
S0-T	0.487	3.8	62.9	0.651	0.634	245
S0-B	0.500	4.0	60.0	0.638	0.674	248
S1-T	0.622	6.8	24.7	0.499	0.609	237
S1-B	0.641	7.7	23.5	0.468	0.579	241
S2-T	0.640	7.1	21.1	0.447	0.586	251
S2-B	0.660	7.6	22.5	0.430	0.596	249
S3-T	0.493	3.6	28.6	0.665	0.670	250
S3-B	0.510	3.9	22.6	0.648	0.664	243

Table 6. Result of the top surface freezing test (Series 2)

Symbol	Time (days)	Symbol	Time (days)
S0-T	20 45	S3-T	60 no crack
S0-B	30 45	S3-B	no crack
S1-T	no crack	S2-T	no crack
S1-B	no crack	S2-T	no crack

(a) Result of freezing and thawing test

regression equation :
$$Y_1 = -4.250X_1 - 7.725X_2 + 7.393 \quad (R=0.847)$$

Y_1 : Omega-transformed value of $Y_1{}'$
$\quad Y_1 = \log(Y_1{}'/(100-Y_1{}'))$
$Y_1{}'$: Volume change(%)
X_1 : Specific gravity
X_2 : Air void size (mm)

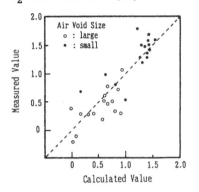

(b) Result of top surface freezing test

regression equation :
$$Y_2 = -1.515X_3 + 108.9 \qquad (R=0.819)$$

Y_2 : Time of crack formation (days)
X_3 : Water absorption (%wt)

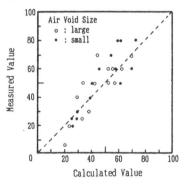

Fig.5. Comparisons of the measured and calculated value

repellent showed no crack for the 60 days of the testing period.

Fig.6 shows the volume change for 50 freezing and thawing test cycles in series 2. and Fig.7 shows the deterioration of specimens at the end of the test, 50 cycle. The volume loss of S0 is 10% to 13% and that of S3, the water repellent type on the market, is 15% to 24% at the 50th

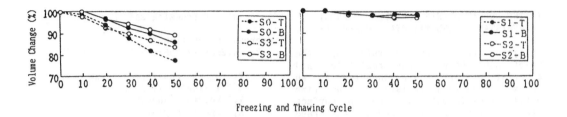

Fig.6. Volume change on freezing and thawing test (series 2)

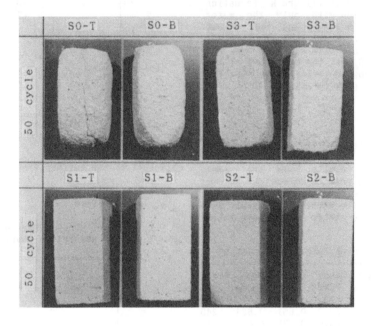

Fig.7. Deterioration of specimens (series 2)

freezing and thawing cycle. In S1 and S2, for increased density AAC which was manufactured for trial purposes, the volume loss is 2% to 3% and the frost resistance is improved.

3.3 SERIES 3

Fig.8 shows the arrangement of AAC panels and surface coatings and the result of observation after 131 days from the test start. Table 7 shows estimation of the result of the exposure test.

Table 7. Result of exposure test

| AAC | Surface coating | | | | | | | | |
| | South | | | | | | | North | |
	A	B	C	D	E	F	G	B	F
S0	▲	○	○	○	○	○	▲	▲	X
S1	○	○	○	○	○	○	○	○	▲
S2	○	○	○	○	○	○	○	○	○
S3	▲	▲	○	○	X	X	X	▲	X

○ :No damage
▲ :Deterioration of surface coating
X :Deterioration of AAC

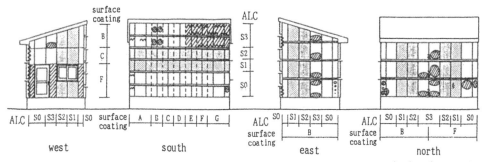

surface
coating

ALC
S3
S2
S1
SO

ALC | SO |S3|S2|S1| |SO   surface| A |B|C|D|E|F| G   ALC SO |S1|S2|S3| SO   ALC | SO |S1|S2| S3 |S2|S1| SO
coating   surface | B   surface | B | F
                                    coating   coating

west   south   east   north

~~~ : crack of surface coating
⬭⬭⬭ : expand of surface coating

Fig.8. Deteriorative conditions of model building after 131 days (series 3)

Frost deterioration of surface coatings or AAC was observed in the case of a combination of SO which was a normal type on the market, with surface coatings A, B, F and G and in the case of S3 which was a water repellent type on the market with all surface coatings. Cracking such as was observed on the top surface freezing test was not observed in this test. The reasons for that are considered to be the short period that the freezing position of panels is maintained and that the water concentration at the freezing point is low in this test.

S1 and S2 which had increased density, showed no damage in the case of all surface coatings. It was confirmed that increased density AAC was more resistant to frost deteriorations.

4. CONCLUSIONS

The conclusion of this study can be summarized as follows:
1) In both the case of laboratory and factory production, increased density AAC with water repellence was comparatively more resistant to frost deteriorations.
2) The relation was obtained between the frost resistance and the material properties. The time until crack formation has a correlation with only the water absorption for 24 hours, and the volume change has a correlation with the specific gravity and the air void size.
3) The result of exposure test was in agreement with that of the other test in a laboratory and it was possible to prevent frost deteriorations by adequate selection of combinations AAC and surface coatings.

REFERENCES

Kamada,E. 1989. Manual of finish coating plans and works for ALC external walls (in Japanese) : Koubunsha
Kamada,E. Koh,Y. and Tabata,M. 1984. Frost deterioration of cellular concrete : The third international conference on the durability of building materials and components, volume 3, 372-382

Advances in Autoclaved Aerated Concrete, Wittmann (ed.)© 1992 Taylor & Francis. ISBN 90 5410 086 9

Chemical resistance of AAC

Gerhard Spicker
YTONG AG, R+D-Centre, Schrobenhausen, Germany

ABSTRACT: AAC cubes, quality category G2/0.5 were stored for 14 days in salt solutions, acids, alkaline solutions and organic solvents. The chemical resistance of AAC to these substances has been determined by means of compressive strength measurements. According to these findings, AAC resists esters, liquid acyclic hydrocarbons, simple aromatic hydrocarbons and carbonhydrates. Furthermore, AAC shows a good resistance to concentrated alkaline solutions as well as to salt solutions with equally strong acidalkaline couples or with low salt solubility. AAC is attacked by acidic salt solutions (by hydrolysis) as well as inorganic and organic acids. For such cases, the acidic attack increases with the rising degree of dissociation as well as with increasing solubility of the calcium salt of the corresponding acid.

1. INTRODUCTION

Due to the numerous fields of application for AAC as a building material, I would like to give a general view over the chemical resistance of AAC to many solutions. An impression of the various fields of application of the building material AAC can be taken form table 1. The widely different industries involved have been listed as well as the chemicals which occur commonly. It is necessary to take the state of the attacking substance into account, i.e. whether the chemical which affects the AAC is in aequeous solution or is a vapour. The corrosiveness of chemicals which only affect AAC in the solid form depends on their ability to form a solution. This is determined by their hygroscopic behaviour and the relative humidity of the surroundings. Consequently, I will restrict myself to the chemical resistance of AAC to solutions in the course of this lecture.

2. EXPERIMENTAL CONDITIONS AND TESTS

Table 2 lists all chemicals whose corrosive effect on AAC has been tested. The individual chemicals have been grouped into aqueous solutions of inorganic and organic salts, inorganic and organic acids and alkaline and organic solvents. All storage tests were performed with AAC material, quality category G2/0,5, which was taken from the current production of the YTONG plant. Two cubes at the same fermentation level were cut out of the hardened block material of a cast form. All AAC cubes were ground with flat surfaces and an edge length of 10 cm and dried at a temperature of 60 °C. Afterwards, the partly-covered cubes were stored for a fortnight in the solutions, listed in table 2 at a temperature of 23 °C ± 2 °C. The storage boxes were shut airtightly for the entire testing period, so that for each system the equilibrium vapour pressure at a given temperature was reached. The direction of fermentation inside each cube was always vertical to the surface of the solution. The control samples were stored under identical conditions; not in a aqueous solution, but in their residual moisture after a 60 °C drying process.

All sample cubes were redried at 60 °C for the compressive strength test according to German standard DIN 4165. The 10x10x10 cm

Table 1: Fields of application for AAC and chemicals that occur there

Industry, Field of application	Prominent chemical		
	as a soild matter	as a solution	as a gas
Fertilizers Agriculture	Nitrates, Phosphates, Sulphates	Organic acid (Manure)	Carbon dioxide, Methane, Ammonia
Paint industry	-	Different organic solvents	Solvent vapours
Tanneries	Arsenide sulfate ("Steep glass"), Aluminium sulfate, Quick lime	Chromium potassium alum, Chromium-(III)-sulphate	-
Pottery	Oxydes, silicates	-	-
Metal upgrading Steel mills	-	Inorganic acids	Acidic vapours
e.g. Aluminium casting	-	Hexachloroethane, Carbon tetra-chloride	Chlorine
Crude oil industry	-	Petrol, Fuel oil, Acyclic hydrocarbons	-
Paper industry	e. g. Alkali silicate salt	Aqueous solutions of alkali silicate salt	-
Dye-Works	Reducing agents	Lyes, Ferrous sulphate hydrate	-
Vehicle workshops	-	Petrol, Engine oil	CO_2, CO, NO_x
Plastic industry		Inorganic + Organic acids	
Food industry		Organic acids, Lactic acid, Acetic acid, Citric acid, Malic acid	CO_2, Fermentation gases Flue gases

AAC cubes were loaded in the fermentation direction and the test pressure was increased at constant speed respectively. The compressive strengths have been calculated as follows:

$$\sigma_p = p / B*L \qquad N/mm^2$$

where:

σ_p = maximum load in N
B = breadth of the sample in mm
L = length of the sample in mm

3. RESULTS

The compressive strengths in N/mm^2 of all sample cubes were determined from double measurements. The corresponding average value was standardized to the average value of the control sample which was set at 100 %. The extent of the control measurement error is given by the size of the symbols in the figures. Fig. 1 shows the relative compressive strengths for the solution stored sample cubes, listed according to increasing degree of chemical attack. The chemicals which attack AAC are summerized here again in the groups of salt solutions, inorganic and organic acids, alkaline solutions and forms of aliphatic and aromatic hydrocarbons.

The material strengths of the AAC cubes have hardly diminished after the storage in the salt solution for 14 days. The not hydrolised as well as the alkaline hydrolised salt solution do not cause a decrease in the material strength.

Table 2: List of the investigated chemical solutions

Type of chemical solution	Concentration in Mole/l (M)
dest. water	pure
tap water	pure
sodium sulphate, Na_2SO_4	1 M
sodium carbonate, Na_2CO_3	1 M
sodium chloride, NaCl	1 M
Lithium fluoride, LiF	1 M
Potassium carbonate, K_2CO_3	2 M
Magnesium sulphate, $MgSO_4$	1 M
Ammonium acetate, NH_4COOCH_3	1 M
Oxalic acid, $H_2C_2O_4$	2 M
Acetic acid, CH_3COOH	1 M, 2 M
Tartraric acid	1 M
Sulphuric acid, H_2SO_4	2 M
Nitric acid, HNO_3	1 M, 2 M
Hydrochloric acid, HCl	2 M
Sodiumhydroxide, NaOH	1 M, 4 M
Acetic acid-n-butyl-ester	1 M
Xylene	pure
Petrol	pure
Fuel oil	pure
D(+)-Glucose	1 M

The acidic hydrolised salt solutions do act acidicly on the material strength of the AAC: Later on I will discuss in detail the effect of acids.

Like all soluble sulphates, MgSO4 should also attack AAC by creating CaSO4. Within the testing period of 14 days, however, there is only an indication of damage to the micro-structure owing to CaSO4 crystallization. The reduced material strength after storage in an ammonium acetate solution is due to the reaction of the remaining calcium hydroxide with the ammonium salt. The products of this reaction are water soluble calcium acetate and volatile ammonia gas. AAC has also proved resistant to the organic solvent acetic acid-n-butyl-ester, petrol and fuel oil. Apart from that, the AAC-material resisted action of the aromatic compound xylene.

Strong alkaline solutions, for which caustic soda solution served as an example in the examination, have little affect. on AAC.

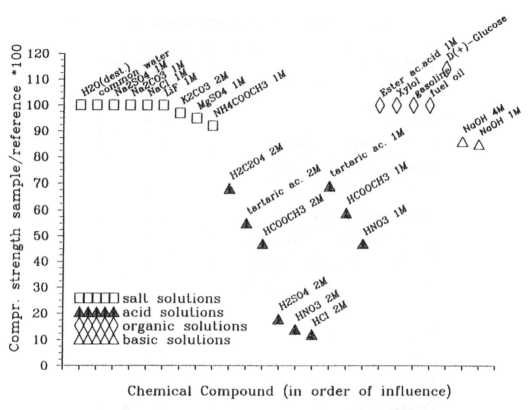

Figure 1: Relative compressive strength of AAC after 14 d storage in different solutions

This is at least valid for concentrations ranging up to 4 moles (14 wt.%).

Noticeable, however, was the increase in material strength after storage in glycose solution. This effect has also been described, independent of this examination, in a study on chemical resistance at the University of Erlangen carried out by Dr. M"rtel's working group. The AAC attacking effect of organic as well as of inorganic acids over the period of examination was significant. At a two molar concentration, the inorganic acids such as sulphuric acid, nitric acid and hydrochloric acid were the most corrosive. The correspondingly lower, unimolar concentrations of inorganic acid (nitric acid) as well as of organic acid (acetic acid) acted less severely. Consequently, for the same dilution the corrosive effect of the more concentrated acid decreases more than with the weaker acid.

The relationship between the extent of chemical attack on AAC and the corresponding acid concentration, is expressed as acidity constants, can be seen clearly in figure 2. With decreasing pk(a)-values of the acids, the relative compressive strength of the acid-stored sample cubes decreases as well. Lower acid concentrations cause a lower decrease in the material strength. The considerable deviations of the relative compressive strength values around the broken trend line can be explained very easily by the additional effect of the different solubilites of the calcium salts which correspond to different acids. Their solubilities are also shown in fig. 2. Higher solubilities therefore enhance the decrease in the material strength. Thus, the salts with high water solubilitiy cause due to the corresponding acids relative material strengths which are below the trend line, while the scarcely soluable salts show relative strengths which lie above the trend line and consequently attack AAC less.

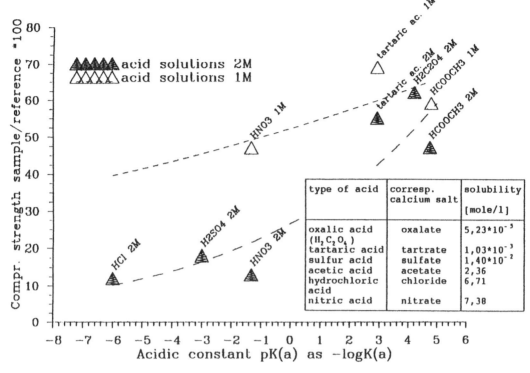

Figure 2: Relationship of chemical resistance (rel.compressive strength) of acids and their acid constants

4. CONCLUSIONS

Regarding the chemical resistance of AAC to aqueous solutions and liquids, a distinction has to be made between the following solvents.

Salt solutions

AAC is resistant
- to aqueous salt solutions which contain ions occuring instrong acids and bases (e.g. NaCl) concentrated acid-alkaline couples (e.g. NaCl)
- to aqueous salt solutions whose corresponding base is stronger than calcium hydroxide (e.g. Na$_2$CO$_3$)

AAC has limited resistance
- to sulphate solutions
- to salt solutions whose corresponding base is weaker than calcium hydroxide (e. g. ammonium acetate)

- to salt solutions whose corresponding acids are more concentrated than bases (e. g. MgCl$_2$)

The chemical resistance of AAC decreases in hydrolised salt solutions; the weaker the corresponding base is or the stronger the corresponding acid is.

Inorganic and organic acids

AAC has limited resistance
- to weak acids whose salts show little water solubility (e.g. oxalic acid)

AAC is not resistant
- to weak acids whose salt show a high water solubility (e.g. calcium acetate)
- to strong acids (e.g. HCL, HNO$_3$)

Alkaline solutions

AAC is resitant
  -to strong lyes (e.g. NaOH)

Organic liquids of cyclic (aromatic) and ali-phatic compounds

AAC is resistant
  - to esters (e.g. acetic acid-n-butyl-ester)
  - to acyclic hydrocarbons (e.g. petrol, fuel oil)
  - to aromatic hydrocarbons (e. g. xylene)
  - to carbonhydrates (e.g. glycol) here is even an increase in material strength.

The examination of the chemical resistance of AAC to solutions shows that far-reaching conclusions can already be drawn from the present observations. Furthermore, it is necessary to record in detail changes in the chemistry and the microstructure of AAC subjected to solutions over long periods of time. Therefore, the work by Dr. M"rtel's group at the Institute for Material Science in Erlangen, is very productive and helpful.

5 Reinforced components

Advances in Autoclaved Aerated Concrete, Wittmann (ed.) © 1992 Taylor & Francis. ISBN 90 5410 086 9

Researches for the design of reinforced aerated concrete beams

K. Janovic & E. Grasser
Technische Universität München, Germany

ABSTRACT: This report deals with experimental and theoretical investigations on reinforced aerated concrete. Several research programs were performed at the Institute for Reinforced Concrete, Technische Universität München. The results of these studies — of course taking into consideration relevant research results of other authors also — should allow to derivate basic dimensioning rules on physically sound models. For practical application simplifications are necessary. The proposals can be considered as a contribution to the actual efforts to unificate international codes of practice.

1 DESIGN RESISTANCE TO AXIAL ACTION EFFECTS

1.1 General considerations

From the extensive studies on the ultimate limit states of resistance to axial stress states (bending moments, axial forces) for reinforced concrete members made of normal concrete it is well-known that one of the most difficult model elements is the behaviour of concrete in the compression zone. Therefore the main attention was directed on the problem of the concrete compression stress distribution in the ultimate limit state and on the relevant compressive strain at the most highly stressed compression fibre. Finally an idealised stress distribution is proposed.

1.2 Compression force in the concrete compression zone

For aerated concrete, a relatively brittle material, the distribution of the compressive stresses will follow a monotonous function up to maximum strain $\epsilon_{c,max}$ at the edge of the section. A descending branche of the stress distribution is not to be expected. For the normal case "neutral axis within the section" the compression force of the compression zone can be represented by the mean value of the compression stresses (σ_m) related to the maximum stress (σ_{max}). For short term loading σ_{max} is equal the strength of a prism f_{prism}. The relationship between σ_m and σ_{max}, the so-called block coefficient α (Völligkeitsgrad) was derived from test results of excentrically loaded concrete prisms, reported by Rüsch and Sell (1961), and of beams, tested by Rüsch et al (1967). In these reports also the cube strength f_{cube} is given. Janovic, Grasser and Kupfer (1975) have shown that the block coefficient α_{cube},

related to the cube strength, is influenced significantly by the cube strength f_{cube}. We derived from the prism tests

$$\alpha_{cube} = 0.63 \quad \text{for} \quad f_{cube} \approx 5.0 \text{ N/mm}^2$$
$$\alpha_{cube} = 0.83 \quad \text{for} \quad f_{cube} \approx 8.0 \text{ N/mm}^2$$

and from the beam tests

$$\alpha_{cube} = 0.60 \quad \text{for} \quad f_{cube} = 5.0 \text{ N/mm}^2$$
$$\alpha_{cube} = 0.80 \quad \text{for} \quad f_{cube} = 7.6 \text{ N/mm}^2$$

This influence can be explained by the interaction of porosity, strength and brittleness of aerated concrete.

More results we gained from a broad research program including 37 tests on beams with the strength of $f_{cube} = 5$ N/mm^2. The reinforcement in 9 of these beams was greater than the reinforcement at the balance point, that means that the concrete in the compression zone crushes at ultimate load before yielding of the reinforcement. From the measured strains of the reinforcement and of the top fibre of the beam the relative depth ζ of the compression zone and the compression force could be derived. In this calculation the eigenstresses due to steam curing of the material were considered as prestressing of the reinforcement and they were taken into account. The block coefficient from these tests proofed to be $\alpha_{cube} = 0.56$.

The mean value of the block coefficient, determined from all before-mentioned research reports is $\alpha_{cube} \approx 0.6$. In relation to the prism strength f_{prism} and assuming a relationship $f_{prism} = 0.95 \, f_{cube}$ as a mean value the proposal for the block coefficient

$$\alpha_{prism} = 0.60/0.95 = 0.63$$

seems to be well based for short term loading.

1.3 Reference strength of concrete

As mentioned before all the tests have been performed under short term loading. Sell and Zelger (1969) found a difference of the strength of aerated concrete between short term tests and tests under sustained load of about 15 %. The reference strength of aerated concrete which is simoultanously the maximum of the stresses at the upper fibre of the compression zone of a section under bending moments in the ultimate limit state is therefore:

$$f_{ref} = 0.85\ f_{prism} = 0.85 \cdot 0.95\ f_{cube} \approx 0.8\ f_{cube}$$

1.4 Concrete strain in the ultimate limit state

Similiar to the behaviour of normal concrete there is a significant strain difference in the ultimate limit state between the most compressed fibre under bending moments (ϵ_{max}) and centrically loaded prisms (ϵ_{centr}). From the before—mentioned research reports mean values of $\epsilon_{max} = -3.0$ ‰ and $\epsilon_{centr} = -2.5$ ‰ were derived. Taking into account some uncertainties and to produce the block coefficient $\alpha_{prism} = 0.63$ with a simplified stress distribution (see Figure 1) it is recommended, to reduce ϵ_{centr} to -2.2 ‰.

1.5 Allowable steel stress

In a new dimensioning concept the behaviour of the different materials (concrete, steel) should be introduced with its real properties in a consistent way. For the reinforcing steel this would mean that the relevant yield strength or — after division by a partial safety factor — the design value could be used irrespective of the strength. But dimensioning a member we have to fulfil not only the safety requirements in the ultimate limit state but also the conditions in the service limit state. Janovic et al have deduced the necessity to limit the yield strength to 420 N/mm^2 in order to avoid untolerable crack width in the service limit state.

1.6 Proposed rules for calculation the resistance to axial load effects

To summarize and to complete the considerations above the calculation of the resistance to axial load effects (bending moments, axial forces) can be based on the following assumptions:

— Idealized stress distribution for aerated concrete (see Figure 1) with the following main features:
  · Maximum strain at the most compressed fibre $\epsilon_{max} = -3.0$ ‰.
  · Strain at contact point between triangle and rectangle stress distribution, also valid for centrically loaded sections: $\epsilon_{centr} = -2.2$ ‰.
  · Reference strength: $f_{ref} = 0.8\ f_{cube}$.

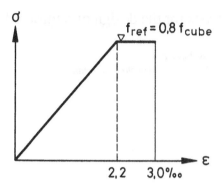

Fig. 1 Proposal for the distribution of compressive stresses in the compression zone of an aerated concrete beam

  · Block coefficient: $\alpha_{prism} \approx 0.63$.

— Possible strain diagrams at the ultimate limit state comparable to the assumptions for normal concrete (e.g. see Eurocode 2, Part 1, Figure 4.11).
— Idealized stress—strain diagrams for steel as bilinear diagrams with yield strength f_y ($\leq$ 420 N/mm^2).

All other necessary assumptions are in principle the same as for normal concrete:
— Plane sections remain plane.
— The reinforcement is subjected to the same variations in strain as the adjacent concrete.
— The tensile strength of concrete is neglected.

A comparison of calculated moments in the ultimate limit state with results from the relevant 9 beam tests showed with calc M_u/obs $M_u = 1.03$ a very good accordance.

2 ANCHORAGE OF REINFORCING BARS

2.1 General remarks

For reinforcement in aerated concrete structures normally smooth bars or wires are used, often coat—protected against corrosion. The remaining bond strength must be neglected. The reinforcement (longitudinal bars, stirrups) is anchored by welded tansverse reinforcement. A safe anchorage is a presupposition for gaining the calculated bearing capacity of a reinforced member made of aerated concrete. The very important problems of the anchorage zone are discussed in the following.

2.2 Distribution of the pressure under the transverse bars

The distribution of the concrete pressure along the transverse bars is studied by Janovic et al (1975) on the basis of the bedding value theory (normally used in soil mechanics; the bedding value is the ratio of soil pressure

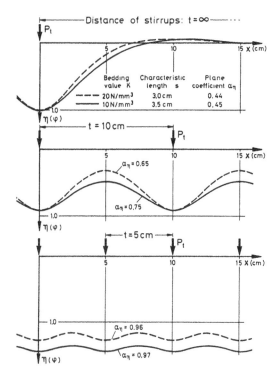

Distance of stirrups: t = ∞ · · · · · · ·

Bedding value K	Characteristic length s	Plane coefficient α_η
– – – 20 N/mm³	3,0 cm	0,44
—— 10 N/mm³	3,5 cm	0,45

Fig. 2 Distribution of the concrete pressure p(x) under a reinforcing bar or the displacement function $\eta(\varphi)$ resp. along the elastically bedded bar with one or more single transverse loads

— in our problem concrete pressure — to settlement). The accuracy of this procedure is sufficient only if either the differences of the pressure along the elastically bedded bar are not too great within the charcteristic length s or either the compressible layer under the bar is thin. The last condition needs some consideration.

The matrix of aerated concrete is a cell structure. The normally very thin walls of the pores create the bearing structure. The compression strength of aerated concrete depends on the stability of the cell walls. After crushing of single cell walls under a reinforcing bar enlarged areas of concrete are activated up to point where crushing of more cell walls is stopped. Obviously the area with crushed cell walls increases with increasing pressure under the bar.

The layer in which crushing of cell walls can occur remains thin. Displacements in this region are great relativ to the thickness of this layer and therefore the stiffness is small compared with undisturbed layers beneath.

So the assumption that a reinforcing bar with a small diameter lies in a thin weak bedding layer on a stiff subgrade is reasonable and therefore the application of the bedding value theory is allowed.

The bedding value theory is based on the connection of the pressure p(x) under the bar and its settlement y(x):

$$p(x) = K \cdot y(x)$$

For aerated concrete the bedding value K was derived by Janovic et al (1975) on the basis of test results from Sell and Zelger (1969) to:

$$K \approx 10 \text{ to } 20 \text{ N/mm}^3$$

For the bedding value theory the well—known equations are used:

$$EI \cdot y(x) = \frac{P \cdot s^3}{8} \cdot \eta(\varphi)$$

$$p(x) = \frac{P}{2s\,\phi} \cdot \eta(\varphi)$$

with:

$$\varphi = \frac{x}{s}$$

$$s = \sqrt[4]{\frac{4EI}{\phi\,K}} \quad \text{(characteristic length)}$$

$$\eta(\varphi) = e^{-\varphi}(\cos\varphi + \sin\varphi)$$

As an example a longtitudinal bar of infinite length with diameter $\phi = 7$ mm, bedded elastically and loaded by single forces — representing the stirrups — at distances of t was calculated. With the bedding values K = 10 and 20 N/mm³ resp. the characteristic length is s = 3.5 and 3.0 cm. Figure 2 shows the settlement $\eta(\varphi)$ for one (upper diagram) or more loads.

If in this case the distance of the loads is greater than about 15 cm the overlapping influence of neighboured loads is neglible. The influence of the K—value on the characteristic length s and on the function $\eta(\varphi)$ or the distribution of the pressure p (x) along the bar is small.

2.3 Spalling of the concrete cover of the transverse bar

The pressure σ_t under a transverse bar produces in the aerated concrete splitting tensile stresses. Figure 3 shows how these splitting tensile forces can be calculated.

A representative test specimen according to Fig. 3 is loaded on one side by the pressure of a transverse bar. Due to Hanecka (1969) the splitting tensile strength of a test specimen loaded with a concentrated line load only at one side is about 15 % higher than the normal splitting strength:

$$f'_{ct,sp} = 1.15\ f_{ct,sp} = 1.15\ \frac{\sigma_t}{e}\ \frac{\phi_t}{\pi} \tag{1}$$

Equation 1 is valid if the pressure along the transverse bar is constant. Figure 2 shows that this distribution is

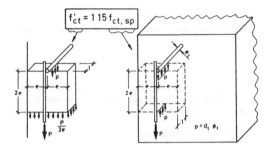

Fig. 3 Splitting of the concrete cover of a transverse bar. Representative tensile splitting test specimen loaded at one side (left) or at both sides (right)

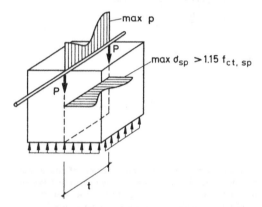

Fig. 4 The interdependance between the non uniform distributed concrete pressure under a reinforcing bar and the tensile splitting stresses acting in a perpendicular plane in a tensile splitting test specimen loaded at one side with two single loads

approximately constant only for small distances of the single loads. Similiar to normal concrete the tensile strength is increasing with increasing gradient of the bearing pressure.

A non uniform distribution of the bearing pressure will produce also a non constant distribution of the splitting tensile stresses perpendicular to the splitting plane (Figure 4). The maximum splitting tensile stress in the ultimate limit state (spalling) at the point of the transverse bar may become even greater than 1.15 $f_{ct,sp}$.

2.4 Relation between bearing pressure and splitting tensile strength

Hanecka (1969) reported, that the splitting tensile strength ($f_{ct,sp}$) of an aerated concrete test specimen under the pressure (σ_t) in a small stripe (transverse bar) increases with the relation of the stripe width (a = ϕ_t) to the width of the cube (b). He distinguished the two characteristic domains a/b:

$$0 < a/b \leq 0.2$$
$$0.2 < a/b \leq 1.0$$

In the first domain a strong settlement of the pressed stripe into the concrete connected with substantially crushing of pores occurs, in the second domain a normal tensile splitting failure is observed. For a/b = 0.2 the compressive strength under transverse bars was about 1.7 f_{cube}.

Sell (1970) investigated the tensile splitting strength with cubes of different sizes (b = 5 to 20 cm). The relative width of the pressure stripe was kept constant (a/b = 0.08). The tensile splitting strength was $f_{ct,sp} = f_{cube}/12$ in all cases without influence of the size of the cube. From this can be concluded that no crushing of the pores of aerated concrete and also no splitting of the concrete cover occurs if for the domain e/$\phi_t \geq 6$ Equation 2 is fulfilled.

$$\sigma_t = 1.15 \, \pi \, \frac{e}{\phi_t} \, f_{ct,sp}$$
$$= 1.15 \, \pi \, 6 \, \frac{f_{cube}}{12} \leq 1.8 \, f_{cube} \quad (2)$$

Janovic and Grasser (1987) derived from the results for the domain e/$\phi_t < 6$ the following formula:

$$\sigma_t = 1.15 \, \pi \, A \, \left(\frac{e}{\phi_t}\right)^{0,5} \quad (3)$$

The factor A can be determined from control test specimens for the determination of the tensile splitting strength with cubes (2e = cube size; a = ϕ_t width of pressure stripe).

$$A = \frac{f_{ct,sp}}{f_{cube}} \, \left(\frac{e}{a}\right)^{0,5} \quad (4)$$

The non uniform distribution of the pressure under the transverse bar and consequently of the tensile splitting stresses too can be taken into account by a correction factor k. This factor is the product of the plane coefficient of the pressure distribution under transverse bars (α_η in Fig. 2) and of an increasing factor which takes into consideration the influence of the tensile splitting stress gradient on the relevant strength. Janovic et al (1975 and 1977) proposed this factor k to be a function of the distances t of the reinforcing bars:

$$\begin{aligned} k &= 1.0 & \text{for} \quad t \leq 12 \, \phi_t \\ k &= 0.7 \cdot 2.0 \, \frac{\phi_t}{t} & \text{for} \quad t \geq 20 \, \phi_t \end{aligned} \quad (5)$$

Values between can be interpolated.

2.5 The tolerable displacements of the transverse bars in aerated concrete

To avoid crushing of the aerated concrete under the transverse bars in chapter 2.4 a limit for the bearing pressure of $\sigma_t < 1,8 \, f_{cube}$ was deduced. With this value, bedding values of K = 20 or 10 N/mm³ resp. (the latter one yielding greater displacements) and a strength of the aerated concrete $f_{cube} = 5$ N/mm^2 the average displacements (settlement) of the transverse bars are

$$\Delta = \frac{\sigma_t}{K} = \frac{1.8 \cdot 5}{20} \text{ or } \frac{1.8 \cdot 5}{10} \approx 0.5 \text{ or } 1.0 \text{ mm resp.}.$$

Anchorage slip of the reinforcement of this magnitude in the ultimate limit state seams to be tolerable.

2.6 Influence of transverse pressure in the anchorage zone

The influence of a transverse pressure σ_{supp} (e.g. pressure at supports) can be taken into account approximately by a fictitious magnification of the splitting tensile strength by the amount of σ_{supp}, as proposed by Janovic and Grasser (1977).

$$f'_{ct,sp} = 1,15\ f_{ct,sp} + \sigma_{supp} \qquad (6a)$$

$$\sigma'_t = \pi \frac{e}{\phi_t} f'_{ct,sp} = \sigma_t + \pi \frac{e}{\phi_t} \sigma_{supp} \leq 2,2\ f_{cube} \quad (6b)$$

In most practical cases the influence of transverse pressure at supports can be taken into account by an enlarging factor to the calculated stress under the transverse bar:

$$\sigma'_t = 1,3\ \sigma_t \leq 2,2\ f_{cube} \qquad (7)$$

The limitation is necessary because the strength increasing of aerated concrete under hydrostatic pressure is relatively small, as unpublished tests by Aschl (1977) have shown (e.g. for $\sigma_1 = \sigma_2 = \sigma_3 = 2,4\ f_{cube}$).

2.7 Number of transverse bars

At some test specimens for pull—out tests (see chapter 2.8.1) with 4 transverse bars in the anchorage zone the steel strains were measured by strain gauges. It proofed that from the whole force in the longitudinal steel bar about 30 % were anchored with the first transverse bar, the other transverse bars anchored 25 to 20 % each. From this it can be deduced that with 4 transverse bars about 80 % of that force can be anchored which can be carried by one transverse bar times four. A representative number n′ of transverse bars which should be considered dimensioning an anchorage may be estimated with

$$n' = n^{0,85} \qquad (8)$$

2.8 Pull—out tests

2.8.1 Research program

To develop, check and complete the dimensioning proposals explained in the chapters 2.2 to 2.7 for the anchorage strength of longitudinal bars with welded transverse bars Janovic and Grasser (1987) performed a research program with pull—out tests at the Institute of Reinforced Concrete of the Technical University Munich. The

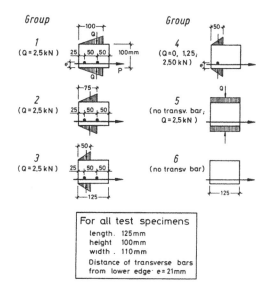

Fig. 5 Dimensions and loading arrangements of the pull—out test specimens with two, one and none transverse bars

following parameters have been taken into consideration:
  concrete cover
  diameter of transverse bars
  number of transversive bars
  transverse pressure at supports

The test program was divided in two categories:
  test specimens with one or two transverse bars (Figure 5)
  test specimens with 4 transverse bars (Figure 6 and Table 1)

Each group of the test specimen in Fig. 5 consisted of 6 pull—out tests, thus 36 all together.

The transverse force (Q = 2.5 kN) was the same for the test specimen with two transverse bars (group 1, 2 and 3), while the length of the support aerea (100, 75 and 50 mm) and consequently the transverse pressure σ_{supp} differed. For the test specimen with only one transverse bar (group 4) the length of the support area was kept constant (50 mm) while the transverse force was different (Q = 0; 1.25 and 2.5 kN).

With test specimens shown in Figure 6 for every series 3 pull—out tests were performed, hence 15 tests together. The following influencing parameters have been studied (see Table 1):

— relation between diameter of transverse to longitudinal bar $\phi_t/\phi_\ell = 6/8$ and 8/10 mm (series I, II, I$_Q$ and III, III$_Q$; index Q means with transverse pressure)

— concrete cover of the longitudinal bar (transverse bar inside) c = 10 and 25 mm (series II)

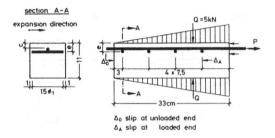

section A-A

expansion direction

Q = 5kN

P

Δ₀

Δ_A

3   4 x 7,5

Q

15 ⌀t

33cm

Δ₀ slip at unloaded end
Δ_A slip at   loaded end

Fig. 6   Dimensions and loading arrangement of the pull—out test specimens with four transverse bars

To avoid a slip fastener effect the distance of the transverse bars was chosen $s_t \geq 2e$ always. In all tests with transverse pressure the resultant force acted in 1/3 of the support length.

2.8.2 Material and measurement of the tests

The test specimens were produced by the Hebel factory in Emmering and precisely cut to the proposed dimensions.

The compressive strength (f_{cube}) and the tensile splitting strength ($f_{ct,sp}$) of the air—dried material was for the test specimens of Fig. 5 (10 cm cubes)

$f_{cube} = 5.28$ N/mm²
$f_{ct,sp} = 0.59$ N/mm² (width of loading stripe: 8 mm)

for the test specimens of Fig. 6 (7 cm cubes)

$f_{cube} = 4.30$ N/mm²
$f_{ct,sp} = 0.45$ N/mm² (width of loading stripe: 8 mm)

For the reinforcement the steel grade S500 was used; the welded ladders were coat—protected with bituminous material.

The pull—out force has been increased displacement—controlled. The slip of the reinforcing bar at the unloaded end was measured by inductivity. The pull—out force of the test specimens of Fig. 6 was controlled by the slip Δ_A at the first transverse bar.

2.8.3 The test results

a) Bond strength (Fig. 5, Group 5 and 6)

The bond strength of bars without welded transverse bar and without transverse pressure (Group 6) was very low ($f_b = 0.026$ to 0.073 N/mm²; mean value 0.050 N/mm²).

The bond strength of bars also without transverse bar but with transverse pressure (Group 5) was with a mean value of 0.050 N/mm² practically the same.

b) Anchorage with one transverse bar (Fig. 5, Group 4)

The development of the pull—out force P and of the end slip Δ_0 was tested with different transverse pressure (Q = 0; 1.25 and 2.5 kN) with two test specimens in each case. The mean values of the transverse pressure were $\overline{\sigma}_Q = Q/(50 \cdot 110) = 0$; 0.23 and 0.46 N/mm². For related test specimens the comparison of the forces P showed a very small scatter.

Test specimens without transverse pressure broke suddenly after reaching the maximum force by concrete splitting (max $\Delta_0 = 0.4$ and 0.9 mm). With transverse pressure the end slip could be increased to more than 2 mm without rupture.

From these results can be concluded that transverse pressure increases deformability and hence the safety of an anchorage with welded transverse bars considerably.

c) Anchorage with two transverse bars
   (Fig. 5, Group 1, 2 and 3)

The resultant force of the transverse pressure (Q = 2.5 kN) was the same in all tests of these three groups. With the different length of the support (100; 75 and 50 mm) mean values of the transverse pressure resulted in 0.23; 0.30 and 0.46 N/mm².

The scatter of the pull—out forces P of each group again was very small. Also the mean values of the three groups differed scarcely. Therefore the results of these groups could be presented in only one diagram (Figure 7).

It is of interest that the doubled values of the pull—out forces of Group 4 (one transverse bar) are not very different from all — single — values of Group 1, 2 and 3 (two transverse bars).

The influence of the number of transverse bars corresponds approximately with Equation 8.

d) Anchorage with four transverse bars
   (Fig. 6 and Table 1)

Figure 6 showes the interdependence between the mean values of the pull—out forces $\overline{P}$ of each series and the end slip Δ_0.

The scatter within every series was small. Up to an end slip $\Delta_0 \approx 0.5$ mm no influence of the concrete cover could be stated. But for greater values of the end slip sudden rupture by splitting occured for small concrete cover (c = 10 mm), while for a greater concrete cover (25 mm) the pull—out force and the end slip could be increased considerably. A transverse pressure produced about the same influence as an increasement of the concrete cover.

2.8.4 Comparison of calculated and observed pull—out forces

For the calculation of the pull—out force the following formula was used:

$$\text{cal } P = n' \ \sigma'_t \ \phi_t \ t \ k \tag{9}$$

with

$n' = n^{0,85}$ (Equ. 8)

n   number of transverse bars

σ'_t pressure under transverse bar (Equ. 6 and 3)

k   factor for the influence of the transverse pressure distribution (Equ. 5)

The relationship between the observed pull—out forces obs P of all tests shown in Fig. 5 and Table 1 and the relevant calculated values cal P according to Equation 9 varied from obs P/cal P = 0.86 to 1.07 with a mean value of 0.96.

Schäffler (1960) mentioned already the unfavourable influence of increasing diameters of transverse bars. Also Eligehausen (1986) reported this effect. Our tests showed for transverse bar diameter of 8 mm (Series III and III_Q in Fig. 8) results at the lower border of the scattering area.

Dimensioning rules for the anchorage of longitudinal reinforcing bars have to take into consideration that all results reported here are related to the ultimate limit state, that an anchorage failure occurs always suddenly without warning and, finally, that all tests were carried out under short term loading. The strength decrease under long duration of the load must be considered.

3 DESIGN RESISTANCE TO SHEAR

In our research reports the problems of the shear resistance of aerated concrete members have been studied extensively. For a thourough discussion of these problems and our test results the space within this paper is too limited. A publication in the near future is intended. In the following only some remarks on special problems connected with aerated concrete are made.

The shear resistance of members without shear reinforcement—can not be well described by the so—called tooth—model — as it is used for normal concrete — without substantial limitations. The bond strength of the reinforcing bars is very small, therefore the longitudinal force in the reinforcement is anchored only by welded transverse bars, sometimes with great distances in the inner part of the span. Also on "aggregate interlock" actions in cracks and on dowel actions can—not be relied upon because of the porosity of aerated concrete, the small diameters of the reinforcement and the diminishing of the compression zone height at the ends of the cracks.

Table 1:   Research program of the test specimens with 4 transverse bars acc. to Fig. 6

Series	Diameter of transverse bar ϕ_t (mm)	Distance of transverse bar to edge e (mm)	Mean value of transverse pressure $\overline{\sigma}_t$ (N/mm²)
I	6	21	—
II	6	36	—
I_Q	6	21	0,14
III	8	24	—
III_Q	8	24	0,11

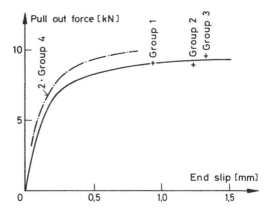

Fig. 7 Mean values $\overline{P}$ of the pull—out forces of test specimens with two transverse bars (acc. to Fig. 5, Group 1, 2 and 3). The dotted line represents the doubled mean values of the test specimens with one transverse bar (Group 4)

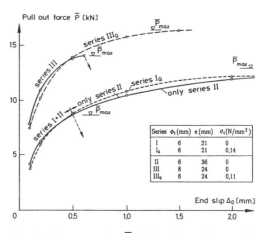

Fig. 8 The mean values $\overline{P}$ of the pull—out forces and the end slip Δ_0 of the test specimens with four transverse bars (acc. to Fig. 6)

179

Tests carried out by Janovic, Grasser and Kupfer
(1975) have shown that the reinforcement has to be
anchored by welded transverse bars. The safety of such
an anchoring system can be checked by rules given in
chapter 2 (e.g. Equation 9). For this the knowledge of
adequate shifting rules is a presupposition. This problem
is discussed by Janovic et al (1975, 1987). One more
research project to study especially these problems is
intended.

REFERENCES

Berger, W. 1967. Werksbericht über die Bemessung von
bewehrten Ytong—Stürzen aus GSB50, Goslar
Ytongwerk, SZ—Watenstedt.

Eligehausen, R. 1985. Stellungnahme vom 4.2.1986
zum Vorschlag für ein Versuchsprogramm von
E. Grasser vom 18.7.1985 mit Ergänzung vom
1.8.1985.

Hanecka, K. 1969. Some remarks on problems dis—
cussed by the session of the Twelfth CEB Committee
"Lightweight Concrete" held at Scheveningen.

Janovic, K., Grasser, E.and Kupfer, H. 1975. Versuche
zur Biege— und Schubtragfähigkeit von bewehrten
Gasbetonbauteilen, Institut für Massivbau,
TU München, Bericht—Nr. 2232/Ja/K.

Janovic, K. and Grasser, E. 1977. Bemessung der End—
verankerung der Längsbewehrung in auf Biegung und
Schub beanspruchten Gasbetonbauteilen, Institut für
Massivbau der TU München, Bericht Nr. 1142/Ja/mb.

Janovic, K. and Grasser, E. 1987. Grundlagen für die
Schub— und Endverankerungsbemessung bewehrter
Gasbetonbauteile (für DIN 4223), Institut für
Massivbau der TU München, AIF—Forschungsvor—
haben Nr. 5475.

Rüsch, H. and SELL, R. 1961. Festigkeit der
Biegedruckzone, DAfStb, Heft 143, Berlin.

Rüsch, H., Kordina, K. and Stöckl, S. 1967. Festigkeit
der Biegedruckzone, Vergleich von Prismen— und
Balkenversuchen, DAfStb, Heft 190, Berlin.

Schäffler, H. 1960. Versuche über die Verankerung der
Bewehrung in Gasbeton, DAfStb, Heft 136, Berlin.

Sell, R. and Zelger, C. 1969. Versuche zur Dauer—
standfestigkeit von Leichtbeton, DAfStb, Heft 207,
Berlin.

Sell, R. 1970. Festigkeit und Verformung von Gasbeton
unter zweiaxialer Druck—Zug—Beanspruchung, DAfStb,
Heft 209, Berlin.

Advances in Autoclaved Aerated Concrete, Wittmann (ed.) © 1992 Taylor & Francis. ISBN 90 5410 086 9

The bond stresses of AAC and slips of reinforcement bars

K. Hanecka

Institute of Construction and Architecture, Slovak Academy of Sciences, Czechoslovakia

ABSTRACT: Bond strength and bond stress-slip working diagram are both important properties for the design of reinforced AAC elements and they been investigated on specimens of size 150*150*150 mm. Three kinds of anti-corrosion coatings applicated on reinforcement bars representing low, mean, and high quality of bond have shown that the distribution of bond have shown that the distribution of bond strresses at given slips of free end of embeded bars otained from push out tests are not symmetrical. Statistical evaluation of the test results has shown that standard deviation as well as skewness depend on the level of bond stress or on slip respectivelly. Results of an analysis give the possibility to determine the characteristic bond strength and to design idealised bond stress-slip working diagrams.

1. INTRODUCTION

It is generally known that reinforcement applicated in aerated concrete elements has to be protected against corrosion.

A number of various corrosive protestions has been used in many countries producing reinforced AAC up to now. Experiences with the use of different kinds of anti-corrosion coatings of reinforcement applied on the same kind of steel and interface have shown different properties related to bond. In the RILEM Model Code for AAC structural members are designed with respect to bond and anchorage by cross welded bars. The bond properties and their consequencies on behaviour of reinforced elements are not sufficiently known in spite of extensive research which has been done in many countries.

The best design for bond and anchorage of reinforcement embeded in AAC can be carried out assuming many laws concerning both these phenomena are known.

The relationship between bond stress and slip of reinforced bar, so-called working diagram, is one

Table 1. Review of tested samples.

Kind of AAC	Mark of serie	Sample size	Diameter of reinforced bar in mm	Mark of corrosive protection	Bond quality of protection
Siporex	S 36	105	8	36	High
	S 37	91	8	37	High
	S 63	37	6,3	V2022/1	Mean
	S 71	38	7,1	V2022/1	Mean
	S 81	64	8	V2022	Low
Calsilox	C 36	104	8	36	High
	C 37	80	8	37	High
	C 81	55	8	V2022	Low

of the important laws of bond.

This working diagram is possible to be obtained in the similar way as, for example, stress-strain diagram for concrete in compression or in tension. It is very important to have stiff system of loading in testing, i.e to load at constant increasing strains at stress-strain or bond-slip diagram.

This paper is devoted to a several aspects related to bond-slip working diagrams and statistical analysis of push out tests carried out on samples made from AAC with reinforced bar.

2 DESCRIPTION OF TESTS

Eight of series of specimens was produced from AAC by technology Siporex and Calsilox as well.

The specimens i.e cubes with size 150 mm were performed by cutting out from slabs 3000*600*150 mm reinforced by four longitudinal smooth steel bars protected against corrosion. The review of the series is in Table 1.

Three kinds of anti-corrosive coatings were used for testing of specimens:

Table 2. Review of compressive strenghts and yield-strenghts of bond in series.

Serie	Compressive strength			Yield strength of bond f_{by} (bond stress at slip s=0,1mm)			
N_o	Sample size n	Mean values f_{cm} (MPa)	Varia-tions δ_c (%)	Sample size n	Mean values f_{bym} (MPa)	Standart deviations s_{by} (MPa)	Skew-ness α_{by}
S 36	18	3,21	9,97	105	1,4455	0,2326	-1,289
S 37	18	3,21	9,97	91	1,3635	0,3532	-1,117
S 63	37	3,16	11,53	37	0,5732	0,6220	-0,827
S 71	38	2,91	13,33	38	0,5427	0,4199	-1,122
S 81	24	3,07	11,77	64	0,2813	0,1859	-0,429
C 36	18	3,42	10,2	104	1,4665	0,3442	-0.989
C 37	18	3,42	10,2	80	1,2790	0,4719	-1,016
C 81	20	3,55	12,27	55	0,5244	0,4783	-0,091

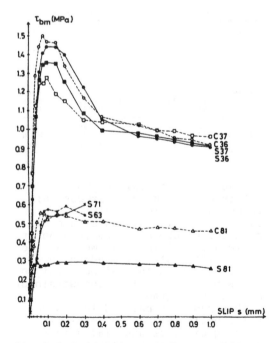

Fig.1 Relationships between slips s and mean bond stresses τ_{bm}

-the coating with lower quality of bond i.e.coating marked as V 2022. This protection consists of high anti-corrosive dispersed colour,

-the coating with medium quality of bond based on the same substance as the first one but with admixture in order to increase the quality of bond,

-the coating with high quality of bond. This kind of protection contains dispersion colour as binder and some admixtures which subsequently improves bond properties.

All these three protections are, according to Edgren (1990) independed of temperature.

To control AAC quality the cylinder samples with diameter 100 mm and hight 100 mm were taken for each test serie. The samples were cuted out from the same slabs as specimens to bond tests.

Results of statistical analysis of AAC on compressive strength are in Table 2.

3 METHOD OF TESTS

The tests were carried out in accordance with Czechoslovakian Standard i.e CSN 73 2036 by Hanecka (1989). Push out tests were carried out on facility recommended by the same standard.

The stiff system of loading was used at all series with exeption of series S 63 and S 71. These two series were tested by soft system of loading.

4 EVALUATION OF TESTS

Development of mean bond stresses, variations and skewnesses in relation with slips of free end of embeded reinforced bar are drawed in figures 1,2 and 3.

From Fig.1 it can be seen that in all tested series the bond stresses increase linearly almost up to slip s=0,1mm. Long-term tests of reinforced elements subjected to bending carried out by Hanecka (1983) confirmed that they were always failur in anchorage if the slip had reached value 0,1mm at any time. Descending part of the diagram drops more rapidly in a higher quality of bond.

Observations of the descending part of the diagram is impossible in case the soft system of loading is used in the testing.

From Fig. 2 it is evident that at the beginning the variations are decreasing with s, and then increasing up to maximum of bond stress.

Also skewness at the beginning increases with s similarly as the variations but after reaching the maximum bond stress, they keep approximately constant.

Relations between bond failures and their corresponding slips are in Table 3. Statistical evalutions of individual serie show that skewness at bond failure (i.e. at maximum of bond stress) is negative and the absolute value increases with mean values of bond failure. On the other hand skewness of the slips corresponding to the bond failures is positive.

The distributions of random values cannot be consider as normal distributions due to significant values of skewness.

In this case the log-normal distribution described by three or only two parameters can be applied as recommended by ISO document by Holicky (1992).

Log-normal distributions of random values i.e yield stredgth of bond (Fig.4) corresponding to the slip s=0.1mm and failure bond strength (Fig.5) of all tested series are drawn in Fig. 4 and Fig.5.

Table 3. Review of bond failures and slips.

Serie	Sample size	Bond failure f_{bu} (maximum bond stress)			Slip s_u corresponding to the bond failure		
		Mean values	Standard deviations	Skewnesses	Mean values	Standard deviations	Skewnesses
N_o	n	f_{bum} (MPa)	s_{bu} (MPa)	α_{bu}	s_{um} (mm)	s_s	α_s
S 36	105	1,4818	0,2259	-1,351	0,1622	0,0708	2,6681
S 37	91	1,3986	0,3264	-1,076	0,1487	0,0668	1,5836
S 63	37	0,5966	0,1324	-0,830	0,0995	0,0698	2,7931
S 71	38	0,6232	0,1187	-0,399	0,1384	0,0871	0,9672
S 81	64	0,3506	0,1558	-0,676	0,3078	0,2686	1,0580
C 36	104	1,5361	0,3264	-0,855	0,1233	0,1205	5,4572
C 37	80	1,4070	0,3929	-0,759	0,1874	0,2582	2,4363
C 81	55	0,6863	0,3953	-0,325	0,3835	0,3626	0,7905

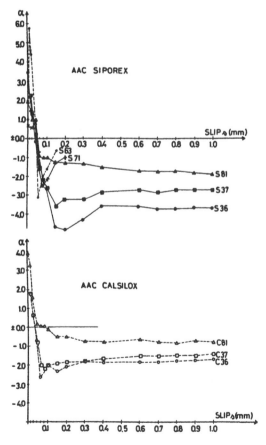

Fig.3 Relationships between slips s and skewness α.

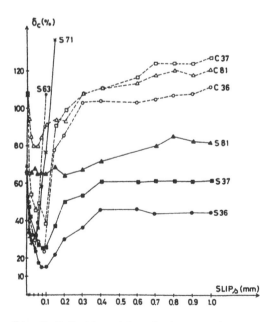

Fig.2 Relationships between slips s and variations δ_c

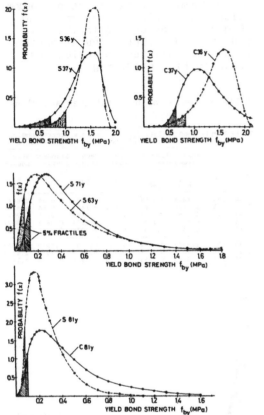

Fig.4 Distributions of yield bond strength of individual series

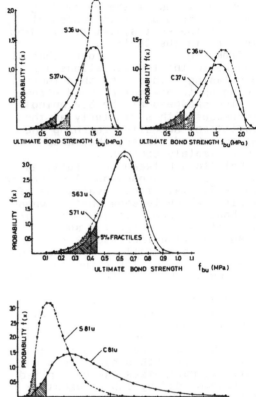

Fig.5 Distributions of ultimate bond strength of individual series

The log-normal distributions have been expressesed by parameters supposing that standard deviation and mean value of the population of a given serie is equal to sample standard deviation and to sample mean value respectivelly.

The two parameters of log-normal distribution were taken account for given serie drawn in Fig.4 and Fig. 5 in the equation (1):

$$(\bar{x}-3s) \quad 0 \qquad (1)$$

where $\bar{x}$ denotes mean value of yield bond strength or failure bond strength,
s is the standard deviation of these random values.

In Fig.4 and Fig. 5 are the shadows represent 5 % of the yield strengths and failure bond strengths calculated from log-normal distributions.

The comparisons of these 5% calculation values from Fig.4 and Fig. 5 for normal and log-normal distributions are listed in Table 4.

From this table it can be seen that normal distribution is not proper for determination of characteristic strength of bond especially at the low bond strength.

The soft system of loading is also not possible for determining characteristic yield strength of bond. Many of spacimens from sample are failured at this system of loading before the yield strength is reached. Therefore the series S 63 and S 71 have been cancelled for further evaluation.

The relationship between ultimate slips s_u and characteristic ultimate bond strength f_{buk} divided by characteristic compressive strength f_{ck} of AAC are drawn in Fig.6. Regression curve in term of hyper-

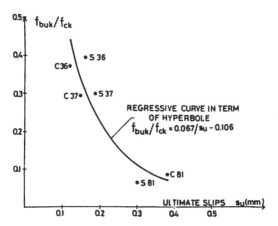

Fig.6 Relative characteristic ultimate bond strengths f_{buk}/f_{ck} in relation with ultimate slips s_u

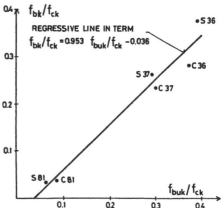

Fig.7 Relative characteristic bond strengths f_{bk}/f_{ck} in relation with relative characteristic ultimate bond strengths f_{buk}/f_{ck}

Table 4. Characteristic values of strenghts.

Serie N_o	Yield strength of bond f_{bk}(MPa)		Ultimate bond strength f_{buk}(MPa)		Compressive strength f_{ck}(MPa)
	AT	DISTRIBUTION			
	Normal	Log-normal	Normal	Log-normal	Normal
S 36	1,0640	1,0110	1,1113	1,0588	2,685
S 37	0,7842	0,7084	0,8633	0,7946	2,685
S 63	-0,4469	0,0914	0,3794	0,4065	2,562
S 71	0,1459	0,1396	0,4285	0,4492	2,273
S 81	-0,0236	0,0874	0,0951	0,1597	2,477
C 36	0,9020	0,8327	1,0008	1,0645	2,847
C 37	0,7626	0,6679	0,7626	0,8647	2,847
C 81	0,0380	0,1081	0,0380	0,2472	2,835

bole is given by formula (2):

$$f_{buk}/f_{ck}=0.067/s_u-0.106 \qquad (2)$$

In Fig.7 there is evaluated relations between characteristic ultimate bond strengths f_{buk} and yield strengths of bond f_{bk}:
The both values are divided by characteristic dompressive strength of AAC. The characteristic strength values are taken from Table 4.
Regression line can be expressed in term

$$f_{bk}/f_{ck}=0.953f_{buk}/f_{ck}-0.036 \qquad (3)$$

The ultimate slip s_u can be determined by using equations (2) and (3).

For a given relative characteristic bond strength f_{bk}/f_{ck}, the ultimate slip is expressed by formula (4):

$$s_u=0.067/((f_{bk}/f_{ck}-0.953)/(-0.036)+0.106)$$

A simplified bi-linear bond-slip diagram can be obtained if it is taken account the previous statements i.e.:
-the bond stresses are linearly increased from zero up to characteristic bond strength corresponding to critical slip s=0,1mm,
-after reaching characteristic bond strength the bond stresses remain constant up to ultimate slip s_u given by formula (4) namely from long time loading effect.
In Fig.8 the simplified bond-slip working diagrams are drawn for given relative characteristic bond strength values i.e. f_{bk}/f_{ck} equal 0.1; 0.2; 0.3 and 0.4 as well.

5 CONCLUSIONS

Pull out tests of bond of anticorrosive coated reinforcing bars have shown that bond stresses up to a slip s=0,1mm of the free end of the

185

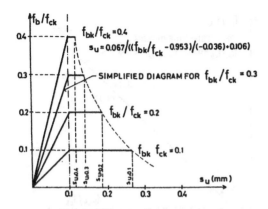

Fig.8 Simplified bond stress-slip working diagrams

bar are almost linear.This slip can be considered as yield bond strength.

The yield bond strength is very difficult to determine if a soft system of loading is used for testing.

The failure of tested elements due to bond at short time test usually starts after slip s=0,1mm if the stiff system of loading is applied at testing. The bond stress at failure can be considered as ultimate bond strength and slip corresponding to this strength as ultimate slip.

Results of these tests have shown that the ultimate slip decreases with increasing ultimate bond strength.

Statistical evaluation has shown that distributions of both yield bond and ultimate bond strength are not symmetrical.

Skewness of all tested series is significantly negative. The dispersions of bond stresses have shown to be a minimum at ultimate bond strength.

Simplified bond stress-slip working diagrams can be considered to be bi-linear,as it is drawn in Fig. 8.

REFERENCES

Edgreen,N. 1990. Proposal of RILEM TC 78 MCA,part 3.9.1 AAC/Reinforcement interface.

Hanecka,K.1983.Reinforcement anchorage in aerated concrete elements subjected to sustained load effects.Stavebnicky časopis VEDA 31:533-541.
Hanecka,K.1989.CSN 73 2036 Test of bond of reinforcement with cellular concrete and their action on each other.Czechoslovakian standard.
Mrazik,A.1987.Teoria spolahlivosti ocelových konstrukcii. (Reliability theory of steel structures.) Vydavatelstvo SAV,Bratislava
Holicky,M.1992.Statistical methods for quality control of building materials and components. The first draft of document ISO/TC 98/SC 2/ WG 3.

Advances in Autoclaved Aerated Concrete, Wittmann (ed.) © 1992 Taylor & Francis. ISBN 90 5410 086 9

Initial steel stresses in reinforced AAC units

A. Koponen & J. Nieminen
Lohja Corporation, Finland

ABSTRACT: Reinforced AAC units are influenced by prestresses generated during autoclaving. Because of these prestresses there will be tensile stresses in reinforcing steel and compression stresses in AAC after autoclaving. In this paper it is discussed how these stresses are generated and what influence they have on the properties of reinforced AAC units. As a result of research in Finland a method is presented by which the initial steel stresses can be determined in a reliable way and evaluated so that they can be taken into account in design calculations. Results from using this method in research and quality control are presented.

1 INTRODUCTION

Prestresses in reinforced AAC units is a subject which is discussed only in a few published papers /1/,/2/.

The magnitude of prestress is different in different factories. In many factories the prestresses decrease rapidly after autoclaving so that they can be ignored. In other factories, particularly those where cement

based anti-corrosion coating is used, the stresses are so significant that they should not be ignored.

In Ikaalinen Siporex factory, where cement based anti-corrosion coating is used, the initial steel stress after autoclaving varies from 80 to 140 MPa depending on the type of unit and the amount of reinforcement. Though these stresses are small compared with the steel stresses used in prestressed concrete units they should be taken into account when designing the reinforcement of AAC units.

2 GENERATION OF PRESTRESSES DURING AUTOCLAVING

The prestresses are due to the different deformation properties of reinforcing steel and AAC during autoclaving when full pressure is 11 atmosphere and maximum temperature over 180 °C.

The main reason for these differences is the fact that AAC has a smaller thermal expansion coefficient than steel as shown in Figure 1. In addition, the deformations

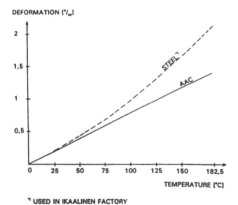

DEFORMATION [°/₀₀]

STEEL"

AAC

TEMPERATURE [°C]

" USED IN IKAALINEN FACTORY

Fig.1 Typical thermal expansion of steel and AAC

of AAC are also influenced by the curing conditions and the specific properties of the raw materials used.

Deformation of AAC during autoclaving in the Ikaalinen factory is presented in Figure 2. These results were obtained by using a special equipment developed by Internationella Siporex AB Central Laboratory Sweden. This equipment consists of a quartz rod attached to AAC, a special displacement transducer also attached to AAC which measures the distance to the free end of the quartz rod and a data acquisition unit.

For a case where the deformations of free steel, free AAC and a reinforced AAC unit are those presented in Figure 3 the generation of prestresses can be explained as follows.

At the beginning of autoclaving the temperature of the AAC mass is about 50 °C. The deformation of free steel is only influenced by its thermal expansion which means that expansion and contraction follow the same line. The deformation of AAC is smaller than that of steel and follows the curve shown in Figure 2. The resulting deformation of reinforced AAC unit is between these two lines.

At the beginning of autoclaving the deformations of the materials don't generate any stresses because there is no sufficient bond between reinforcing steel and AAC and the steel bars can slip inside AAC. At some point before the full pressure is reached the bond increases so much that compression stresses are generated in steel and tensile stresses in AAC. This is due to the fact that steel expands more than AAC. During the full pressure period these stresses increase because AAC shrinks, see Figure 2. If the bond develops too early the tensile stresses of AAC can be so high that transversal cracks can occur in the reinforced AAC unit during autoclaving.

During the descent of the autoclave pressure the compression stresses of steel and the tensile stresses of AAC will gradually

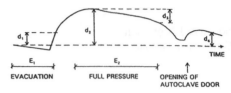

TYPICAL DEFORMATION CURVE

EVACUATION   FULL PRESSURE   OPENING OF AUTOCLAVE DOOR

TYPICAL DEFORMATION VALUES

	d_1 ($^o/_{oo}$)	d_2 ($^o/_{oo}$)	d_3 ($^o/_{oo}$)	d_4 ($^o/_{oo}$)
Ikaalinen factory (1983)	0	1,1	0,2	0,9

Fig.2 Typical deformation curve of AAC during autoclaving

fade and some time before opening the autoclave door there are hardly any prestresses in the reinforced AAC unit. After opening the autoclave door AAC expands rapidly and this means that there will be tensile stresses in reinforcing steel and compression stresses in AAC. During this phase the temperature of AAC unit is about 100 °C from which it will cool to ambient air temperature. During the cooling period steel contracts more than AAC and prestresses will increase in the reinforced unit; the lower the final ambient temperature the greater the prestresses.

Finally, the reinforcing steel has a tensile stress of 80 to 140 MPa corresponding to a compressive stress in AAC of 0.5 to 0.7 MPa. If the steel and AAC were now allowed to move freely relative to each other the relative length difference would be 0.8 to 1.1 mm/m.

3 INFLUENCE OF PRESTRESSES

Because of prestress AAC is under compression which improves the performance in serviceability limit state as the cracking resistance in bending is increased. On the other hand, prestress of this magnitude has no significant influence on the properties in ultimate limit state. Basic guidelines for taking prestress into account in design calculations are given in Chapters 5.2 and 5.3.

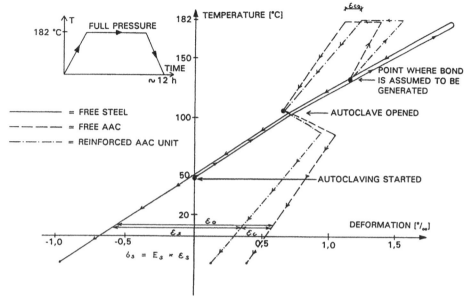

Where:

ϵ_o = relative length difference between AAC and steel

ϵ_s = elongation of reinforcing steel

ϵ_c = compression of AAC in reinforced unit

ϵ_{∞} = elongation of AAC in reinforced unit

during full pressure

E_s = modulus of elasticity of steel

σ_s = initial steel stress

Fig.3 Deformations of free steel, AAC and reinforced AAC unit as a function of autoclaving temperature after bond has generated between steel and AAC

As mentioned above too great deformation difference can cause transversal cracks during autoclaving.

Another problem that can arise if the prestresses are very high is that end cracks can occur due to splitting forces near the ends of a prestressed unit. If the splitting stresses are greater than the tensile strength of AAC, cracks will appear. The location of splitting forces and two typical types of end cracks are presented in Figure 4. End cracks at the level of reinforcement are more dangerous than end cracks in the middle because they can extend far deeper into the unit.

End cracks can be avoided by using transversal reinforcement near the ends of unit. This reinforcement can be hooks, welded bars or special 'ladder' reinforcement, see Figure 5.

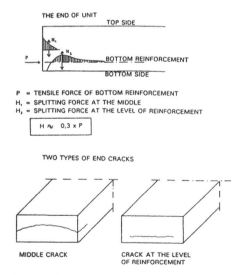

Fig.4 Splitting forces at the end of unit and two types of end cracks

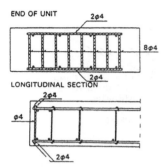

Fig.5 Ladder reinforcement

4 INITIAL STEEL STRESS MEASURE-MENTS IN QUALITY CONTROL

To keep the problems caused by too great prestresses under control and to justify taking prestress into account in design calculations it is necessary to measure the level of initial steel stress in a reliable way.

In the Ikaalinen factory the measurement of steel stress is done by using strain gauges glued to the surface of the reinforcing bar.

The measurement is done in the following way:

1. The AAC unit is placed on its edge as shown in Figure 6 so that stresses due to own weight are insignificant.

2. Normally the steel stresses shall be measured on three longitudinal bars (tension reinforcement) representing the top, middle and bottom positions during casting. The bars shall be uncovered at mid-span, see Figure 6.

3. After the removal of the anti-corrosion coating two strain gauges are glued onto each bar, on opposite sides.

4. When the glue has hardened the first readings of the strain gauge values are recorded.

5. Then the steel bars are cut so that they are free of stresses and the second set of readings is recorded.

The steel stress (f_{so}) is calculated from the strain values as follows:

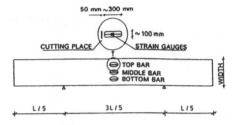

Fig.6 The test arrangement of initial steel stress measurement

$$f_{so} = (\epsilon_1 - \epsilon_2) \times E_s$$

where:
ϵ_1 = mean value for the two gauges from the first readings
ϵ_2 = mean value for the two gauges from the second readings
E_s = modulus of elasticity of steel bar

The mean steel stress value (f_{som}) of the three measured steel bars is used for the evaluation of test results. The modulus of elasticity of AAC (E_c) should also be determined and it is useful to measure the camber of the unloaded unit at mid-span; it can be compared with calculated camber.

This test method is now under preparation in CEN/TC 177/WG 3 with the title 'Determination of steel stresses in unloaded autoclaved aerated concrete components'.

5 EVALUATION OF TEST RESULTS

5.1 Relative length difference

When prestress is used in design by calculation, it is not possible to use a measured steel stress value directly because that value is significant only for the type of unit tested. For example if two AAC units are produced in the same mould, one with smaller amount of reinforcement and the other with more reinforcement, the first unit has bigger initial steel stress because the compression of AAC is smaller in this unit.

The value of relative length difference

190

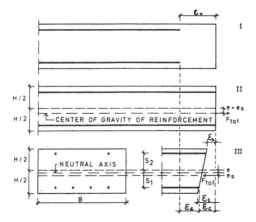

Phase I: Relative movements are free and stresses are zero.

Phase II: The length difference is neutralized by means of an external force (F_{tot}) affecting the reinforcement only.

Phase III: Interaction between AAC and reinforcement is established and external counterforce (F_{tot}) which affects the whole cross section is introduced.

The total of external forces is now zero.

Symbols:

A_1 = area of bottom (tension) reinforcement

A_2 = area of top (compression) reinforcement

E_c = modulus of elasticity of AAC

E_s = modulus of elasticity of steel

n = E_s/E_c

Fig.7 The analysis of prestress by means of an assumed external force

already shown in Figure 3 is not depending on cross section values and can therefore be used in design. It is defined as the difference in length between AAC and reinforcing steel when relative movements are not prevented and can therefore be considered as a material property which depends on raw materials, type of anti-corrosion coating and autoclaving process.

5.2 Calculation of relative length difference

The calculation of relative length difference

(ϵ_o) and initial camber (a_o) can be made with the help of Figure 7 when the cross-section values of the AAC unit and the modulus of elasticity of AAC and steel are known.

A. Cross section values of AAC unit

Moment of inertia (J):

$$J = B \times H^3/12 + n \times (A_1 \times S_1^2 + A_2 \times S_2^2) - A_t \times e^2$$

Where:

$A_t = B \times H + n \times (A_1 + A_2)$

$e = n \times (A_1 \times S_1 - A_2 \times S_2)/A_t$

Bottom section modulus (W_1) and top section modulus (W_2):

$$W_1 = J/(H/2 - e); \quad W_2 = J/(H/2 + e)$$

The distance between the center of gravity of reinforcement and neutral axis (e_s):

$$e_s = \frac{A_1 \times (H/2 + S_1) + A_2 \times (H/2 - S_2)}{A_1 + A_2}$$

$$- e - H/2$$

B. Calculation of relative length difference

The external force (F_{tot}) can be calculated using following formula:

$$F_{tot} = \epsilon_o \times (A_1 + A_2) \times E_s$$

The compression strain at bottom and top fibre of AAC:

$$\epsilon_1 = F_{tot} \times (1/A_t + e_s/W_1)/E_c$$

$$\epsilon_2 = F_{tot} \times (1/A_t - e_s/W_2)/E_c$$

and the tensile strain in steel:

$$\epsilon_s = \epsilon_o - \epsilon_c$$

Where:

$$\epsilon_c = \epsilon_1 - (\epsilon_1 - \epsilon_2) \times (H/2 - S_1)/H$$

The relative length difference (ϵ_o) can now be calculated by resolving the equation:

$$f_{som} = \epsilon_s \times E_s$$

Where:
f_{som} = the measured mean steel stress

C. Calculation of initial camber

Initial camber (a_o) can be calculated using the following formula:

$$a_o = \frac{\epsilon_o \times (A_1 + A_2) \times E_s \times e_s \times L^2}{8 \times E_c \times J}$$

Where:
L = the length of the AAC unit

5.3 Design values

As mentioned previously it is sufficient to take prestress of this magnitude into account only in serviceability limit state design, where they can have a significant influence on cracking moment and camber. Prestress is introduced in the calculation by the parameter 'relative length difference' which is derived from measured steel stress values as described in Chapter 5.2.

Normally the steel stresses are determined after a short stabilizing period of about one week. If the variation of test results is reasonable the mean relative length difference (ϵ_{om}), evaluated from such tests, therefore represents the short-term properties. The value to be used in design should represent the long-term properties and should therefore take into account the loss of prestress due to shrinkage and creep. In most cases it is sufficient to take shrinkage into account simply by subtracting an estimated shrinkage value (normally 0.2-0.25 mm/m) from the short-term ϵ_{om} - value. Creep will be automatically taken into account by using a reduced value of the modulus of elasticity of AAC. Normally a creep factor of 1.0 is used for AAC.

Depending on the type of anti-corrosion

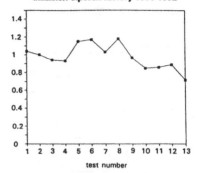

Fig.8 Relative length difference ϵ_o - values (mm/m) in the Ikaalinen factory

coating loss of prestress can also take place due to slip of the reinforcement. This has to be verified by long-term bond tests.

The estimated value of the long-term prestress can be verified by measuring the steel stresses in old AAC units. It should be kept in mind, however, that great variations have to be expected in such tests due to different and usually un-known load history of old units.

6 TEST RESULTS

6.1 Quality control test results in the Ikaalinen factory

Initial steel stress measurements have been made in the Ikaalinen factory since 1982. Nowadays it is part of the normal quality control which is supervised by the authorities and made with the frequency of at least two tests per year and when changes are made to raw materials or autoclaving process. Normally tests are made more often as can be seen in Figure 8 where the relative length difference values calculated from the measured steel stresses are presented. Until 1991 the calculation of relative length difference has been done by using nominal modulus of elasticity of AAC and nominal cross section values of AAC unit.

To ensure efficiency and reliability the strain gauge readings are directly recorded

Table 1. Reinforcement area of tested AAC units

	Bottom A_1 (mm^2)	Top A_2 (mm^2)
Unit 1	393	95
Unit 2	393	95
Unit 3	251	95
Unit 4	251	95
Unit 5	143	95
Unit 6	143	95

Table 2. The modulus of elasticity of AAC (E_c)

	E_c from bending test (MPa)	E_c from strain gauges (MPa)	E_c from Demec gauge (MPa)
Unit 1	1678	1680-1765	1540-1765
Unit 2	1664		
Unit 3	1682		
Unit 4	1679		
Unit 5	1791		
Unit 6	1566	1640-1840	

Table 3. Measured initial steel stresses and calculated relative length difference values

	Mean measured steel stress f_{som} (MPa)	Relative length diff. ϵ_0 (mm/m)
Unit 1	88.4	0.97
Unit 2	88.5	0.98
Unit 3	107.8	0.95
Unit 4	105.1	0.93
Unit 5	125	0.88

The average of ϵ_0 is 0.94 mm/m and the standard deviation is 0.04 mm/m.

on a data logger. The resulting ϵ_0 - value is evaluated by a computer program developed for the purpose.

6.2 Test series made by Technical Research Centre of Finland (VTT)

In connection with the current work in CEN/TC/177 on European norms for reinforced AAC products the initial steel stresses were determined on a test series of 6 reinforced units manufactured at the Ikaalinen Siporex factory in Finland. The tests were carried out under the supervision of Technical Research Centre of Finland /3/. Another objective was to make the calculation of relative length difference more accurate by using measured modulus of elasticity of AAC and measured cross section values.

The verification was done by comparing the relative length difference values of AAC units with different reinforcement (produced in the same mould) and comparing the calculated initial camber values of AAC units with measured ones.

6.2.1 Test units

The AAC units were cast on January 1991 all in the same mould and were taken out of the autoclave on Jan 21st. Dimensions of all the units were 200×600×5980 mm^3. The nominal dry density of AAC was 500 kg/m^3. Reinforcement of units is presented in Table 1.

6.2.2 Cross section values of AAC units

Cross section values like dimensions of units and location of reinforcing bars were measured after the steel stress measurement. Afterwards it was found out that the measured distance between centre of steel bars and surface of unit could not represent the real situation and thus nominal values were used in the analysis.

6.2.3 Modulus of elasticity of AAC

The modulus of elasticity of AAC was determined with bending test on 24th of January. Load was applied as a line load at midspan and increased stepwise up to 50 % of the nominal load. Deflection was measured at each step. The modulus of elasticity of AAC was calculated from measured deflection curve with the help of

Table 4. Measured and calculated initial camber

	Measured mean values a_o (mm)		Calculated values a_o (mm)	
	0 day	10 days	0 day	10 days
Unit 1	12.1	13.9	12.5	14.2
Unit 2	10.2	14.4	12.7	14.4
Unit 3	8.4	9.4	8.7	9.9
Unit 4	7.7	9.4	8.6	9.7
Unit 5	3.0	3.5	4.0	4.6
Unit 6	2.9	4.0		

regression analysis. The modulus of elasticity of AAC was also measured with the prism test by using strain gauges and Demec gauge. Results show good agreement, see Table 2.

Because the initial steel stresses were measured at the age of 10 days, measured short term E_c (from bending test) had to be multiplyed with a reduction factor in the analysis to take into account the effect of creep during 10 days. The used reduction factor was 0.86 which based on Swedish creep tests /4/. When calculating the initial camber 0 days after autoclaving short term E_c was used.

6.2.4 Initial steel stress and relative length difference

The initial steel stress measurements were carried out on 31st of January, using strain gauges according to the method presented in Chapter 4. One of the six strain gauges of unit 5 failed and all results of the unit 6 were lost during cutting of steel bars. The calculation of relative length difference was done according to Chapter 5.2. Results are presented in Table 3.

6.2.5 Initial camber

Initial camber was measured several times when the units were unloaded on their storing position placed on edge as shown in Figure 6. The calculation of initial camber

was done according to Chapter 5.2, using ϵ_o - values given in Table 3 (also when calculating the camber after 0 days which is not exactly correct). The camber values measured 0 days (21.1) and 10 days (31.1) after autoclaving were compared with the calculated values. The results show good agreement, see Table 4.

7 CONCLUSIONS

In the opinion of the authors of this paper, the test results presented here allow the following conclusions to be drawn:

- The test method developed in Finland and presented in this paper offers a reliable method for determining the initial steel stresses.

- The research tests at the Technical Research Centre as well as the quality control tests confirm that the prestresses in products manufactured at the Ikaalinen factory are consistent and not subjected to uncontrollable variations.

- As a result of the conclusions above it is justified to take prestresses into account in the design calculations for reinforced AAC units in the serviceability limit state.

REFERENCES

/1/ Schäffler, H. 1960. Eigenspannungen in bewehrten platten aus dampfgehärtetem Gas- und Schaumbeton. Proceedings of the Rilem Symposium Lightweight Concrete held in Göterborg 1960: 213-224.

/2/ Cividini, B. 1981. Long term deflection of aerated reinforced concrete slabs. The International Journal of Cement Composites, Volume 3, Number 3: 213-221.

/3/ Technical Research Centre of Finland, Research Report No. RTT 1999/91.

/4/ Nielsen, A. 1983. Shrinkage and creep - Deformation parameters of aerated, autoclaved concrete. Autoclaved Aerated Concrete, Moisture and Properties, edited by F.H. Wittmann, Amsterdam: 189-205.

Advances in Autoclaved Aerated Concrete, Wittmann (ed.) © 1992   Taylor & Francis. ISBN 90 5410 086 9

Prestressed concrete bars as reinforcement for AAC

Hans-Gert Kessler

Hebel AG, Munich, Germany

ABSTRACT: The use of high-strength concrete prestressed bars (so-called stressbars) as droopy working reinforcement has, by explainable causes, not proved efficient for normal concrete. However, utilized in lightweight concrete, prestressed bars give real additional advantages; therefore, in this range reinforcing with stressbars should be taken into consideration. This affirmation can be confirmed for example, by using stressbars in autoclaved aerated concrete (AAC).

1 INTRODUCTION

As known, lightweight concretes — with the exception of expanded slate concrete – are hardly

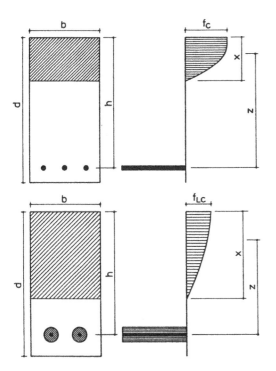

Fig. 1: Steel reinforced normal concrete (above) and stressbar reinforced lightweight concrete (below)

suitable for applying the idea of prestressing on account of their lower resistances and especially because their often unfavourable rheological properties (creep & shrinkage) result in inadmissible prestress losses.

The well-known advantages of prestressed concrete, namely
-- activation of the whole cross-section,
-- absence of cracks to a great extend, and
-- considerable steel savings
may be cedet, at least partially (approximately on the level of the so-called "incomplete prestressing"), to lightweight concrete by using prestressed concrete bar-shaped inserts as reinforcement, for which can be proposed the term of "stressbars".

These prestressed bars which are made from high-strength concrete represent a reinforcement alternative — initially thought for normal concrete elements with usual density and normal strength – which replaces the usual reinforcing steel. The element itself may not be considered as prestressed; it works only together with the stressbars using them as an embedded droopy reinforcement (Figure 1). At the best, it may be called a partial prestressing by which even the prestressed "parts" are not of lightweight concrete. Especially AAC is and remains an absolutely non-prestressable material.

2 ADVANTAGES OF STRESS BAR REINFORCEMENT

There exist however some differences with regard to a steel reinforced (normal or lightweight) concrete, as the behaviour of a stressbar reinforced cross-section becomes similar to that of an incomplete prestressed section.

This behaviour is based on the fact that the steel in the bar exists in an extended state, even before the element works statically. Therefore, the tensile force which is to be taken over by the bar appears fully at relatively slight strains due to the fact that strains are additional for the steel. There results a shift of the neutral axis to the reinforcement, thus creating
- a reduced crack danger (of course not a complete absence of cracks as in the case of full prestressing),
-- an enlarged share of statically working (pressed) concrete, reducing in this way the inactive part which only increases the load.

Besides, the concrete strain f which is necessary for obtaining the required bending compressive force, can be smaller at increased x -- in compliance with the strength properties of most of the lightweight concretes (especially AAC). While the advantages mentioned above -- the crack reducement and the better static concrete utilization -- are valid also for normal concrete, this represents a further advantage when using stressbars in lightweight concrete.

Last not least, we must as well mention the great economical advantage of prestressed concrete which is given also by using prestressed bars. The employed highstrength steels (useless for reinforced concrete on account of their great strain) are more expensive, but not at the same degree as the more increased strength leads to steel savings.

3 DISADVANTAGES OF STRESS BAR REINFORCEMENT

The fact for which a reinforcing of normal concrete with stressbars has not been accepted by the praxis is due to some important reasons, i.e.

1. the real intention to obtain the statical advantages of prestressed concrete without its technological troubles could hardly be achieved. For example, it is still necessary to use either casting forms linked to a prestress bed or moulds of a stiff kind (and therefore heavy) enabling the taking over of the prestress forces until the concrete is hardened. This necessity is only shifted from the prefabrication of the elements to the preceding manufacture of the bars. Thus, the hardening processes and times as well as the necessary production areas are approximately doubled; besides, there appears also the problematic handling of the

bars during their transportation in the intervening time.

2. The stressbars themselves, clumsy-looking as reinforcing members, are nevertheless very delicate as prestressed products. Their manufacture requires greatest care (e.g. in respect of the right centric position of the wire strand in order to assure the straight form also after the transfer of the prestress) and, as mentioned above, a careful handling. Besides, the manufacture of the bars gives (in comparison to other prestressed concrete elements) a reduced productivity -- if the latter is defined, for example, in a cu.m output per sq.m production area and time unit.

3. The reinforcing can only be provided in a straightline shape; not even the possibility to splay the bars like wires in a prestress bed is given.

The decisive point of view for a renouncement to the reinforcing of normal concrete with stressbars was the consideration that it is not very useful to reinforce an element in the roundabout way over previously prestressed bars, when the element itself can be prestressed.

4 SPECIAL ADVANTAGES BY USING STRESSBARS IN AAC

Exactly this consideration becomes irrelevant in the case of an lightweight concrete element - because of its usual non-prestressability (exception the already mentioned expanded slate concrete). The manufacturers of lightweight concrete elements have always regretted to renounce to the advantages of prestressing, therefore the use of prestressed bars is really indicated to get at least partially the advantages of prestressed concrete.

A further advantage, especially for lightweight concretes, has already been shown in chapter 2 viz. the possibility to ensure the necessary bending compressive force with a small f as a result of the augmented x (Figure 1; of course the latter also reduces at the same time the lever arm z but this represents a lesser evil).

Especially when using stressbars in AAC two further advantages result, i.e.

1. The direct sheathing of the wire cord by a completely crackless highstrength concrete (B 55 and more) allows the renouncement to a corrosion protection, which else is indispensable in AAC.

2. Above all, the relatively great perimeter of the stressbars (their clumsy-look has already been mentioned) allows the transfer of internal forces from the bars to the surrounding AAC by means of very reduced shearing stresses, which undoubtedly

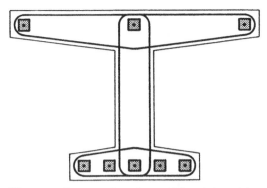

Fig. 2: Cross-section of a prefab reinforced by stressbars (longitudinal) and stirrups (transverse)

can be transmitted to AAC. As known, the anchoring of the reinforcing members in AAC is one of the most difficult problems. By using stressbars, this problem is settled in an elegant manner. Besides, the production of stressbars offers the possibility to profile them without trouble.

Having in view the special advantages when using stressbars in AAC, it is indicated to revert to the disadvantages stipulated under chapter 3.

The double production effort - first the manufacture of stressbars from highstrength normal concrete in a prestressed concrete factory, and secondly the production of the AAC elements with this lightweight material fabricator - cannot be avoided; just on them are based all the other ad-

vantages. The same is also valid in connection with the difficult transportation of the bars.

The possible reluctance of the prestressed concrete producer against the manufacture of a product like stressbars - inefficient for him -- can be removed by a corresponding price setting. The important steel savings could thus allow compromises, satisfying both parts.

The straightness of the reinforcing guidance is usually given with most of the reinforced products of the AAC industry.

5 THE PRACTICAL REINFORCING WITH STRESSBARS, ESPECIALLY OF AAC

A further point of view which recommends the use of stressbars particullary for AAC, is the following:

As already mentioned, stressbars as reinforcing members have a considerable cross section in spite of their delicacy as concrete elements; aprox. 50 sq.cm are required not only for the taking over of the prestress force, but also for the dangerless handling of the bars up to their incorporating. On account of this, they are less suitable for the employment within delicate cross-sections (Figure 2). On the other hand AAC elements are characterized not only by simple cross-sections which are rectangles in their basic form (Figure 3, a...d) but also by the usual renouncement to the shearing reinforcement in plates. Nevertheless, on the other hand, if used in prefabricated elements made of

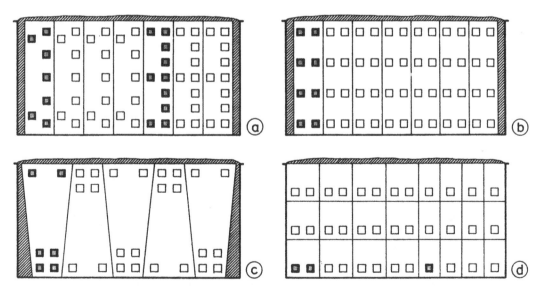

Fig. 3: Stressbar reinforced AAC elements cast in blockform: a. roof and ceiling slabs, b. wall plates, c. girders, d. lintels

other lightweight concretes, it is possible to utilize a complex wickerwork formed alongside by stressbars and crosswise by reinforcing steel, as shown in Figure 2.

The fixing of the reinforcement in the casting form requires several special considerations – but nevertheless there are different solutions for this problem. Since bar spacers become superfluous by the concrete wrapping of the steel, the flat form casting of the elements by laying the stressbars simply on the bottom of the form, is to be taken into consideration (Figure 4). In this way, a greater width of the plates is possible. It is above all a question of careful handling of the prefabricated elements, if there can be renounced to their upper reinforcement. Tensile strength on the upper side of the element during transportation and intermediate depositing can be avoided by corresponding lifting devices and by a conscientious control of these operations.

6 EXAMPLE FOR APPLICATION

The proposal made in Figure 4 was further devel-

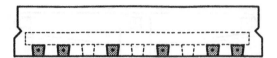

Fig. 4: Cross-section of a stressbar reinforced AAC slab without upper reinforcement, cast in a flatform

opped and applied to a series of seven roof slabs 1,500 mm wide and up to 9,000 mm long (Figure 5); therefore, profiled stressbars made of B 55 concrete and prestressed by a wire strand of 9.3 mm nominal diameter were designed. With the same geometry (nominal cross-section 7·7 = 49 sq.cm) the bars would furnish further possibilities if cast from B 75 or B 100 and prestressed with 11.0 and 12.5 mm diameter wire cord respectively.

The established steel consumption contained in the last column of the table included in Figure 5 shows values which can be considered even as spectacular. Certainly, this is due not only to the utilization of highstrength prestress steel, but also to the complete renounciation to an upper reinforcement; this renounciation is based on the mas-

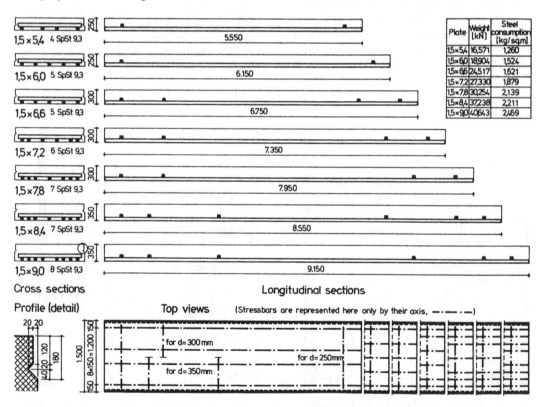

Fig. 5: A series of stressbar reinforced AAC roof slabs (1.5 m width) without upper reinforcement.

siveness of the elements.

7 CONCLUSIONS

This paper wants to give a reason for thinking
about a new way to reinforce lightweight concrete,
especially AAC.

The advantages and disadvantages of the employment of highstrength concrete prestressed bars
(so-called stressbars) have been compared and
their special advantages in lightweight concrete
(e.g. AAC) emphasized. The high efficiency of
stressbar reinforcement has been elucidated by
means of an example.

Advances in Autoclaved Aerated Concrete, Wittmann (ed.) © 1992   Taylor & Francis. ISBN 90 5410 086 9

Aerated concrete used in composite action with ordinary concrete From blocks to elements

B.G. Hellers
Royal Institute of Technology, Stockholm, Sweden

O. Lundvall
Swedish University of Agriculture, Uppsala, Sweden

ABSTRACT: An interaction of aerated autoclaved concrete (aac) and prestressed reinforcement is realised by a combination with ordinary portland cement concrete (pcc). Blocks of light weight concrete are put together and prestressed into floor elements, wall elements and into elements spanning over doors and window openings. Tests on floor members confirm the composite action. The capacity of interaction between materials and the bonding of prestressing wires are critical. So, in practice the quality control must be concentrated on the interface conditions between materials. In case blocks are made into elements on the building site, which is quite possible, this is a decisive requirement. Design is demonstrated for floor members of up to 9 m span and 0.3 m depth. Thanks to an initial prestress walls can remain fully compressed even after deflection under an eccentric floor load. Internal stiffness distribution helps avoid large torsional deformations of trimmer beams caused by the eccentric load from the floor elements. The prestress of the beam gives a considerable increase to its shear capacity.

1 INTRODUCTION

Aerated concrete is produced to standard in unreinforced blocks and reinforced products, such as wall, roof and floor elements. The reinforcement is unstressed and covered with cement slurry for corrosion protection and improved adherence to the covering aac. The combination of technologies, well integrated in the production, leads to a complex situation with storage and delivery of a large number of standardised items. There are two answers to this problem - one is to produce only to order, which requires a booming demand or a program economy, the other is to be flexible and produce only blocks , in a limited variety of sizes. They can later be combined by prestressing and concrete grouting to any element configuration in want according to a specific project order.

1.1 *Material properties*

The structural combination of prefabricated aac blocks and pcc calls for well defined materials. The aac is taken with the general density 5 kN/m^3, for which most properties are known.

Table 1. Material properties (MPa).

	E	f_{cc}	f_{ct}	f_v
aac	1800	1.30	-	0.25 (0.1)
pcc	32000	17.5	1.20	2.5 (1.0)

The possible technique for an element production to be established in situ will put a practical limit to the pcc quality. On a building site in Sweden the top quality concrete is normally K45 (Swedish standard). In most cases pcc is not taken above K40. This strength level is used here. The elastic constants and design stresses are shown in table 1. They are characteristic values taken according to the Swedish regulations BBK 79, a national version of the CEB-FIP international model code.

Regular partial coefficients are used with a medium safety classification (γ_n = 1.1). No tension stress on aac is accepted. The design shear stress, either from torsion (pure shear) or bending, is highly influenced by the effects of prestress. Figures in brackets are without prestress while the plain figures reflect a prestress level of 45% of the ultimate

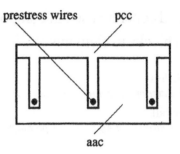

Figure 2. General section of a floor member.

Figure 1. Traces in aac blocks before casting.

compression in design. The nominal relation between elasticities in compression is 1 to 18, while the relation between design stresses is a little less, 1 to 13.5. Still, the condition of perfect interaction, that a composite of aac and pcc can function as one homogenous material, is at hand. The model will require, however, a complete adherence at the interface between the two materials in order to produce a continuous strain flow across the intersection. In pouring concrete over aac, the light weight material, which is very porous, must be soaked in advance in order to avoid capillary dehydration of the green concrete.

1.2 *Prestressing equipment*

To prestress is a matter of special mechanical equipment which must be readily available on site. This requirement can now be satisfied by reasonably cost standardised products, such as a portable prestressed bed, mini stressing jacks, round wire or strands up to 16 mm, grips and releasing pieces. The availability almost anywhere of prestressing equipment has highly contributed to the progress of the technology, its frequency in application and the quality of its performance. In this case only two dimensions, ø5, ø7 of dented round steel wire are considered. The pretension steel is from Halmstad Jernverk, Hjulsbro (Linköping). It is marked 1570/1330, indicating the significant values of ultimate and 0.2 (yield)-stresses.

$f_{us} = 1570$ MPa

$f_{ys} = 1330$ MPa

The characteristic values which are defined at the 0.05 - fractile level are 5% higher.

$f_{uk} = 1.05f_{us} = 1650$ MPa

$f_{yk} = 1.05f_{ys} = 1400$ MPa

According to the Swedish regulations, Lorentsen (1963), the design value of the steel stress is limited by

- at prestressing $0.8f_{yk}/0.65f_{uk}$

- after prestressing $0.75f_{yk}/0.6f_{uk}$

whichever is the lowest.

Calculations in the following are based on $0.75f_{yk} = 1050$ MPa, where the steel has a strain level of 0.5%. The elastic compression of the composite sections is important. The level is 10% due to the fact that a large part consists of a low modulus material, aac. It has to be compensated for, the way regulations suggest, by increasing the strain at prestressing to 0.55%. The estimated prestress forces after compression and losses are

- ø5: P = 20.6 kN

- ø7: P = 40.4 kN.

In the tests related to the present technology a variety of methods for applying prestress have been used. Early tests were realised with simple mechanical device, a spanner and an open bolt with a nut. Later tests were realised with more sophisticated equipment, such as a Paul mini stressing jack with grips and a simple releasing piece.

2 BLOCKS TO ELEMENTS

Prefabricated aac blocks are put together in close contact leaving cut-open traces for the prestressed reinforcement in which the grouting of pcc will take place. The width of each trace is $3\varnothing$. The picture below illustrates the preparation of a floor member for which the grouting is concluded by forming a top pcc layer. The integration of reinforcement in the pcc splines requires a careful compaction to secure the evolution of bond between steel and concrete. The splines serve as a large structural and stabilising reinforcement of the aac for which it is equally important to secure the adherence to concrete. The critical property in this case is the capillary suction from the aac which must be readily reduced by prior soaking. In practice this means that the aac must be soaked through by putting the blocks in water over night. The moisture from the aac helps the pcc to mature properly. The pcc splines balance the major action from the prestressed rebars. The sideways interaction between elements must be secured by adding a tongue and groove. Such a development is not included in this paper. It is also justified, however, by earlier experimental conclusions, which strongly suggests the mechanical interaction between blocks in an aac wall structure, see Hellers, Sahlin (1972).

2.1 Floor member

A general section of a floor member is shown in figure 2.

The calculations in this paper are restricted to the span variation between 6 and 9 m. The maximum relation between span and depth is 30. Depths vary from 0.2 to 0.3 m. All members are 0.6 m wide which coincides with the present Swedish standard. Combinations of member dimensions appear in the tables 3 and 4. The total load in calculating the actual load effects has three components, the gravity weight load, the permanent load from flooring and secondary walls, estimated to 1.0 kN/m^2, and the variable load. Two different variable load levels are considered, 2.5 kN/m^2 and 4.0 kN/m^2. The partial coefficients to be used on the variable load are, at the failure load state 1.3 and at ordinary load state 0.5 according to the Swedish code of practice, NR 1 (1989). The design is primarily based on failure load while the

Figure 3. Crack within an aac block.

deformations are checked at ordinary load. Two parameters decide the capacity of the critical section, the thickness of the pcc layer, d_1, and the eccentricity of the prestressing force, e. No adherence can exist between the aac blocks. A crack in the individual block confirms a good adherence between the aac and pcc materials. The crack behaviour in the pcc spline within the element is difficult to register, however, and would require indirect sensing.

2.1.1 Calculation model

At failure load it is assumed that the cracked section is fully open up to the top pcc layer. Using the symbol P to denote the prestressing force, the conditions of equilibrium require

$$d_1 = \frac{P}{f_{cc}b}; \quad (b = 0.6\ m)$$

$$e = \frac{M_d}{P}; \quad (e < d)$$

M_d = design bending moment.

Up to the ordinary load level the condition is that all sections remain uncracked. The combined modulus of elasticity in the region with aac and pcc splines, is calculated according to a parallel model

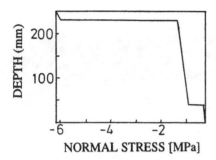

Figure 4. Stress diagram at ordinary load.

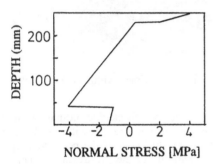

Figure 5. Stress diagram at prestress.

Table 2a. Variable load 2.5 kN/m^2.

Span (m)	P (kN)	Rebar ø5mm	Rebar ø7mm	d_1 (mm)	e (mm)
6	121.2	-	3	12	132
7.5	161.6	-	4	16	170
9	182.2	1	4	18	233

Table 2b. Variable load 4.0 kN/m^2.

Span (m)	P (kN)	Rebar ø5mm	Rebar ø7mm	d_1 (mm)	e (mm)
6	182.2	1	4	18	120
7.5	202.0	-	5	20	181
9	242.4	-	6	23	233

$$v = \frac{A_{pcc}}{A_{tot}}$$

$$E = vE_{pcc} + (1-v)E_{aac}$$

Relations between the moduli of elasticity are established, n, where the lowest value equals unity, $n_3 = 1$. The position of the neutral axis is calculated with the formula

$$\bar{z} = \frac{\sum n_i A_i z_i}{\sum n_i A_i}$$

where z_i marks the position of the neutral axis of the individual layer from the bottom. When the position of the neutral axis is established, the next step is to calculate the bending moment of the eccentric prestress load in combination with permanent and variable loads up to the ordinary level. The stresses are calculated using Navier's formula

$$\sigma = \frac{N}{A} + \frac{M}{I}\zeta$$

where ζ denotes the coordinate with respect to the neutral axis. Considering the variation of stiffnesses across the section the formula is modified into

$$\sigma_i = n_i \left(\frac{P}{\sum n_i A_i} + \frac{M}{\sum n_i I_i} \zeta \right)$$

The form indicates non-continuous stress variations in the transition from one region to another. Calculated stresses are related to design values. The option is that all sections of the floor member keep their flexural stiffnesses unreduced from prestressing to a state of ordinary load. The full compression capacity of the aac is used at the upper load level, as illustrated in figure 4.

The required prestress forces in design are relatively large. One consequence is that the stress levels in connection with only prestressing are considerable. A stress distribution at prestress is illustrated in figure 5. The compression capacity of the aac is exhausted while the tension stress of the pcc top layer calls for a supplementary reinforcement. An appropriate steel mesh will supply the pcc layer with the necessary tension strength.

2.1.2 Design results

The design process involves the choice of a total prestress force. Material economy requires that all rebars are used up to their design values. So, the amount of prestress reinforcement is a step function which is reflected in tables 2a and 2b, along with the structural geometry. The span/depth relation is constant, 30.

The member deflects upwards under prestress. The negative deflection is of the order 1/300 of the span. The deflection is counteracted by the permanent gravity weight. At the ordinary load level, the deflection is only slightly positive, the total deflection from the initial negative value being of the order 1/250 of the span. Above the ordinary load level the member starts cracking up. The calculation of deflections becomes more complex in that the flexural stiffness is reduced by cracking. An iterative computer program can handle this phenomenon, which is illustrated in figure 6, up to an asymptotic approach to the failure load, which is very close to the design load effect.

Tests indicate that an imperfect adherence between the concrete materials strongly increases the deflection. Assuming linear elastic conditions of both materials and no tension capacity, neither of aac nor of pcc, a computer calculation shows the following stress distribution at the mid-section of the floor member at the design and failure load.

The stress distribution rectifies the calculation model in design, assuming that the pcc stresses of the top layer are equalized, reflecting a plastic behaviour. The sensitivity of the design in varying the variable load and the span/depth relations is demonstrated in tables 3a and 3b. Varying the span/depth relations from 25 to 30 implies a considerable increase of the pcc/aac relation and the amount of prestressed reinforcement.

2.2 Wall member

In the traditional elastic theory the full joint interaction between a floor member and a masonry wall without tensile strength is outlined. In the failure state the joint may plasticize which helps bring the reaction towards the core. Anyway, the wall is influenced by a considerable reduction of flexural stiffness in the vicinity of the upper and the lower joint, which radically lowers the bearing capacity of the wall, in particular endangering its stability. By prestressing wires along the wall element it is possible to keep all sections entirely compressed. No top layer is needed. If the rebars are placed coinciding with the neutral layer, the initial deflection of the wall is omitted.

Stability problems are suppressed. The load in

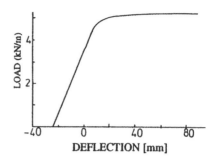

Figure 6. Deflection curve for a floor member.

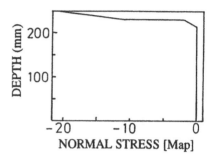

Figure 7. Stress diagram at failure load.

Table 3a. Spans and heights (2.5 kN/m^2 load).

Span (m)	Depth (mm)	Span/ depth	pcc/ aac	$\mu \cdot 10^3$
9	300	30	0.229	0.964
7.5	300	25	0.139	0.641
7.5	250	30	0.198	1.00
6	250	24	0.125	0.644
6	200	30	0.151	0.962
4.5	200	22.5	0.091	0.491

Table 3b. Spans and heights (4.0 kN/m^2 load).

Span (m)	Depth (mm)	Span/ depth	pcc/ aac	$\mu \cdot 10^3$
9	300	30	0.322	1.30
7.5	300	25	0.199	0.855
7.5	250	30	0.268	1.30
6	250	24	0.149	0.768
6	200	30	0.258	1.40
4.5	200	22.5	0.102	0.641

calculating the wall structure is taken from the design load effects on the floor members. The gravity weight of the wall itself is neglected.

It is possible to use walls of depth dimensions

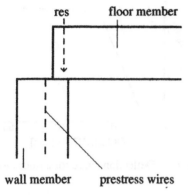

Figure 8. Wall member, joint and floor member.

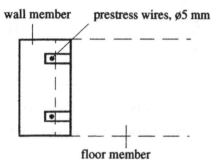

Figure 9. Section of a reinforced wall member.

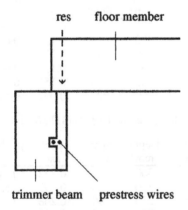

trimmer beam   prestress wires

Figure 10. Trimmer beam.

300 or 250 mm in case of up to two extra storeys, using spans of up to 9 m. In case of a 200 mm wall the span must be reduced to 7.5 m. The total prestress force used in these calculations is 41.2 kN in each element, which requires two ø5 mm wires. The conclusion is

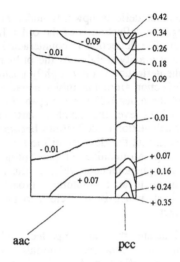

Figure 11a. Shear component τ_{xy}.

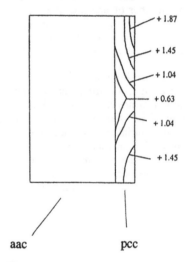

Figure 11b. Shear component τ_{xz}.

that by prestressing it is possible to build 3-storey structures with considerable load bearing spans and flexibility.

2.3 Trimmer beam

Trimmer beams are needed to span over openings, such as window and door openings. The beam carries an eccentric load from the adjacent floor member. It is possible to reduce the value of the twisting moment by a stiffness distribution which decreases the eccentricity, or by a displacement of the neutral layer closer to the load action of the resultant from the floor member. In a 300 mm deep wall the eccentricity reaches a minimum when the depth of the pcc layer is 59 mm (figure 10). Then the eccentricity is approximately 30 mm. Taking a span of 9 m and a variable load of 4.0 kN/m^2 the resulting force acting on the beam is 41.7 kN/m. In case the span is restricted to 2.4m a beam height of 450 mm is required.

To get hold of the appearing shear stresses, which are critical at the beam ends, a finite element analysis is tried, where the distribution of material properties is considered. The shear capacity of the trimmer beam is influenced favourably by the prestress. The pcc layer will need, in addition, mesh reinforcement to secure the required strength of pcc in bending and shear. The shear components τ_{xy} and τ_{xz} in the trimmer beam at the clamped end sections, are visualized in the figures 11a and 11b.

3. CONCLUSIONS

Blocks of aac can be combined to elements by prestressing and pcc grouting. Floor members of up to 9 m span have a high load bearing capacity. They are stiff but sensible to imperfect adherence between materials. Wall members of 3 m height have a load bearing capacity of up to 3 storeys when they are compressed to avoid stiffness reduction. Trimmer beams of up to 2.4 m span have a considerable load bearing capacity and torsional strength, thanks to a favourable stiffness distribution and the strengthening effect of prestress.

REFERENCES

BBK.1988.Bestämmelser för betong-
      konstruktioner. utgåva 2 (in Swedish)
      Statens betongkommitte. Stockholm.
Hellers, B. G. Sahlin, S. 1972. Loading test on
      foundation walls unsupported at the top
      edge. Nat Swedish Building Research,
      R36. Stockholm.
Lorentsen, M. 1963. Spännbetong (in Swedish)
      SVR:s Förlag, Stockholm.
NR 1. 1989. Nybyggnadsregler (in Swedish)
      BFS 1988:18, Boverket, Stockholm.

Advances in Autoclaved Aerated Concrete, Wittmann (ed.)© 1992   Taylor & Francis. ISBN 90 5410 086 9

Survey on the European standardization work of CEN/TC 177: Intended EN 'Prefabricated reinforced AAC components'

D. Bertram
Ministerium für Bauen und Wohnen, Düsseldorf, Germany

N. Fichtner
DIN, Deutsches Institut für Normung e.V., Berlin, Germany

ABSTRACT: The main task of the Construction Products Directive is the removal of technical barriers in the field of building and civil engineering. That means that harmonized standards for the construction products should be established as far and as quickly as possible. In 1989 CEN decided to standardize "Prefabricated reinforced AAC components" within its TC 177. A European standard shall take account of different levels of the essential requirements within the Member States so that a classification of the products should be included. Under this aspect the following survey on the current European standardization work of CEN/TC 177 is given.

1 INTRODUCTION

At the end of 1988 DIN, supported by the Standards Institutes of Denmark (DS), the Netherlands (NNI), Sweden (SIS) and Italy (UNI) submitted to CEN an application for a new European standardization project. In March 1989 the Technical Bureau (BT) of CEN decided to accept this proposal and to create a new TC 177. The secretariat was allocated to DIN.

At the first meeting of the TC in Berlin in October 1989 three Working Groups (WGs) were set up:

WG 1   Autoclaved aerated concrete (AAC) components
WG 2   Components of lightweight aggregate concrete with open structure (LAC)
WG 3   Test Methods (for both materials)

The following countries take an active part in the standardization work up to now: A, B, D, DK, F, NL, N, S, SF, UK.

CEN/TC 177 unanimously agreed to prepare a harmonized European Standard for the AAC-components in accordance with the Construction Products Directive (CPD) (1). That means in particular that it is tried

- to develop ENs according to the performance concept;
- to consider the essential requirements;
- to consider the attestation of conformity;
- to consider marking and labelling (CE-mark).

Perfomance concept:
Performance levels and requirements to be fulfilled by products in future in the EC-(EFTA-) Member States shall be laid down in classes in the interpretative documents and in the harmonized standards in order to take account of different levels of essential requirements for certain works and of different conditions prevailing in the Member States. Harmonized standards should include classifications that allow construction products which meet the essential requirements and which are produced and used lawfully in accordance with technical traditions warranted by local climatological and other conditions to continue to be placed on the market.

Essential requirements:
The products must be suitable for construction works which (as a whole and in their separate parts) are fit for their intended use account being taken of economy, and in this connection satisfy the following essential requirements where the works are subject to regulations containing such requirements.

1. Mechanical resistance and stability
2. Safety in case of fire
3. Hygiene, health and the environment
4. Safety in use
5. Protection against noise
6. Energy and heat retention

Attestation of conformity:

The conformity of products with the harmonized standards is ensured by means of production by manufacturers and of supervision, testing, assessment and certification by independent qualified third parties, or by the manufacturer himself.

CE-mark:

A product is presumed fit for use if it conforms to a harmonized standard. Products thus considered fit for use are easily recognizable by the CE-mark; they must be allowed free movement and free use for their intended purpose throughout the Community.

The intended standard is a product standard for prefabricated reinforced AAC components. The components are intended for use in different types of loadbearing and non-loadbearing applications. The standard will cover components intended to be used in building constructions such as

- Roof anf floor components
- Wall components
- Lintels, beams and piers
- Other prefabricated components.

The intended content of the product standard is given in table 1.

Table 1. Contents

1. Scope
2. Normative References
3. Definitions, Symbols and Abbreviations
4. Materials Design and Manufacture
5. Requirements for Components
6. Classification
7. Attestation of Conformity
8. Description of the Components
9. Marking, Labelling and Packaging

Annex I: Design by Calculation
Annex II: Design by Testing

2 MATERIAL DESIGN

The main properties of AAC laid down in the standard are compressive strength, dry density and thermal conductivity:

- the mean compressive strength should not be less than 1,5 N/mm²

- the dry density is usually in the range of 300 to 800 kg/m³

- the declared thermal conductivity

value is normally within the range of 0,08 to 0,17 W/mK and is directly related to the dry density.

The following strength and density classes are proposed for the time being

Strength classes (N/mm²)

1,5	3	4,5	7
2	3,5	5	
2,5	4	6	

Density classes (kg/m³)

300	450	600	800
350	500	650	
400	550	700	

The compressive strength and the dry density are obtained from tests carried out on specimens sampled in accordance with the intended standard and measured in accordance with the test standards EN 679 (2) and EN 678 (3) which are in preparation in TC 177. The characteristic compressive strength fck (5 % fractile) and the characteristic dry density ρ dry (95 % fractile) shall be stated by the manufacturer for the purpose of design by calculation and of thermal/sound insulation. The test method for thermal conductivity is specified by CEN/TC 89/WG 8 and is based on ISO 8302 (Guarded hot plate apparatus) (4).

The reinforcement placed in AAC components is generally protected with coatings applied during the manufacturing process. AAC itself does not provide sufficient corrosion protection to the steel reinforcement due to its high vapour permeability and low ph-value. The efficiency of the corrosion protection shall be determined in accordance with the test method (5) prepared by TC 177.

In the manufacturing process the following raw materials combined with additives and agents, where appropriate, may be used in general:

Siliceous material: Natural or ground
                       sand
                       Pulverized fuel ash
                       Ground granulated
                       blast furnace slag

Calcareous material: Cement
                       Lime

Water

Cell generating material.

3 REQUIREMENTS FOR COMPONENTS

In this standard as far as possible account is taken of the essential requirements of the CPD. The Standard mainly deals with the first and the last item of the essential requirements, namely

- Mechanical resistance and stability
- Energy economy and heat retention

but reference is made to the others.

The dimensions of the components are not covered by the standard but permissible dimensional tolerances to the size are given.

As already mentioned AAC is characterized by its characteristic compressive strength and by its characteristic dry density. These values shall be declared by the manufacturer, they may be given as strength and density classes according to a classification scheme. Some of the countries taking part in the standardization work are not yet familiar with a classification scheme, therefore the TC allows the alternative that the manufacturer states the values by himself.

The declared thermal conductivity value of the component must be stated, the value may be either determined from test measurements carried out on specimens or from tables.

4 DESIGN OF THE COMPONENTS

Due to the different historical development of standardization there are two possibilities for the design of components, namely

- design by calculation (Annex I)
- design by testing (Annex II)

Annex I, design by calculation, is based on the principles of Eurocode 2 (6). Two limit states are taken into account

- the ultimate limit state
- the serviceability limit state.

The Standard specifies the essential requirements on the materials AAC and steel reinforcement and also rules for the de-

sign, for instance for

- bending and longitudinal force
- shear
- bond and anchorage
- buckling.

The design stress-strain diagrams are standardized; for AAC see Figure 1 and for reinforcing steel see Figure 2.

In these figures γ_C and γ_S are the partial safety factors for AAC and for the reinforcement. The following values are proposed:

$\gamma_C = 1,3$

$\gamma_S = 1,1$

In Figure 2 a horizontal top branch may be used as an alternative. In both cases the steel strain is limited to 0,01.

Due to a number of assumptions the strain distributions in the ultimate limit state for bending and longitudinal force are possible as shown in Figure 3. According to Eurocode 2 (6) the tensile strength of the reinforced components is ignored.

The method for shear design is based on three values of design shear resistance

V_{Rd1}: design shear resistance without shear reinforcement,
V_{Rd2}: maximum design shear force that can be carried without crushing of the notional concrete compressive struts, and
V_{Rd3}: design shear force that can be carried by a section with shear reinforcement.

The values taken as a basis comply with test results for AAC, the formulae used correspond to the structure of the formulae for normal weight concrete given in Eurocode 2. The formulae are based on a truss-model made of struts and ties.

A main problem of the AAC components are bond and anchorage of the reinforcement bars. The longitudinal tensile and compressive reinforcement shall be anchored by welded transversal bars. In this context it must be discussed, too, whether or not bond may be taken into account and under which relevant conditions.

The standard gives also some formulae for the ultimate limit states induced by structural deformation (buckling), but the discussion on this subject still continues.

Finally it can be remarked that the standard complies with all relevant structural requirements.

In Annex II, Design by testing, it is assumed that the manufacturer states the characteristic loadbearing capacity of the components, either based on results from full scale type testing or performance testing of full scale components. The design by testing method can replace design by calculation for one or several of the component properties, but the equivalence of the two methods must be given.

5 ATTESTATION OF CONFORMITY

The standard will incorporate a chapter for "Attestation of Conformity". According to the CPD, Annex III (1) the choice and combination of the four methods given depend on the requirements on the particular product. The Standing Committee to the CPD is responsible for the decision on the system to be chosen for the attestation of conformity, the TC can only give a recommendation. The final decision will be included in the mandate issued to CEN. At present the level "ii) first possibility" is proposed, that means that the declaration of conformity of the product is based on tasks for the manufacturer and tasks for the approved body. The conformity control should only refer to the finished product and the factory production control, basic products should not be included.

6 FURTHER TASKS

At the moment European Standards on fire resistance and behaviour are not available. With regard to the importance of safety in case of fire (an essential requirement of the CPD) in the design of the products, the standard gives the following two alternatives based on a resolution of the CEN/BT (the highest decision committee of CEN):

- a normative "National Annex"

- an informative Annex III which may be used to comply with national requirements in the field of fire resistance.

By this procedure the users of the EN are enabled to actually apply these Standards.

Due to the different properties of the materials between normal concrete and AAC particular technical specifications are necessary. The work in CEN/TC 250/SC2 on Eurocode 2 relates only to normal concrete and excludes AAC, so that a special standardization work should be done in close contact with CEN/TC 250 experts.

The objectives of structural design rules will be e.g.

- structural stability and stiffness
- accidental eccentricity
- floor plates and wall plates
- joints
- connections

Within TC 177/WG3 (Convenor: Professor van Nieuwenburg) a number of test standards are prepared for determination of the various requirements of AAC, e. g. compressive strength, dry density, drying shrinkage, modulus of elasticity, strength of welded joints, corrosion protection, bond strength between AAC and reinforcing steel etc. At present the aforementioned first three Draft European Standards are submitted to the CEN members for CEN enquiry (see (2), (3), (7)). If these drafts will become European Standards, CEN members are bound to give these European Standards the status of national standards without any alteration.

REFERENCES

(1) Directive relating to Construction Products; Council Directive of 21 December 1988 (89/106/EEC)

(2) EN 679 Determination of the compressive strength of autoclaved aerated concrete; 1992 (At present at the draft stage)

(3) EN 678 Determination of the dry density of autoclaved aerated concrete; 1992 (At present at the draft stage)

(4) ISO 8302: 1992 Thermal insulation; Determination of steady-state thermal resistance and related properties; Guarded hote plate apparatus

(5) EN ... Test methods for verification of corrosion protection of reinforcement in autoclaved aerated concrete and in lightweight aggregate concrete with open structure; 1992 (At present at the draft stage and intended for publication)

(6) ENV 1992-1-1:1991 Eurocode 2: Design of concrete structures - Part 1: General rules and rules for buildings

(7) EN 680 Determination of the drying shrinkage of autoclaved aerated concrete; 1992
(At present at the draft stage)

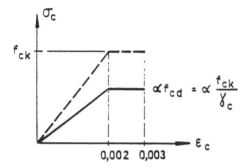

Fig. 1 Bi-linear stress-strain diagramme of AAC

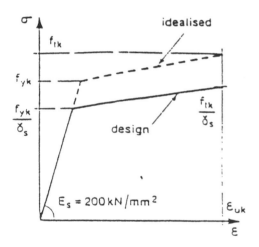

Fig. 2 Design stress-strain diagramme for reinforcing steel

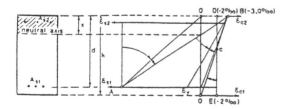

Fig. 3 Strain distributions in the ultimate limit state

213

Advances in Autoclaved Aerated Concrete, Wittmann (ed.) © 1992 Taylor & Francis. ISBN 90 5410 086 9

Intended European Standards (EN) 'Test methods for autoclaved aerated concrete' in CEN/TC177: An overview

D. Van Nieuwenburg
Magnel Laboratory for Reinforced Concrete, Gent University, Belgium

S. Karl
Institut für Massivbau, Technische Hochschule Darmstadt, Germany

ABSTRACT: A series of test standards has been developed for the determination and verification of the main properties and performances of AAC and prefabricated reinforced components made of this material. These test standards are referred to in the relevant product standard "Prefabricated components of autoclaved aerated concrete". The scope, the principle, and some interesting details of these test standards are presented in the following report. More detailed informations to the determination of the drying shrinkage and the determination of the bond strength are given in the paper "Some new ideas and proposals" of D. Van Nieuwenburg and B. De Blaere.

1 INTRODUCTION

The intended European product standard from CEN/TC 177 for prefabricated reinforced AAC components is presented in the report of Dr. Bertram and Mr Fichtner. The conception of this EN for AAC components is based on the possibility of verification of the essential properties required in the product standard by means of unified standardized European test methods.

The following test methods are treated in the different ENs:

- Compressive strength;
- Flexural strength;
- Dry density;
- Moisture content;
- Drying shrinkage;
- Static modulus of elasticity;
- Creep strains under compression;
- Bond behaviour between reinforcing steel and AAC by means of the "push-out" test;
- Bond strength - Beam test;
- Verification of corrosion protection of reinforcement;
- Dimensions of components;
- Performance test for components under transversal load;
- Performance test for vertical components;
- Steel stresses in unloaded AAC components.

In the following, the main topics of the present drafts or draft proposals, respectively, are presented.

2 PROPERTIES OF THE AAC

2.1 Compressive strength (EN 679)

The compressive strength is determined on test specimens, normally cubes with an edge length of 100 mm, taken from prefabricated reinforced components. The test specimens shall be conditioned prior to the test until the moisture content ranges from (4 to 10)% by mass.

2.2 Flexural strength

The flexural strength is determined

by applying an uniform bending moment by means of third-point loading in the middle third of the span of a simply supported prismatic test specimen, taken from a prefabricated component.

The method of centre-point loading may also be used and is specified in an annex.

As for the determination of the compressive strength, the test specimens, normally prisms with a section of 50 by 50 mm and a minimum length of 200 mm, shall be conditioned until their moisture content ranges from (4 to 10) % by mass.

2.3 Dry density

The dry density is determined on test specimens taken from prefabricated components by oven-drying to constant mass at $(105 \pm 5)\,°C$.

2.4 Moisture content

The moisture content of the AAC in a prefabricated component is determined by cutting test specimens from the component without disturbing the moisture content and subsequent oven-drying to constant mass, as it is done for the determination of the dry density.

2.5 Drying shrinkage

The drying shrinkage is determined on prisms with the dimensions 40 x 40 x 160 mm taken from prefabricated components. Prior to the test, the prisms are saturated in water until their moisture content exceeds 30 % by mass. After a waiting period of at least 24 hours, enabling sufficiently uniform moisture distribution within the AAC, a zero reading of the gauge length is taken and the test specimens are allowed to dry in air, at a relative humidity not less than 45 %, until the moisture content is below 4 % by mass. During this period at least four measurements of gauge length and mass of the prisms are executed. Subsequently, the prisms are dried to constant mass

at $(105 \pm 5)\,°C$ in order to enable calculation of the moisture content at the time of the individual length measurements. The strain versus moisture content curve is drawn and from this the conventional value of drying shrinkage is derived as the relative length variation corresponding to a change in moisture content from (30 to 6) % by mass.

2.6 Static modulus of elasticity

This test method is specified in a unified paper which also deals with the determination of the E-modulus of LAC (Lightweight Aggregate Concrete with open structure).

The modulus of elasticity is determined on prismatic test specimens (normally 100 x 100 x 300 mm) taken from reinforced components. It is calculated from the difference of longitudinal compressive strains corresponding to the increase of longitudinal compressive stress from a basic test stress σ_b (approximately 5 % of the compressive strength of the concrete) to the upper test stress σ_a (in general one-third of the compressive strength of the concrete). The test specimens shall be taken in such a way that their longitudinal axis is perpendicular to the rise of the mass during the manufacture, and they shall be conditioned with respect to their moisture content as described for the determination of the compressive strength. The gauges for determining the longitudinal deformations or strains shall be attached to at least 2 (better 4) opposite longitudinal surfaces of the test specimens. If measurements are performed at two surfaces only, these surfaces should be parallel to the direction of rise.

2.7 Creep strains under compression

Creep strains are determined on test specimens taken from prefabricated components. The creep strain at a given age is defined as the total strain under compression at that age, less the instantaneous strain under the same stress occurring during the application of the load, less the strains due to shrinkage from the time of loading until the considered age.

The strains due to shrinkage are determined on unloaded control specimens equal to those used in the creep test and stored under the same conditions.

Prior to the test, the creep specimens and the control specimens shall be conditioned, at a temperature not exceeding 60°C, until a moisture content μ_m between 4 and 10 % by mass has been obtained. At least 72 h before the beginning of the creep test until completion of the test the specimens shall be stored in air at a temperature of (20 ± 1)°C and a relative humidity of (55 ± 5) %.

If not specified otherwise, the creep specimens shall initially be loaded at an age of 28 days. The applied load shall be sustained constant, in general for a period of at least one year.

2.8 Bond behaviour "Push-out" test

This is a rather simple test for determining the bond behaviour between reinforcing bars and AAC which can be used for quality control purposes. Values to be used in design cannot be derived from the test results.

Test specimens are taken from prefabricated reinforced components by cutting them in transverse direction. After conditioning to a moisture content of the AAC between 6 and 10 % by mass, a compressive force is applied onto the cut end of a reinforcing bar while the opposite surface of the test specimen is supported in such a way that pushing-out of the loaded bar is not hindered. The force is increased at a prescribed steady rate until the bond fails and the bar is pushed out. The slip of the bar relative to the concrete is measured at the free end of the bar,

and the load-slip curve is recorded.

2.9 Bond test for reinforcing steel – "Beam test"

This is a new test method which has been developed at the Magnel Laboratory for Reinforced Concrete of the University of Gent.

Beams with an embedded longitudinal reinforcing bar are cut from prefabricated components. In the middle of the span a hinge is provided by removing part of the AAC in the tensile zone of the beam. The beam is simply supported and loaded by two vertical loads in the middle third of the span. The elongation of the steel is measured in the central part where the AAC has been removed, in order to determine the tensile stress in the bar. In addition, the slip of the steel relative to the AAC is measured at both ends of the beam, and the curve representing the slip as a function of the elongation of the steel is recorded. From this the interesting bond parameters for design can be derived. For more detailed informations see the paper "Some new ideas and proposals" of D. Van Nieuwenburg and B. De Blaere.

2.10 Verification of corrosion protection of reinforcement

This test standard covers the procedures for verification of the effectiveness of the corrosion protection of reinforcing steel embedded in AAC components (and also in LAC components).

Three different short-term tests and one long-term test are provided. The corrosion protective system is considered to be suitable for reinforced components according to the relevant product standards, if it passes at least one of the three short-term tests, the choice of which is free, or the long-term test. In cases of doubt, the results of the long-term test, which is the reference test method, are decisive. The test methods can be used to examine the fundamental suitability of a certain corrosion protection provision and for current quality control purposes.

The specification of three different short-term test methods is based upon historical reasons and established procedures rather than on technical needs. Since all the short-term tests are more severe than the reference long-term test, the manufacturer is free to chose the one to which he is the most familiar or which seems to be the most convenient or practical to him. Experience has shown that protective systems which pass one of the short-term tests will always fulfill the requirements of the long-term test too.

Principle: Test specimens consisting of AAC with embedded reinforcing bars are cut from prefabricated components and exposed to a defined corrosive environment for a specified series of cycles (short-term tests) or for a specified period of time (long-term test). Immediately after the end of the exposure the concrete and the protective coating are removed from the bars, and the steel surface is examined visually for corrosion. The amount of rust present on the steel bars is recorded and compared with admissible limits.

Method 1 is a short-term test consisting of 10 cycles of 2 h wetting in sodium chloride solution with a concentration of 3 % and 70 h drying in air with specified evaporation rate. The test is based on the method specified in the German AAC Standard DIN 4223 (version July 1958).

Method 2 is a short-term test with 30 alternating cycles of 3 h wetting with water and subsequent 21 h drying in warm air at (40 ± 5)°C, ventilated to achieve a specified evaporation rate. This test is similar to the short-term test in the draft of the German AAC Standard DIN 4223 from August 1978.

Method 3 is a short-term test with 4 x 28 alternating temperature cycles between 25 and 55°C in moisture saturated air (4 cycles per day). The test is similar to that specified in the Danish Standard DS 420.1 (September 1967), which served also as a basis for the corresponding RILEM Recommendation.

Method 4 is a long-term test, consisting of an exposure to moisture saturated air at room temperature for a period of 1 year. As mentioned before, this is the reference test method. It has been taken, with only slight modifications, from the German AAC Standard DIN 4223 (July 1958).

2.11 Dimensions of components

The length, width, thickness and, where required, other essential dimensions of the components are measured. In addition, for nominally rectangular components, the squareness is checked.

2.12 Performance test for components under transversal load

This standard covers the procedure for the determination of the mechanical performances of prefabricated reinforced components under flexural and/or out-of-plane shear load. The component is simply supported at its ends in horizontal position and loaded by two vertical line loads in the outer quarter points of the span. In special cases an other arangement of loading may be chosen, e.g. for the determination of the shear capacity.

The load is applied in steps, and midspan deflections are measured. The load corresponding to the formation of the first crack is determined, and crack width in the serviceability limit state is measured. Finally, the loadbearing capacity is determined.

2.13 Performance test for vertical components

This paper is at present in a very early stage. It will be based, to a far extent, on Scandinavian experience.

2.14 Steel stresses in unloaded AAC components

Due to differences in the coefficients of thermal expansion of AAC and steel and other influences steel bars embedded in AAC components may be under tension after the autoclaving process, before any external load has been applied. These tensile stresses may have an influence on the deflections and on the cracking behaviour.

In order to determine these steel stresses, the component is placed on one of its longitudinal edges in vertical position and supported in a way that the steel is free from external stresses. The AAC is removed from longitudinal bars in the central part over a length of approximately 300 mm, and strain gauges are glued on two opposite sides of the bars. After hardening of the glue a zero reading is taken, and then the steel bars are cut in a distance of approximately 200 mm from the strain gauges in order to release the steel stresses. The initial steel stresses in the bars are calculated from the difference of strain readings and the modulus of elasticity of the steel.

The possible use of the results of this test has to be further discussed within the working group for the product standard.

Advances in Autoclaved Aerated Concrete, Wittmann (ed.)© 1992   Taylor & Francis. ISBN 90 5410 086 9

Intended European Standards (EN) 'Test methods for autoclaved aerated concrete' in CEN/TC177: Some new ideas and proposals

D. Van Nieuwenburg & B. De Blaere

Magnel Laboratory for Reinforced Concrete, Gent University, Belgium

ABSTRACT : This paper deals with some new ideas and proposals for the determination of typical characteristics of autoclaved aerated concrete (A.A.C.). Firstly a test method for the determination of a conventional value of the drying shrinkage of the unreinforced material is presented. Secondly two methods are proposed for the determination of the bond characteristics between the reinforcement bars and A.A.C. The "Beam-test" aims at the derivation of design values for the determination of anchorage systems. The "Push-out test" is essentially meant for quality control purposes or for verification of the bond quality of new coatings for reinforcement bars.

1. INTRODUCTION

In view of European normalisation of lightweight concrete products a continuous effort is made to improve existing and to develop new test methods for determination of typical characteristics necessary for design and quality control.

For both lightweight aggregate concrete and autoclaved aerated concrete, a number of elaborated test methods is available. The following table lists a number of existing drafts for CEN test methods on A.A.C.

Tests on A.A.C. itself.
- Determination of the compressive strength of A.A.C.
- Determination of the dry density of A.A.C.
- Determination of drying shrinkage of A.A.C.
- Determination of moisture content.
- Determination of bending tensile strength.
- Determination of modulus of elasticity in compression of A.A.C.
- Determination of creep deformations under compression.

Tests on reinforcement steel in A.A.C.
- Determination of steel properties by the tensile test.
- Determination of the strength of cross-welded bars by shear testing.
- Determination of bond strength between reinforcing steel and A.A.C. by the push-out test.

- Determination of bond strength in A.A.C. by the "Beam-test".
- Test methods for verification of corrosion protection.
- Determination of steel stresses in A.A.C elements.

Tests on entire reinforced components.
- Performance test for reinforced components of A.A.C. (transversal loading).

In this paper major attention is only paid to the rather new aspects in the determination of drying shrinkage and the determination of bond strength by beam-testing and push-out-testing.

2. TEST METHODS

2.1. Determination of the drying shrinkage of autoclaved aerated concrete

Just after autoclaving A.A.C. components may have a moisture content of about 30 %. The equilibrium moisture content in normal indoor conditions, reached after a rather long period, is about 6 %. Shrinkage caused by drying can cause a considerable decrease of dimensions and subsequent damage to building parts with great dimensions like walls and roof surfaces.

In annex I of the product standard "Design by calculation" a shrinkage value of 0.25 ‰ was considered as being realistic. Hereafter a short description for the determination of a "conventional shrinkage value" is given. This value can be useful for verifying possible evolutions in production control and to derive a specific value for design purposes which could be indicated in the technical documentation of the manufacturer.

The test is executed on three prisms with length axis perpendicular to the rise direction, respectively taken from the upper third, the middle and the lower third part of the product in rise direction.

Prisms (cross section : 40 mm x 40 mm) with a minimum length of 160 mm are dried in constant climate conditions from a moisture content of over 30 % to less than 4 %. Mass and gauge length evolution in time are registered.

Initial moisture content is reached by wetting in water. To ensure an initial moisture content of 30% it is important to know the dry mass of the specimen. This dry mass can be determined approximately by previous determination of the dry density of a comparable specimen taken from the same sample. The real initial moisture content is determined at the end of the test procedure by drying the prism.

After reaching the initial moisture content the specimens are sealed in a plastic foil or similar material in order to prevent loss of moisture and stored at 20°C for at least 24 h in order to homoginize the moisture content.

After removal of the sealing material the prisms with attached gauge plugs are conserved in a climatized room (20°C and > 45 % relative humidity). Reference reading of gauge length and mass is executed at the start of this conservation period. At least four measurements of the gauge length and of the mass are executed at suitable time intervals until the moisture content is less than 4 %.

After this procedure specimens are dried to constant mass in a ventilated oven (105 ± 5°C) in order to calculate the real moisture content (μ_t) at the different measurement intervals.

Relative length variation ($\Delta\ell$) is determined as a function of moisture content in a graph, as presented in fig. 1. Moisture content is defined as

$$\mu_t = 100 \ \frac{\Delta m}{m}$$

with Δm : loss of mass of the specimen, when drying from the moist state under consideration to dry condition.

m : mass after drying at 105°C ± 5°C.

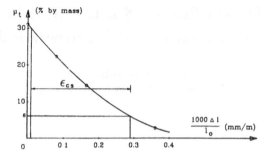

Figure 1.

Conventional drying shrinkage ε_{cs} is calculated as

$$\varepsilon_{cs} = \frac{1000 \ \Delta\ell}{\ell_0} \qquad (mm/m)$$

with $\Delta\ell$: variation in gauge length between 30 % and 6 % moisture content (mm).

ℓ_0 : initial gauge length (mm).

2.2. Evaluation of bond characteristics

Anchorage of longitudinal reinforcement bars in AAC is conventionally assured by several cross-welded transversal bars. Several mathematical models are presented for the design of the anchorage system, more specifically the diameter and the distance between the transversal bars. Of course a minimum number of them is necessary for the assembling and the correct positioning of the longitudinal reinforcement.

The contribution of reliable bond characteristics to the anchorage could lead to an economy in reinforcement design. Since in most cases smooth steel, coated with a variety of products is used, the bond performance is not an obvious reliable phenomenon.

In order to check bond behaviour as a whole and individual performance of coating types, adequate test methods are developed. A first test method, based on a RILEM recommendation for bond characteristics determination in ordinary reinforced concrete, aims at the measurement of the real conventional bond strength and should lead thus to the determination of design values.

A second test method, consisting of a simple push-out test permits a quick, efficient check of the reliability of the bond characteristics and can

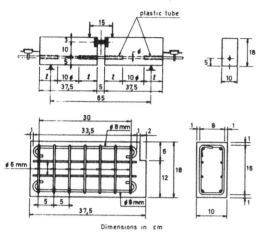

BEAM-TEST TYPE 1 $\varnothing < 16$mm

plastic tube

Dimensions in cm

Figure 2.

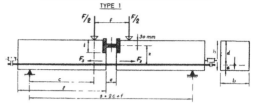

Figure 3.

be useful as a statistical verification method for production and quality control. Both methods are the subject of an important test program, set-up at the Magnel Laboratory for Reinforced Concrete of the University of Gent.

2.2.1. Beam test

In order to approach the situation where the anchorage of the longitudinal bars is essential namely that of a transversally loaded partially cracked horizontal slab, a beam shaped test specimen is chosen.

In contrast with the specimens developed for normal concrete bond determination, test beams for AAC must be cut out of production slabs. This way of manufacturing limits the free choice of specimen dimensions considerably. Fig. 2 shows the design of the test beam for normal concrete and bars with a diameter smaller than 16 mm. The beam comprises two parallelepipedal reinforced concrete blocks, interconnected at the bottom by the test bar and at the top by a steel hinge. The beam is loaded in simple flexure by two forces of equal magnitude, disposed symmetrically with regard to the mid-span section of the beam. The blocks are provided with an important transversal reinforcement, thus avoiding early bond rupture through cracking of the concrete due to shear. The bond length in each block is limited to ten times the diameter of the bar by means of soft plastic tubes, placed around the bar.

For specimens of A.A.C. a quite different and more delicate choice of dimensions has to be made to optimize specimen behaviour. A very important consideration concerns that the bond length equals the block length. It is practically impossible to diminish the bond length through cutting along the bar. Mainly due to this fact the block length is submitted to contradictory requirements. Increasing the block length leads to an increased total bond force and a higher steel stress in the test bar. Increasing the block length to specimen height ratio diminishes the shear span and the shear force resistance.

Due to the absence of the shear reinforcement system cracking of the concrete blocks is very likely. Early rupture of the block before slip of the bar occurs, prohibits the calculation of reliable conventional bond strength values.

Envisaging these basic data a primary design has been developed. Bond length is chosen at 50 $\varnothing$ ($\varnothing$ = bar diameter). The conventional mean bond stress (τ_{bm}) and the steel stress in the mid section are related as follows

$$F_s = \pi \cdot \ell \cdot \varnothing \cdot \tau_{bm} = \frac{\sigma_s \cdot \pi \cdot \varnothing^2}{4}$$

$$\sigma_s = 200 \, \tau_{bm}$$

For an estimated value of 2 N/mm^2 for τ_{bm} a steel stress of 400 N/mm^2 is reached. The conventional shear stress $\tau_c = V/bd$ in the concrete is limited to about 0.3 N/mm^2 through an adequate choice of the cross-section dimensions of the beam (b, d, h).

Fig. 3 shows the design scheme of the test beam and the load configuration. Table 1 gives the dimensions of the different beams as a function of the tested bar diameter.

The manufacturing of the specimens departed from a series of slabs with thickness 200 mm cast in one mould. The concrete is characterised by the following data

223

TABLE 1
Dimensions of beam test specimens type 1

Bar Diameter	Bond length ℓ (mm)	Block width b (mm)	Specimen depth h (mm)	d (mm)	f (mm)	e (mm)	Span s (mm)
5,5	275	125	125 200	100 175	150	40	480
7	350	150	150 200	120 170	200	60	620
8	400	170	170 200	140 170	200	60	720
10	500	180	200	170	200	60	920

TABLE 2
Summary of test results beam test type 1 (average values)

Cement coated bars	Rupture type	Number of results	τ_{bm} (N/mm^2) Lowest	Average	Highest	σ_s (N/mm^2) Average
Ø 5,5 mm	slip	3	1,62	1,80	2,07	360
	cracking	5				
Ø 7 mm	slip	4	1,24	1,61	1,90	322
	cracking	4				
Ø 8 mm	slip	4	1,34	1,53	1,78	306
	cracking	4				
Ø 10 mm	slip	-	1,29	1,36	1,51	272
	cracking	8				
Untreated bars						
Ø 5,5 mm	slip	1	1,98	2,11	2,42	423
	cracking	7				
Ø 7 mm	slip	-	1,63	1,72	1,79	344
	cracking	8				
Ø 8 mm	slip	1	1,60	1,65	1,73	330
	cracking	7				
Ø 10 mm	slip	-	1,29	1,35	1,46	270
	cracking	8				

TABLE 3
Dimensions of beam test specimens type 2

Bar diameter	Bond length ℓ (mm)	Block width b (mm)	Specimen depth		f (mm)	e (mm)	t (mm)	Span s (mm)
			h (mm)	d (mm)				
5,5	220	150	200	175	200	221	100	570
7	280	150	200	170	200	200	100	690
8	320	170	200	170	200	200	100	770
10	400	170	200	170	200	200	100	930

TABLE 4
Summary of test results beam test type 2

Cement coated bars	Rupture type	Number of results	τ_{bm} (N/mm^2)			σ_s (N/mm^2)
			Lowest	Average	Highest	Average
Ø 5,5 mm	slip	3	1,45	2,36	4,57	378
	cracking	5*				
Ø 7 mm	slip	7	1,24	1,75	2,08	281
	cracking	1				
Ø 8 mm	slip	5	1,42	1,63	1,82	260
	cracking	3				
Ø 10 mm	slip	5	1,40	1,45	1,48	231
	cracking	3				

* Concrete rupture in compression zone

$f_c \sim 4,6$ N/mm^2
$\rho_c \sim 490$ kg/^{3}

Mainly modified cement coated bars were tested. As a reference some uncoated bars completed the test program. The tested diameters were 5.5 mm, 7 mm, 8 mm and 10 mm.

An important advantage of the use of an hinge in the compression area at mid-span is the constant load-steel-stress ratio. This advantage how-ever is weakened by the important and delicate effort, necessary to manufacture the specimen.

During the deformation controlled loading procedure the relation slip-load is registered at both bar ends. When failure occurs at one end through slip or cracking the failing bar end is fixed and the procedure is continued until the remaining bar end fails as well.

In table 2 a summary of test results is given. In most cases rupture through cracking of the blocks cannot be avoided. The crack pattern consists mostly of one rather inclined crack near the centre of the beam and a nearly horizontal crack along the bar reaching the support see fig. 4. At the bottom surface a splitting crack along the bar is observed. All cracks appear at rupture load. Even with a deformation controlled loading procedure crack growth is difficult to observe. In a minority

TABLE 5
Summary of push-out test

Coating	Bar diameter	Number of results	τ_{bm} (N/mm²)			
			Lowest	Average	Highest	Spread
Cement-production type	5,5	80	0,9	2,17	3,5	0,73
	7	27	0,7	1,89	2,8	0,71
	8	26	1,1	1,91	3,5	0,51
	10	18	1,0	2,13	3,1	0,51
Manganese phosphate	5, 6, 7, 8	56	0,35	1,00	2,10	0,32
Zinc phosphate	5, 6, 7, 8	42	0,50	1,65	2,70	0,43
Cement-prototype	5	14	1,30	2,30	2,95	0,46
	6	14	1,85	2,45	2,90	0,72
	7	14	1,84	2,35	2,70	0,34
	8	14	1,60	2,15	2,65	0,35
Rust-binder	5	14	0,34	0,58	1,18	0,27
	6	14	0,09	0,72	1,18	0,29
	7	14	0,34	1,10	1,78	0,46
	8	14	0,27	0,59	1,35	0,34
Bitumen	5	14	0,17	0,30	0,45	0,09
	8	14	0,42	0,66	1,07	0,20
Paint	5	14	0,65	1,22	1,59	0,33
	6	14	0,68	0,95	1,31	0,17
	7	14	0,56	0,85	1,17	0,18
	8	14	0,10	0,67	1,15	0,31
Untreated	4	14	1,75	2,85	3,75	0,50
	5	14	1,68	2,45	3,15	0,49
	7	14	1,05	1,75	2,15	0,29
	8	14	1,25	1,60	2,10	0,27

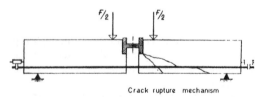

Crack rupture mechanism

Figure 4.

of cases rupture through pure slip is achieved, thus permitting a reliable bond stress calculation. The conventional mean bond strength τ_{bm}, defined as

$$\tau_{bm} = \frac{F_s}{\pi . \varnothing . \ell}$$

reaches values of 1.4 to 1.7 N/mm² in the case of true slip rupture. When rupture with cracking occurs higher values are registered (up to 2 N/mm²). Real slip rupture occurs almost only with cement coated reinforcement bars. Untreated bars do not slip and present higher bond strength values. Specimens higher positioned in rise direction present about 10 % higher bond strength values than lower placed, higher density specimens.

The mean value of τ_{bm} decreases with increasing bar diameter. Specimen depth (d) and width (b) are noticed to be insignificant test parameters in this test series. Steel yielding does not appear.

A most important conclusion, drawn out of this first test series is that conventional bond stress could be a reliable and rather low-spread characteristic.

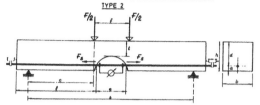

Figure 5.

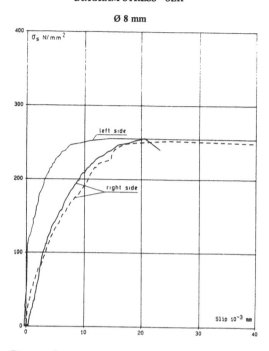

Figure 6.

The study of this first test series lead to a second program where the following aspects were altered (see fig. 5). Omitting the integration of the hinge in the cut-out specimen diminishes the risk to early bond damage due to cracking. Through this choice indirect steel stress determination through elongation measurement becomes necessary. Bond length is diminished to 40 Ø. The mid section shape of the specimens is obtained by diamond coring. The cut surface is similar in shape and localisation to a possible first crack. In this way cracking could be deferred. Cross section dimensions (b, d) are kept constant for all diameters (table 3)

$$b \approx 170 \text{ mm}$$
$$d \approx 170 \text{ mm}$$

The only varying parameter left is the block-length ℓ which is related itself to the span. The thus reached limited range for total load values and shear stress values τ_c facilitates test praxis.

Steel stress is indirectly measured and registered by means of a resistive extensometer, mounted on the bar. Bar displacement is measured at both ends by means of inductive transducers.

The test program comprises 16 tests on cement coated bars (Ø 5.5, 7, 8, 10 mm), taken from the same production mould as the first series. Table 4 shows a summary of calculated test results.

As expected, slip rupture is more frequently achieved, mainly for specimens, situated lower in rise direction. Still a considerable number of specimens shows cracking at maximum load. In all these cases a first crack appears nearly parallel to the cylindrical saw cut at a distance of a few cm. This first crack is followed by a more horizontal crack pointing to the support and a splitting crack visible at the lower surface. Cracked specimens show higher bond stress values just as do the specimens with smaller bar diameters. Bond stresses at rupture are higher than in the first program,

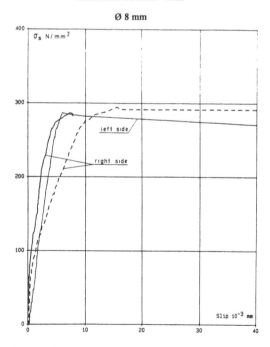

Figure 7.

due to the smaller bond length.

Fig. 6 and 7 show respective diagrams for a slip-rupture test and a crack-rupture test.

At this moment, the intention exists to expand the test program with other coating systems and concrete qualities. Specimen design will be improved to be usable for the entire range of present and future production materials. The achievement of this effort will lead to a standard test method for determination of the basic bond design strength.

2.2.2. Push-out test

Monitoring bond strength in current production requires a low cost, efficient and repeatable standard test procedure. Specimens must be easy to fabricate and test apparatus must be elementary and affordable.

The most easy design is that of a prismatic specimen with standardized length, transversally cut out of a longitudinally reinforced element. Since no bars stick out of this prism a pull-out test cannot be performed. A push-out test though can be realized using a compression test device.

In the Magnel Laboratory for Reinforced Concrete this test method is applied in an extensive test program on the performance of a series of possible steel protection coatings for A.A.C. It is also the subject of studies within European normalisation.

Fig. 8 shows the principal test set-up for 200 mm high prisms. At the top of the bar the concrete is cut away by a circular dry saw over a length of 10 mm, so that a crown-shaped pushing device can be used to apply the load. At the bottom the concrete block rests on a plate leaving a cylindrical hole Ø 20 mm under the bar. At this bar end slip can eventually be measured. The entire set-up is placed in a compression loading system with a capacity of 25 kN (bond length = 190 mm). Tests can be performed either manually or load- or deformation-controlled.

In the Magnel Laboratory eight test series were executed comprising the following coatings :

    A - Cement coating-production type
    B - Manganese phosphate
    C - Zinc phosphate
    D - Cement coating prototype
    E - Rust binder (thin layer)
    F - Bitumen
    G - Paint
    H - Untreated

Different bar diameters (5.5 to 10 mm) have

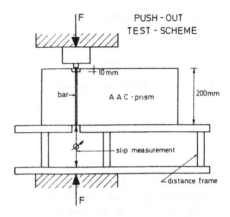

Figure 8.

been tested. Table 5 shows a survey of the test results. Conventional bond strength values listed are calculated as follows

$$\tau_{bm} = \frac{F}{\pi . \varnothing . \ell} \qquad (N/mm^2)$$

ℓ : bond length : 190 mm
F : maximum load (N)

Highest bond values are obtained with cement coated bars and untreated bars.

Paint type and phosphate type coatings perform less. Rust binder treatment gives relatively low τ_{bm} values and bituminous coatings provide neglectible bond. In this last case thixotropic behaviour is noticed. Through this fact still lower bond strength is expected on long duration load.

Bond strength values are generally slightly higher with push-out testing than these resulting from a beam test. This difference can be explained by the different stress distribution along the bar.

In most cases bond strength decreases with increasing diameter. Lower situated bars in risc direction reach lower bond strength values than higher positioned bars. Spread is higher with push-out testing compared with beam testing. Still reliable mean values are measured.

As a general conclusion one could say that push-out testing is an efficient economic test method for bond strength quality control. The obtained values cannot directly be used for design since stress distribution in the test specimen cannot be compared with that in real full scale components.

In the CEN document a slightly different scheme, with registration of the "Bond stress-slip" relation is proposed.

6 Masonry

Advances in Autoclaved Aerated Concrete, Wittmann (ed.) © 1992   Taylor & Francis. ISBN 90 5410 086 9

The practical use of AAC masonry to meet the performance requirements of buildings in Europe

C. A. Fudge & A. H. Riza
Celcon Blocks Limited, Grays, UK

ABSTRACT: With the implementation of the Construction Products Directive by individual Member States of the European Community in 1991/2, construction products have to meet six so called "Essential Requirements", relating to health, safety and energy economy. In most European countries, national regulations and standards already exist relating to the performance of products in use. This paper examines how aac masonry is being used in Europe to meet the Essential Requirements for walls, very often as the sole construction material. Examples are given showing how the combination of structural, thermal and sound performance is achieved in both cavity and solid walls. Aac masonry has an important role to play in the design of the envelope of a building and this paper will examine how the material can be incorporated to produce weather-tight constructions.

1 INTRODUCTION

Throughout the European Community, individual Member States currently impose a wide range of different obligations on the construction industry in the interests of safety and consumer protection. However well justified in themselves, these obligations have resulted in technical barriers to trade because of differing requirements and procedures in various Member States. It was agreed in 1985 that harmonisation of technical requirements and standards should be undertaken to eliminate barriers to trade between Member States to overcome this long-standing problem.

In most European countries, national regulations and standards already exist relating to the performance of products in use. To enable the harmonisation process to begin, the European Commission published the Construction Products Directive (Official Journal of the European Communities, 1989) to be implemented by individual Member States during 1991/1992. The main feature of this Directive is that construction products, when incorporated into permanent works, have to meet six so called "Essential Requirements". These Requirements relate to health, safety and energy economy, depending on the intended use of product. This paper

examines the practical use of aac masonry to meet these performance requirements. Whilst, in the main, European practice is examined, the principles can be applied Worldwide to meet the general requirements for buildings.

2 PERFORMANCE REQUIREMENTS TO BE MET

The "Essential Requirements" of the Construction Products Directive make it necessary for products to be suitable for construction works and fit for their intended use. The requirements generally concern circumstances which are foreseeable.

(a) Mechanical resistance and
      stability
The construction works have to be designed and built in such a way that the loadings that are liable to act on it during its construction and use will not lead to any of the following
- collapse of the whole or part of the work,
- major deformations to an inadmissible degree,
- damage to other parts of the works or to fittings or installed equipment as a result of major deformation of the loadbearing construction, and,
- damage by an event to an extent

disproportionate to the original cause.

(b) Safety in case of fire
The construction works should be designed and built in such a way that in the event of an outbreak of fire,
- the loadbearing capacity of the construction can be assumed for a specific period of time,
- the generation and spread of fire and smoke within the works are limited,
- the spread of fire to neighbouring construction works is limited,
- occupants can leave the works or be rescued by other means and
- the safety of rescue teams is taken into consideration.

(c) Hygiene, health and the environment
The construction work has to be designed and built so that it will not be a threat to the hygiene or health of the occupants or neighbours, in particular as a result of any of the following:
- the giving-off of toxic gas,
- the presence of dangerous particles or gases in the air,
- the emission of dangerous radiation,
- pollution or poisoning of the water or soil, and
- the presence of damp in parts of the works or on surfaces within the works.

(d) Safety in use
The construction work has to be designed and built in such a way that it does not present unacceptable risks of accidents in service or in operation such as slipping, falling, collision, burns, electrocution and injury from explosion.

(e) Protection against noise
The construction works have to be designed and built so that noise perceived by the occupants or people nearby is kept down to a level that will not threaten their health and will allow them to sleep, rest and work in satisfactory conditions.

(f) Energy economy and heat retention
The construction works and its heating, cooling and ventilation installations must be designed and built in such a way that the amount of energy required in use shall be low, having regard to the climatic conditions of the location and the occupants.

3 MEETING THE ESSENTIAL REQUIREMENTS

3.1 Structural requirements

The process of harmonisation of standards across Europe is enabling a better understanding of different design approaches with the substantial experience of some countries in particular forms of construction being recognised by the others. Structural design by simple "rules of thumb", for example, is a prescriptive approach to design of masonry for low rise dwellings. It is a common method of compliance in some nations, even though the masonry built to the maximum permissible limits cannot always be justified by arduous detailed calculation procedures. However, years of practical experience and gradual development have produced a set of rules which are perfectly satisfactory. This is a new design philosophy to many countries but acceptance is growing in the world arena with the imminent publication of a draft international standard (International Standards Organisation) and the RILEM publication " Recommended practice for construction of aac". The latter is an adaptation of the former with the rules being specifically applied to walls constructed of aac.

Generally, for a two storey building with concrete floors and a timber roof, walls can either be at least 150mm thick solid construction or two 100mm leaves with a cavity to meet structural requirements. In either case, aac units can be used. Traditionally, solid wall constructions tend to be around 200mm thick and cavity walls seldom less than two 100mm leaves, with aac being used increasingly as the loadbearing element because of its other inherent characteristics e.g. thermal insulation properties. By reference to the simple perscriptive rules, or other design codes, it can be seen that aac can be perfectly adequate for most domestic buildings and will have no problems in meeting the Essential Requirement for Mechanical Resistance and Stability. Economic wall thicknesses are achievable using aac masonry, whilst higher buildings can also be constructed using higher strength aac masonry units.

Whilst aac masonry may be used in loadbearing walls, it has also been incorporated extensively throughout Europe for infill in domestic framed structures. In seismic areas, economic wall thicknesses are obtained, when designed into, say, a concrete frame.

For non-domestic framed buildings, aac masonry has again been used, taking into account its flexural strength and hence resistance to lateral loading.

One of the major benefits of using aac is its low density compared to other forms of masonry construction. With a reduction in self-weight of up to a third of other types of masonry, aac units are used to provide economic, structurally adequate non-loadbearing walls within common structures. This in turn reduces the sizes of structural members carrying the masonry and therefore the size of foundations.

3.2 Energy conservation requirements

Energy conservation is regulated in many countries by limiting the amount of heat lost through the external elements of a building. This is often done by setting maximum thermal transmittance values (U-values) for each element. The U-value set by each country depends on such factors as climatic conditions, economy and environmental considerations, but, in most situations, the statutory requirements can be met where aac is used as the main structural element without the need for secondary insulation. In the middle European regions, for example, requirements can usually be met using 215mm thick low density aac in solid wall constructions.

In calculating U-values to determine heat losses, steady state conditions are assumed to exist. In practice, however, the thermal conditions within a building tend to be dynamic. A building which has a high structural mass will take longer to be heated by the warm daytime air causing a delay in the rise of the inside air temperature. The result is that before the inside air temperature matches that of the outside, the external air begins to cool down for the evening. Consequently, the variation in internal air temperature is less extreme than would be expected within a lightweight structure. The ability of a material to resist changes in temperature is referred to as its thermal inertia, and, although it is commonly associated with mass, its thermal conductivity and specific heat are also important factors. Despite the fact that aac is a lightweight material, it has a useful thermal inertia because of its favourable combination of all of the above. This has been shown by research (Energy Conservation Committee 1979) where twenty occupied residential units built of different materials namely, aac, timber frame and reinforced concrete were monitored. The results showed that the units built with aac exhibited a lower rate of energy consumption than the other constructions.

The contribution of thermal inertia to energy conservation is being recognised gradually. Some national standards already have this in place and outside of Europe, for example, the California Energy Commission permits higher U-values for walls built using materials with properties equal to that of aac (Aroni 1990). There is no doubt that in the future energy conservation regulations will take into account the dynamic thermal behaviour of buildings enabling the full potential of aac as a structural insulating material to be realised. Aac, having the lowest thermal conductivity of any concrete masonry unit, enables the thermal requirements of any particular Member State to be met using the thinnest wall constructions.

Thus, aac masonry provides a cost effective solution, with the additional benefit of being capable of thin-joint construction. In areas of Europe not subjected to the most onerous requirements for thermal insulation, aac masonry forms the sole insulating material. In solid wall construction, using aac, no external or internally faced thermal insulant is necessary, as with other forms of masonry. Where cavity walls are used, clear cavities can be obtained with an inner leaf of low density (less than 500 kg/m^3) aac in thicknesses of 100 to 150mm. The advantage of clear cavities, as will be discussed elsewhere, is the reduction in risk of rain penetration to the inner surface of the construction.

Aac masonry has also been used in the foundations and to form the basement of a building. Because of its intrinsic thermal properties, a reduction in heat loss is again achieved in the ground floor/basement area.

3.3 Fire resistance requirements

The inherent thermal characteristics of aac, coupled with the fact that it is a non-combustible material, means that it has excellent fire resisting properties. Although the design of buildings against fire spread is covered by national regulations, the requirements in each country are generally the same as the Essential Requirement, i.e. premature failure of the structure should be prevented and the spread of fire inside a building restricted. Provision is normally made by specifying minimum periods of fire resistance for the relevant elements of the structure. In general, there is ample material in a wall construction to prevent the spread of flame externally, except where combustible cladding is used. Aac, being non-combustible, provides a good medium for this construction. One of the major benefits of aac masonry, is its use as an internal wall. It can have a fire resistance at least as good as, if not better, than other types of masonry materials. In addition, aac is also used to clad steel columns and frames to provide the necessary fire resistance. Again, slender, economic solutions are achievable.

3.4 Acoustic requirements

In general, the sound insulation qualities of any type of masonry is dependent on both density and porosity. The mechanism for sound transmission involves a source, direct or airborne, setting up vibrations in the masonry which in turn causes the air beside it to vibrate. It is these new airborne vibrations that are heard. Although density is important in that a heavy material is not easily set into vibration, the air porosity will also have a significant effect on the performance of the wall. The cellular structure of aac gives it very low air porosity and, consequently, a high resistance to the passage of airborne sound relative to density. As such, aac masonry has been found to have its own 'mass law' (Luckin K.R., 1986) and some national standards recognise this fact.

The thickness of aac required depends on the level of insulation required, which is set by national standards or regulations.

Various constructions are used throughout Europe to satisfy national requirements. These range from solid wall constructions, to cavity walls with and without insulants in the cavity. Where a wall separates two different dwellings, solid walls can be used from approximately 200mm in thickness and upwards, depending on the level of insulation needed. Cavity walls, with cavities of around 75 to 100mm, are constructed with aac masonry leaves of some 100mm to 150mm. Aac masonry also provides a useful contributory factor in reducing indirect or flanking sound transmission via paths adjacent to the main separating construction. The thickness of the wall required is again dependent upon national requirements,

but is generally a minimum of 100mm.

This form of masonry is also used in walls within a single occupancy building to reduce the noise from one room to another. It is ideally suited to provide the necessary sound insulation and with its low self weight reduces the loads on floors in dwellings which tend not to be over-engineered.

3.5 Health, hygiene and environmental requirements

The Essential Requirement for Hygiene, Health and the Environment requires not only the exclusion of rain and moisture from the outside, but also the prevention of surface and/or interstitial condensation.

The resistance to rain penetration of masonry walls depends on many factors including surface finishes, workmanship, mortar, thickness of the wall and the presence of a cavity. Rendering can, in many cases, provide adequate resistance to rain penetration, but in certain severe conditions total resistance is provided by use of an impervious external cladding system. Due to its structure, aac has a far superior resistance to water penetration than most other types of masonry. This is recognised in some national codes (British Standards Institution, 1985) which show that external solid walls built of aac may be thinner than those of other materials to meet the same rain resistance criteria. Resistance to rain using relatively thin solutions can be achieved with fairly modest wall thicknesses. As previously discussed, traditionally solid walls tend to be around 200mm or thicker, so the use of a rendered aac wall will give the thinnest solution without the need for additional precautions against rain penetration. In addition, thin-joint masonry reduces the risk of rain passing through the interface between units and traditional mortar joints. Where there is a risk of wind-driven rain for example the North Sea areas of Northern Europe, aac masonry has been used extensively in cavity walls. The design philosophy is that any water passing through the outer leaf will run down the face in the cavity and pass through specially designed 'weep-holes' at the base of the wall.

Condensation and mould growth is also controlled by several factors, including moisture generation, ventilation, thermal insulation and heating. Water

vapour is generally produced in a
building by the normal activities of the
user. In most countries, minimum
ventilation rates are specified by
national regulations to ensure that
most, if not all, of the vapour produced
is removed. Surface condensation can
occur when the vapour which is not
removed comes into contact with a cold
surface. Whilst the walls, roof and
floors may be adequately insulated
overall. there are often localised
areas. such as around windows or doors,
where the insulation standards may be
lower. With aac constructions the risk
of these "cold bridges", and hence
condensation, is greatly reduced by
closing the edges of openings in aac and
by retaining the blockwork across the
lintels to maintain a higher level of
insulation. Aac constructions,
therefore, can help to avoid the need
for additional detailing or insulation
which may be necessary with other
materials. Thin-joint masonry is used
to great advantage to achieve this
requirement.

The need to prevent condensation or
moisture penetration to the internal
surface of a building is twofold. The
primary concern is environmental - i.e.
the prevention of mould growth which can
be hazardous to health, and the second
is the need to prevent moisture damage
to internal finishes. Unlike some other
materials, aac is not under normal
conditions significantly affected by
high moisture levels and has good
freeze/thaw characteristics. This
coupled with its resistance to sulphate
attack, makes it highly suitable for use
below ground level. Although not used
widespread throughout Europe, in the UK,
aac is commonly used for the
construction of walls below ground,
without surface protection, in soils
with ground water soluble sulphate
levels not exceeding 2500 mg/L. An
additional benefit of using aac in this
position is that it provides a useful
level of edge insulation to the floor
slab. As well as reducing the risks of
cold bridging at the floor/wall
junctions, the floors thermal
transmittance value is also
significantly reduced.

4 EUROPEAN WALL CONSTRUCTIONS

Traditional forms of masonry wall
constructions vary across Europe. In
the UK, and most of northern Europe,
cavity walls with two relatively thin
leaves (generally some 100mm each with
50-100mm cavities) predominate, whereas
to the south, walls tend to be solid and
perhaps 250-350mm thick.

Several factors have influenced the
type of construction that has become
popular in each country including
climate, national structural codes and
traditional appearance.

Cavity walls are generally perceived
as offering better resistance to rain
penetration than solid walls. This is
why they predominate in the north since
conditions of exposure tend to be more
severe, particularly around the North
Sea coastal areas. The combination of
wind and rain demands such
construction. To the south, walls tend
to be solid, reflecting milder
conditions. In France or Spain, for
example, 300mm thick solid walls are not
uncommon. This difference is also
reflected in the national structural
codes. In Germany, where walls are
generally solid, the cavity
constructions that have been built tend
to have inner leaves around 200-250mm
thick. The walls are basically designed
as solid loadbearing walls with the
outer leaf acting as self supporting
cladding. Countries with greater
experience of cavity constructions tend
to have design codes which recognise the
contribution of the outer leaf to the
overall performance of the wall. In
Denmark, Holland and the UK, for
example, relatively slender walls
comprising of two 100mm leaves are
commonplace with the outer leaves often
built with facing materials such as clay
or calcium silicate brickwork.
Consequently this can give the buildings
a more pleasant appearance than plain
rendering which, more often than not,
covers the solid wall constructions of
the south.

5 SUMMARY

It has been shown that aac masonry can
fulfill all of the essential
requirements as laid down by the
European Directive. Different Member
States have various levels attached to
the requirements e.g. different U-values
to meet the climatic needs of the
country. However, in the main aac can
be the sole material to meet the
combined requirements. In external
walls, structural, thermal, sound and
rain peneration requirements are met,
depending on location, with either solid
or cavity wall constructions. Solid

walls in the range of 200 to 300mm give the structural characteristics with very often no need to apply additional thermal insulation. In cavity walls, economic wall solutions are possible which again may or may not include additional insulants. The low density of aac compared to other types of masonry unit not only provides savings in the design of supporting structural members and foundations, but also in the time and labour used to construct the walls.

When used internally, the inherent properties of the material allows economic walls to be built providing good sound and fire resistance. Again, being one of the lowest density solid masonry materials, cost savings on structural members are possible.

Other aspects, such as workability, have not been covered by this paper, but these all add to the reasons why this type of masonry unit is used extensively throughout various parts of Europe.

REFERENCES

Aroni, S. 1990. On energy conservation characteristics of autoclaved aerated concrete, Materials and Structures, 1990, 23,68-77.
British Standards Institution, 1985. BS 5628 : Part 3 : Code of Practice for use of masonry materials and components, design and workmanship.
Energy Conservation Committee. Report on Temperature and Energy Consumption, ALC Institute, Japan, 1974.
International Standards Organisation. ISO 9652 : Part 2. Masonry Designed by Simple Rules.
Luckin, K.R., Jones, A.J. and Engledow G, 1986. Sound Insulation Performance of Autoclaved Aerated Concrete, British Masonry Society. Proceedings No 2.
Official Journal of the European Communities, 1989. Council Directive (89/106/EEC), No L40 of 11.2.1982, 12-26.

Advances in Autoclaved Aerated Concrete, Wittmann (ed.) © 1992   Taylor & Francis. ISBN 90 5410 086 9

Compressive strength and modulus of elasticity of AAC masonry

P. Schubert & U. Meyer
Institute for Building Material Research, Aachen University of Technology, Germany

ABSTRACT: The paper presents the latest results of analysis for the compressive strength and modulus of elasticity of autoclaved aerated concrete (AAC) masonry. Regression equations are used to describe the compressive strength of AAC-masonry as a function of the compressive strengths of the masonry units and the mortar respectively for masonry with thin-layer, normal and light-weight mortars. Since the influence of mortar compressive strength is generally negligible (thin-layer mortar and normal mortar) or extremely slight (light-weight mortar), a good approximation of masonry compressive strength can be achieved exclusively by regarding it as a function of masonry unit compressive strength. A comparison of the analyzed results with the permissible stresses specified in DIN 1053 Part 1 revealed additional compressive strength reserves in a number of cases, possibly allowing higher permissible stresses.

The modulus of elasticity of masonry with thin-layer mortar can be described with high precision as a function of masonry compressive strength. For masonry with normal mortar, the reliability of the results is somewhat less. In the case of light-weight mortar masonry, there were too few available results for analysis. According to the analyzed results, the modulus of elasticity values quoted in DIN 1053 Part 2 and by Schubert (1992) are in some cases too high.

This analysis of results is intended as a contribution to the preparation of the final version of EC 6 "Masonry".

1 INTRODUCTION

The fracture mechanism of masonry has long been understood, see Hilsdorf (1965). In masonry under compressive stress at right angles to the bed joints, the masonry units are in a triaxial compression-tension-tension stress state, while for reasons of equilibrium the mortar in the bed joint is in a triaxial compressive stress state. The compressive strength of the masonry is additionally influenced by the usually different transverse deformation properties of the masonry units and the mortar respectively, especially in the fracture zone. If the mortar possesses a higher transverse deformability, additional transverse tensile stresses arise due to the bond between the units and the mortar. These reduce the potential resistance of the masonry units to transverse tensile stress and hence the compressive strength of the masonry. This case is particularly likely to occur if high-strength masonry units with low transverse deformability are combined with low-strength masonry mortars, e.g. with light-weight mortars (high transverse deformability). The permissible basic stresses specified for light-weight mortars in DIN 1053 Part 1 are accordingly much lower than those for normal mortar. Identical compressive strengths for individual units and for the masonry as a whole may be expected only if the transverse deformations of the units and the mortar up to the instant of fracture are always the same, if the head joint influence can be neglected and if test conditions are comparable or have been taken into account. The strength efficiency ratio α_N, i.e. the ratio of masonry compressive strength to unit compressive strength, would then be equal to 1.

In the head joint region, no significant transmission of forces can occur in masonry with normal and light-weight mortars, due to the low adhesive strength between the mortar and the units. This results in a reduction of masonry compressive strength of roughly 10 to 20 %. Because of the high adhesive strength of head joints laid with thin-layer mortar, the head joint has little or (in the case of low-strength units) no effect on masonry compressive strength.

The factor α_N provides a clear indication of the way in which properties of the units and the mortar affect the utilization of unit compressive strength and of the remaining potential for increasing masonry compressive strength by Schubert (1985).

The modulus of elasticity of compressively stressed masonry perpendicular to the bed joints is an important factor in dimensioning masonry (e.g. buckling calculation) and in assessing deformations and crack resistance. Like masonry compressive strength, it is essentially determined by the properties of the units, especially due to the high ratio of masonry units to mortar when large-format blocks are employed.

2 COMPRESSIVE STRENGTH OF AERATED CONCRETE MASONRY
2.1 General Observations on the Analysis

The analysis covers a total of 378 tests on AAC-masonry from 32 sources. Each individual masonry tests has been logged and analyzed. The unit compressive strengths include the shape factors according to the masonry unit standard. The masonry compressive strengths have been converted to a uniform slenderness (height/thickness) $\lambda = h/t = 5$ using the correction factor a' according to Mann (1983):

$$f_{c,ma} = a' \cdot f'_{c,ma}; a' = 0.966 + 0.00136 \cdot (h/t)^2. \tag{1}$$

Regression calculations were performed using the test values and various formulations. Where possible and useful, statistical parameters such as the confidence interval, quantiles, coefficient of correlation R and standard error of the estimate were determined. Measured and calculated strength values were also compared.

2.2 Masonry Compressive Strength and strength effiency factor
2.2.1 Fundamental Considerations

As noted in Section 1, significant variables influencing masonry compressive

strength are the transverse tensile strength of the masonry units together with the transverse deformability properties of the masonry units and the masonry mortar. The ratio of unit tensile strength (along the longitudinal axis of the units or across the unit width) to unit compressive strength is therefore highly significant. The greater this ratio, the higher will be the expected masonry compressive strength. It is not yet possible to describe the transverse deformability properties of the masonry units and the masonry mortar with sufficient accuracy. The transverse strain modulus E_q of the units and the mortar has therefore been used as a property parameter. This is determined as the secant modulus at a compressive stress $\sigma_c = 0.3 \cdot \max \sigma_c$ and the associated transverse strain. The smaller the E_q value, the greater will be the transverse deformation under compressive stress. In view of the factors noted in Section 1, roughly identical transverse deformabilities for mortar and units are desirable. These can, however, be approximated only roughly by means of the transverse strain modulus, since E_q takes no account of transverse deformation properties at higher stresses. Very few research results are currently available for any of the parameters noted above (Schubert (1992)). The assumed average ratio of unit tensile strength to unit compressive strength is 0.10.

In block masonry with thin-layer mortar, no influence of mortar properties on masonry strength is to be anticipated, due to the low bed joint thickness of 2 to 3 mm and the comparatively high mortar compressive strength. In AAC-masonry with normal and light-weight mortar, a significant influence of the mortar on the masonry compressive strength is to be expected only if higher-strength blocks are combined with lower-strength mortars, i.e. where units in strength class 6 and above are combined with group II standard mortar or light-weight mortar.

In recent years, the formulation

$$f_{c,ma} = a \cdot f_{c,u}^b \cdot f_{c,m}^c \tag{2}$$

with

$f_{c,ma}, f_{c,u}, f_{c,m}$: compressive strength of masonry, units, mortar

has proved fundamentally suitable for the quantitative description of masonry compressive strength. The formulation is empirically based and is not valid for

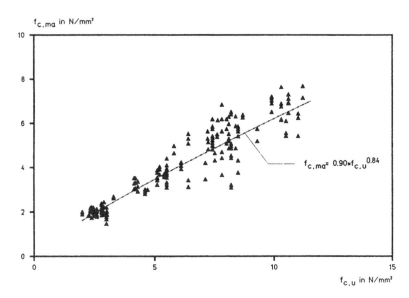

$f_{c,ma}$ in N/mm²

$f_{c,ma} = 0.90 \cdot f_{c,u}^{0.84}$

$f_{c,u}$ in N/mm²

Fig.1   Relationship between the compressive strength of masonry an masonry units-
precision-block units with thin layer mortar

$f_{c,m}$ = 0. In this case, the masonry compressive strength would be zero, which is, however, neither theoretically nor experimentally true. Other approaches, for example in the form

$$f_{c,ma} = a \cdot f_{c,u}^{b} \cdot (1 + c \cdot f_{c,m}^{d}), \qquad (3)$$

which do not include this disadvantage, are difficult to use and fail to yield a more precise description of the test values. The first formulation noted above will therefore be employed in the following analyses.

In assessing the strength efficiency factor α_N in respect to the potential which still exists for increasing masonry compressive strength, the differing slendernesses of the masonry and block specimens respectively must be taken into account. If an approximate slenderness of λ = 1 is assumed for the block, the corrective factor for masonry compressive strength (λ = 5) according to Mann (1983) or Voellmy (1957/58) will be a'= 1.03 or 1.36.

A decrease of only about 3 % in compressive strength when slenderness is increased from 1 to 5 would be at variance with current knowledge in this field. The formulation by Voellmy (1957/58) will therefore be employed in the following discussion. Since the blocks used in the masonry tests had a slenderness h/t = 1

(the influence of block length is similar to that for the masonry specimens), the shape factor of the masonry block should not be taken into account in assessing the factor α_N. Given identical slenderness of the masonry and block specimens, it is thus true to say that:

$$\alpha_N = \alpha'_N \cdot 1.36 \cdot 1.2 = a'_N \cdot 1.6. \qquad (4)$$

This means that, taking into consideration the differing slendernesses of the masonry specimen and the masonry block, the maximum possible load factor will be approximately 0.6.

2.2.2 Precision Block Masonry with Thin-
Layer Mortar

The analysis covers a total of 221 test results from 1983 onwards. As was to be expected, the influence of mortar compressive strength on masonry compressive strength is negligible. The regression equation

$$f_{c,ma} = 0.90 \cdot f_{c,u}^{0.84} \quad (R = 0.94) \qquad (5)$$

then appears as the mean value function. The equation is shown in Fig. 1. The standard error determined for the estimate

239

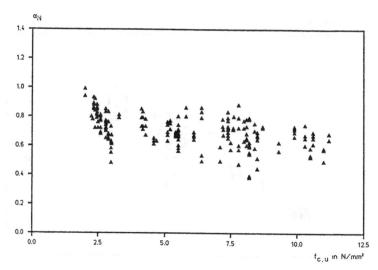

Fig.2 Compressive strength efficiency ratio $\alpha_N = f_{c,ma}/f_{c,u}$ and compressive strength of masonry units-precision block units with thin layer mortar

was 0.046 for the first parameter and 0.026 for the second.

The factor α_N is shown in Fig. 2.

2.2.3 Block Masonry with Normal and Light-Weight Mortars

140 tests were evaluated for masonry with standard mortar. The resulting mean value function (cf. Fig. 3) was:

$$f_{c,ma} = 1.09 \cdot f_{c,u}^{0.68} \cdot f_{c,mo}^{0.02} \quad (R = 0.82).$$

(6)

Here again, the standard error for the parameters a, b, c was extremely small.

Due to the again slight influence of mortar compressive strength, the relationship between the respective compressive strengths of the masonry and the units was also determined:

$$f_{c,ma} = 1.10 \cdot f_{c,u}^{0.69} \quad (R = 0.80).$$

(7)

Only 17 test results were available for masonry with **light-weight mortar**. The reliability of the analyzed results is consequently low. The following relationships were found:

$$f_{c,ma} = 0.89 \cdot f_{c,u}^{0.64} \cdot f_{c,m}^{0.09}$$

(8)

and

$$f_{c,ma} = 1.10 \cdot f_{c,u}^{0.64}.$$

(9)

As expected, the influence of mortar compressive strength is higher with light-weight than with normal mortars, but is still extremely slight by comparison with the influence of unit compressive strength.

Fig. 4 shows the factor α_N.

2.3 Results of Analysis as Compared to Permissible Basic Stresses in DIN 1053 Part 1

For comparison with the calculated mean value curves for AAC-masonry with thin-layer, normal and light-weight mortars, the permissible basic stresses σ_0 in DIN 1053 Part 1 were converted into mean compressive strength values for the same slenderness $\lambda = 5$:

$$f_{c,ma} = \sigma_0 \cdot \frac{2.0}{0.85 \cdot 0.80} \cdot 1.10 = \sigma_0 \cdot 3.24$$

(10)

where 2.0 is the safety factor, 0.85 the fatigue influence, 0.80 the difference between the mean value and the 5 % quantile

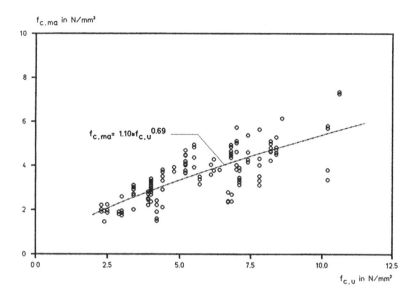

Fig.3 Relationship between the compressive strength of masonry and masonry units-block units with normal mortar

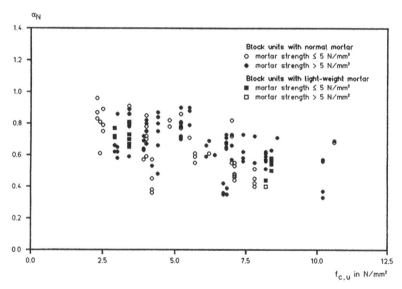

Fig.4 Compressive strength efficiency ratio α_N = $f_{c,ma}/f_{c,u}$ and compressive strength of masonry units-block units

and 1.10 the conversion of slenderness λ = 10 to λ = 5.

Fig. 5 indicates the mean value curves from the regression calculations and the calculated $f_{c,ma}$ values. As will be evident from the figure, analyzed results indicate that higher permissible basic stresses would be possible in some cases.

3 MODULUS OF ELASTICITY
3.1 General Observations on the Analysis

The analysis covered a total of 95 tests. The modulus of elasticity as a secant module at σ_c = 0.3 max σ_c and the associated strain for the first loading were determined. Too few test results for AAC-masonry with light-weight mortar were

241

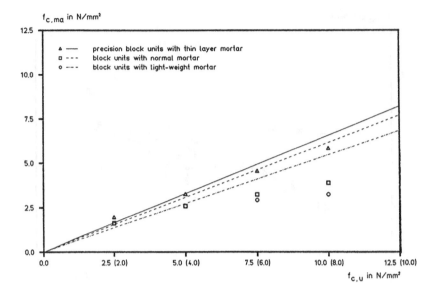

Fig.5 Compressive strength of masonry and masonry units
-evaluated curves and the values from German standard

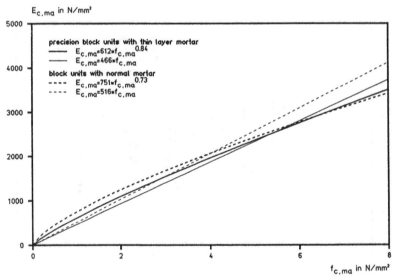

Fig.6 Relationship between the compressive E-Modul and compressive strength of masonry

available, and no analysis could be attempted. The selected formula was the relationship between the modulus of elasticity and the compressive strength of the masonry:

$$E_{c,ma} = a \cdot f_{c,ma}^{b}. \qquad (11)$$

As with the analysis of compressive strength results, statistical parameters were established. The relationship between the modulus of elasticity and the compressive strength of the masonry units was also evaluated. Since only very few test results (n = 7) were available, the analyzed results represent only a coarse approximation.

242

3.2 Moduli of Elasticity of AAC-Blocks and AAC-Masonry

Allowing for the restrictions noted above, the analysis of modulus of elasticity values for **AAC-blocks** revealed the following relationships:

$$E_{c,u} = 901 \cdot f_{c,u}^{0.68} \quad (R = 0.90) \qquad (12)$$

and

$$E_{c,u} = 475 \cdot f_{c,u} \quad (R = 0.82) \qquad (13)$$

The following regression equations were determined for **AAC-masonry** with thin-layer mortar (cf. Fig. 6):

$$E_{c,ma} = 612 \cdot f_{c,ma}^{0.84} \quad (R = 0.96) \qquad (14)$$

and

$$E_{c,ma} = 466 \cdot f_{c,ma} \quad (R = 0.95). \qquad (15)$$

For **AAC-masonry** with normal mortar the equations were (cf. Fig. 6):

$$E_{c,ma} = 751 \cdot f_{c,ma}^{0,73} \quad (R = 0.75) \qquad (16)$$

and

$$E_{c,ma} = 516 \cdot f_{c,ma} \quad (R = 0.69). \qquad (17)$$

The correlation between measured and calculated moduli of elasticity was good in all cases.

A comparison of the analyzed results with the modulus of elasticity values given in DIN 1053 Part 2 and by Schubert (1992) revealed distinctly lower analyzed values in some cases. On this basis, existing modulus of elasticity values may need to be reduced, especially in the range of higher masonry compressive strengths.

REFERENCES

Hilsdorf, H. 1965. Untersuchungen über die Grundlagen der Mauerwerksfestigkeit. Materialprüfungsamt für das Bauwesen, TU München, Bericht Nr. 40.

Mann, W. 1983. Druckfestigkeit von Mauerwerk; eine statistische Auswertung von Versuchsergebnissen in geschlossener Darstellung mit Hilfe von Potenzfunktionen. Berlin : Ernst & Sohn – In: Mauerwerk-Kalender, S. 687-699

Schubert, P. 1985. Zur Druckfestigkeit von Mauerwerk aus künstlichen Steinen. Institut für Bauforschung der RWTH Aachen (Hrsg.) Wiesbaden, Berlin : Bauverlag GmbH – In: Baustoffe '85.

Schubert, P. 1992. Eigenschaftswerte von Mauerwerk, Mauersteinen und Mauermörtel. Berlin: Ernst & Sohn – In: Mauerwerk-Kalender 17, S. 125-135.

Voellmy, A. 1957/58. Festigkeitskontrolle von Betonbelägen, Bd. 3. In: Betonstraßen-Jahrbuch, Fachverband Zement, Köln

Advances in Autoclaved Aerated Concrete, Wittmann (ed.)© 1992 Taylor & Francis. ISBN 90 5410 086 9

From the compressive strength of the units to the compressive strength of masonry built with thin layer mortar

K. Kirtschig
University of Hannover, Germany

ABSTRACT: The revised Eurocode 6 will include masonry structures which are built with thin layer mortar in walls and columns. One of the provisions for this is that the compressive strength of masonry using thin layer mortar is known. This provision is met. The required compressive strength can be described by the compressive strength of the masonry units and the compressive strength of the masonry mortar. There are, however, some details which have to be clarified. They are related to open questions in testing procedures for the units and for the mortar and the masonry itself. This contribution deals with some aspects of the solution of the different points.

1 INTRODUCTION

There exists a report EUR 9888 EN by the Commission of the European Communities entitled "Eurocode No 6, Common unified rules for masonry structures". This report is rather a first draft than a code and at present it is being revised after the Commission has received many comments. Some of the comments deal with the characteristic compressive strength of masonry. One special point here is that in the present draft there is no information on the characteristic compressive strength of masonry when thin layer mortar is used. The reason for this omission is that at the time when the Eurocode was drafted the use of thin layer mortar was not so common that it was neccessary to deal with it in detail. It was enough to mention it only in order not to exclude it. In the meantime the situation has changed: Most of the European countries use thin layer mortar or are at least interested in it. This means that the next edition of Eurocode 6 has to provide more information on masonry constructions using thin layer mortar. How this could be done is at present under discussion in the relevant committee of CEN (CEN TC 250/SC 6). This contribution will outline some details of the present work. Moreover it will also give some general aspects on the relationship between the characteristic compressive strength of masonry on one side and the compressive strength of the used materials, i. e. the units and the mortar, on the other side.

2 GENERAL INFORMATIONS ON THE TREATMENT OF MASONRY WITH THIN LAYER MORTAR

a) The code gives rules for thin layer mortars with compressive strengths of at least 5 N/mm² or more but the compressive strength is not supposed to be greater than 10 N/mm². The reason for these limitations is that there is experience only with thin layer mortar of these strengths. To be more precise there is at present experience only with compressive strength of approximately 10 N/mm². It is, however, generally believed that a 5 N/mm² thin layer mortar could also be included.

b) The units must be suitable for

use with thin layer mortars, i. e. they have to meet certain requirements for the tolerances of the dimensions of the units. There are at present requirements for tolerances in the drafts for calcium silicate and aerated concrete units of CEN TC 125. I. e. only those units may be used. However even if for other units requirements for the tolerances would be included in the CEN TC 125 specifications these units could not be used for this type of masonry, because there is not enough experience with these materials.

c) For the calculation of the characteristic compressive strength of masonry the normalised compressive strength of the units is not supposed to be greater than 40 N/mm². The reason for this restriction, again, is that there is no experience with units of higher strengths. Experience in this context especially means that there are no test results available on the compressive strength of masonry using units with higher compressive strengths.

d) The thickness of the masonry must be equal to the width or length of the units so that there is no longitudinal mortar joint through all or part of the length of the wall. The background of this requirement is that all the tests were carried out with wall specimens without longitudinal mortar joints and that it is known from tests using normal purpose mortar that a longitudinal mortar joint reduces the masonry strength by approximately 15 %.

3 COMPRESSIVE STRENGTH OF MASONRY

3.1 Some theoretical aspects

a) The compressive strength of masonry is usually given as a function of the compressive strenth of the units and the compressive strength of the mortar. Theoretical considerations show, however, that it would be more adequate either to use the tensile strengths of the materials, or even better, their deformation characteristics. As long, however, as the different properties are proportional to each other it is allowed to use the compressive strength. This is very useful because the determination of the compressive strength is comparably easy.

b) The purpose of the mortar is, first, of all to smoothout the inaccuracies of the dimensions of the units and, more specifically, of the bed faces. For this, a certain thickness of the joints is required. In case of general purpose mortar the required thickness is 10 to 20 mm and varies from country to country. Because the mortar has different deformation behaviour than the units, it influences the strength of the masonry. The influence may be positive or negative: If the mortar deforms more than the units the influence is negative, and if the mortar deforms less than the units the influence is positive in this respect that it at least does not reduce the masonry strength. The degree of reduction depends on the thickness of the bed joints: The thicker the bed joints the greater is the strength reduction. To minimise the reduction it is, therefore, useful to reduce the thickness of the bed joints as much as possible, i. e. to such a degree that the inaccuracies of the unit dimensions are still smoothened. This is, as is well known, one of the ideas behind the use of thin layer mortar (the second idea being to increase the thermal resistance, a property which this contribution will not deal with). Because the thickness of the bed joints depends on the inaccuracies of the dimensions of the units, it is necessary to minimise the inaccuracies. This can easily be done in case of aerated concrete units and also in case of calcium silicate units. This is the reason why for those types of units thin layer mortar is of special interest.

c) Aerated concrete units have a rather low strength, especially in comparison with the strength of the thin layer mortar. This means that, from the theoretical point of view, the compressive strength of the units will be best used when they are layed with thin layer mortar. Test results confirm the theoretical consideration.

3.2 The compressive strength of the units and its determination

This clause will be rather extensive because it will deal with some points about which the author feels that there are some misunderstandings or even that they are not understood at all.

a) The background of the shape factor

The strength of any material is a numerically fixed mechanical value, it is rather a figure which depends on the test conditions, the geometry of the specimens etc. In the case of the compressive strength of masonry units and more specifically of aerated concrete units, it means: Units of definetely the same material but of different dimensions have different compressive strengths. If, however, walls are built with these units of different dimensions one reaches the same masonry strength. Consequently it is not sufficient to define the compressive strength of the units by one figure but also by the dimensions of the units. As a result, one has units of exactly the same materials, with different compressive strengths, with which, however, one gets the same compressive strength for the masonry. This must be confusing not only for the user but also for those engineers and architects, who are not very much involved in this field. To get rid of this confusion the shape factor was created. The idea behind this shape factor was to correct the different strengths of units of different dimensions but of exactly the same materials in such a way that all of them get the same unit strength. The provision for this procedure was that one unit was to be chosen as a a standard unit with fixed dimensions and a shape factor equal to 1.00. As such a standard unit, a unit with a height of 100 mm and a width of 100 mm, regardless of the length is now proposed. This unit can be seen in the table where also the shape factors are given for other units as well. The following example may contribute to a better understanding:

Units of exactly the same material were tested in compression. The

Table. Shape factor to allow for the dimensions of the masonry unit

Height mm \ Width mm	90	100	150	200	≥250
50	0.80	0.75	0.70	-	-
65	0.90	0.80	0.75	0.70	0.65
100	1.05	1.00	0.95	0.85	0.75
150	1.25	1.20	1.10	1.00	0.95
200	1.40	1.35	1.30	1.20	1.05
≥250	1.50	1.40	1.35	1.25	1.20
Interpolation is permitted.					

dimensions of the units were: unit A with a width of 150 mm and a height of 150 mm; unit B with a width of 90 mm and a height of 150 mm. The results of the tested strengths were: 19.8 N/mm² for unit A and 17,4 N/mm² for unit B. The shape factors for the units taken from the table are 1.10 for unit A and 1.25 for unit B. The corrected strengths become 1.10 · 19.8 = 21.8 N/mm² and 1.25 · 17.4 = 21.8 N/mm². This strength is now considered to be the strength of a unit of the same material with a height and width of 100 mm for which the equation for the characteristic strength of masonry is valid (see clause 3.4). It is necessary to make a final remark regarding the values of the shape factor as they are given in the table. The values were determined by experiments carried out at several laboratories. Different types of units (different materials, solid and perforated units) were tested. The test results scatter in a wide range. This is very understandable for this sort of research work. The presented shape factors, therefore, are not very accurate and can show only certain tendencies. This, however, is sufficient enough for the solution of the problem in principle. Further research work could be done in this field. The chance to improve the figures are, however, very low. Germany has drawn years ago a conclusion which actually should still be favoured as a more adequate solution to the problem: Since the main influence on the strength is the heigth of the units, the shape factor as defined in the German specifications depends only on the height of the units and only two factors are introduced: 1.1 for units with a height between 175 mm and 239 mm and 1.2 for units with heights of more than 239 mm.

b) The influence of the moisture content of the units when being tested

The strength of masonry and masonry units depends also on the moisture content: The higher the moisture content the lesser the strength. It is generally accepted that in the design of masonry, a masonry strength belonging to an air-dry condition of masonry should be taken. That means that the units should be tested when they are air-dry. To define the term of "air-dry" raises some difficulties. In CEN and also in ISO it is, therefore, agreed to test the units when they are water saturated - this condition being easier to define - and that the obtained strength is afterwards converted to an air-dry condition. For this conversion multipliers are given. Another well - to - define conditioning is oven-dry. Units may also be tested in this condition and the strength obtained in this way may then be obtained strength is also converted to an air-dry condition by use multipliers. It may be mentioned that the figures of the multipliers are different for the different materials of the units. They are given in the drafts for the units specifications.

c) The normalised compressive strength of masonry units

The strength of the units which includes the shape factor and which also takes into account the influence of the moisture content (in air-dry condition) is called the normalised compressive strength. It is that strength which has to be used for the calculation of the characteristic compressive strength of masonry (see clause 3.4).

3.3 The compressive strength of mortar

It is not necessary to deal with this property, because there are, at present no essential changes in CEN

TC 125 and in CEN TC 250/SC 6. One point, however, must be mentioned. It is the way how the compressive strength of the mortar is tested: The specimens are formed in steel moulds so that the water is not absorbed while in a wall the water is absorbed by the units. This considerably affects the strength of the mortar. It must be stated that the tested mortar strength has not much to do with the strength of the mortar in a wall. It is, therefore, not surprising that the correlation between the mortar strength and the masonry strength is very bad. Though the problem is known, there is for the time being, not much interest to solve this problem internationally, for instance in the CEN specifications or ISO standard. The time obviously is not mature to deal with this problem and to work out a solution to it. Instead, people look for better correlations for the compressive strength of masonry depending on the strength of the units and the mortar without taking into account that a better correlation can not be found because testing procedures used to determine the mortar strength are simply wrong.

3.4 The compressive strength of masonry built with thin layer mortar

The dependency between the masonry strength and the strength of the used materials, i. e. the masonry units and the mortar can be described by an equation of the following type:

$$f_k = K \cdot f_b^{\alpha} \cdot f_m^{\beta} \qquad (N/mm^2)$$

where K, α and β are coefficients

f_b is the normalised compressive strength of the units (N/mm^2)

f_m is the compressive strength of the mortar (N/mm^2).

For general purpose mortar and for thin layer mortar the coefficients α and β may be taken to be $\alpha = 0.65$ and $\beta = 0.25$. It is generally agreed that in the design of masonry the characteristic strength of masonry will be used. In this case K may be

chosen as follows, provided that the
thickness of the masonry is equal to
the width or length of the units, so
that there is no longitudinal mortar
joint through all or part of the
length of the wall (only this case
is of interest here):

For general purpose mortar K = 0.60
for Group 1 masonry units and K =
0.55 for Group 2 units. In the case
of thin layer mortar the
corresponding figures are K = 0.70
and k = 0.60. This means that in the
case of thin layer mortar the
compressive strength is
approximately 15 % or 10 % higher
than in the case of general purpose
mortar. It must be added that the
main difference between Group 1 and
Group 2 units is the amount of
vertical holes. They are units less
than 25 % for Group 1 and units
between 25 and 55 % for Group 2.

Advances in Autoclaved Aerated Concrete, Wittmann (ed.) © 1992   Taylor & Francis. ISBN 90 5410 086 9

Mathematical model applied to AAC masonry with thin joints

P. Delmotte & J. D. Merlet
Centre Scientifique et Technique du Bâtiment, Paris, France

ABSTRACT

The knowledge of masonry strength under vertical loading, as well as other structural parameters, such as wall slenderness or excentricity loading, constitutes a fundamental element for design of loadbearing masonry walls.

Calculation is now an essential alternative to experiments.

The current use of AAC units is to execute masonry with thin layer mortars : the model commented here is capable of taking in account the benefit effect of low thickness of mortar joints in that case.

This model is based on hypotheses that it is a direct relation between strength and overall deformation.

It describes the behaviour of a couplet masonry unit and mortar joint under loading, and specially takes in account the overall deformation of couplet.

We are analysing successively the effect of :

. comparative dimensions of the two constituants, masonry units and mortar joints,

. their characteristics of strength and deformation capacity,

. behaviour of masonry under loading during tests and specially surrounding joints.

Others questions in relation to masonry design in case of horizontally loading in and out of plane for AAC masonry are finally taken up in wiew to contribute to an as complete as possible model code.

1 FOREWORD

Knowledge of the masonry strength (f_M) under vertical loading, as well as of the effect of the other structural parameters such as slenderness and eccentricity of the loads applied constitutes an fundamental element in design of loadbearing masonry walls.
Since the experimental determination of f_M cannot be obtained easily because of the size of wallets, of the great care which their construction needs, of the great number of required test units resulting from the high dispersion of test results generally observed and of the time needed for mortar hardening,

the alternative of strength determination by calculation is appeared to us as an evidence.
But as opposed to other building techniques for which the theoretical values of the characteristic strength of the materials used may easily be calculated with an acceptable range of approximation, calculation becomes more difficult in the case of masonry for several reasons the principal of which are the following :
- masonry is a composite material ; its strength is a fonction not only of its components considered separately, but also of certain combined effects (for instance hooping-counter hooping effects between units and mortar

joint) ; failure phenomena may thus be different from those observed on units alone
- depending on the materials used to manufacture the units, values of failure strength have their own specific dispersions which are usualy very high
- "intrinsic" mechanical characteristics, i.e. values which have the physical meaning which their use in behaviour models would require, cannot be obtained accurately owing to the conventional character of experimental modes used to determine f_b and f_m

Our aim in this paper is to provide in the particular case of AAC units associated to thin layer mortar a simple formulation representative enough, after analysis of the phenomenon observed during testing and of the supplementary information obtained from the "finite element" approach.

Such a model is capable to taking in account the benefit effect of low thickness of joints allowed by using thin layer mortars in that case

2 VERTICALLY IN PLANE LOADBEARING WALLS

2.1 General

The aim of calculation is to establish relation between unit test strength and masonry strength, then between masonry strength and wall strength

f test $\longrightarrow$ f masonry $\longrightarrow$ f wall

The model is based on hypothesis that it is a direct relation ship between strength and overall deformation. It describes the behaviour of a couplet masonry unit and mortar joint under loading and specially takes in account the overall deformation of that couplet

2.2 Overall deformability of a unit- mortar couplet

If both unit and mortar are of the same width, the mean stress in the assembly is equal to the following:

$$\sigma_0 = E_b . \Delta h/h \qquad (1)$$
or to $\sigma_0 = E_m \Delta e/e \qquad (2)$
or even to $\sigma_0 = E_M . \Delta (h+e)/(h+e) \qquad (3)$

In addition, we have :
$\Delta (h+e) = \Delta h + \Delta e \qquad (4)$

The following may be derived from (1),(2),(3) and (4):
$(h+e)/E_M = h/E_b + e/E_m \qquad (5)$

Finally, we obtain:
$E_M = E_b . E_m . (h+e)/(h.E_m + e.E_b) \qquad (6)$

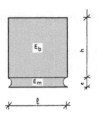

Fig. 1 - Unit-mortar couplet

2.3 Hypothesis linking strength to deformability

Suppose the behaviour of the constituent materials is both elastic and linear. Although this hypothesis generally does not prove true near failure, it nevertheless remains acceptable for AAC masonry whose failure behaviour is of the brittle type.

The following relations may then be written:

$$f_b = E_b . \epsilon_{bu} \qquad (7)$$
$$f_m = E_m . \epsilon_{mu} \qquad (8)$$

Because of the linear behaviour of both unit and mortar, the same then applies to the assembly, which results in the following:
$$f_M = E_M . \epsilon_{Mu} \qquad (9)$$

E_M is given by (6) and ϵ_{Mu} by the same reasoning as the one referred to above, as follows:

$$\epsilon_{Mu} = (h.\epsilon_{bu} + e.\epsilon_{mu}) / (h+e) \qquad (10)$$

2.4 General formulation

The following may be deduced from (6), (7), (8), (9) and (10), supposing e lower than h:

$$f_M = \frac{h.f_b.f_m.\epsilon_{bu}}{h.f_m.\epsilon_{bu} + e.f_b.\epsilon_{mu}}$$

and

$$f_M = f_b . \frac{1}{1 + \dfrac{e}{h} . \dfrac{\epsilon_{bu}}{\epsilon_{mu}} . f_m} \qquad (11)$$

Consequently, the determination of f_M also depends on the knowledge of ϵ_{bu} and ϵ_{mu} values, or E_b and E_m.

The diagrams herewith enclosed (n°2 and 3) show several applications of the formula (11).

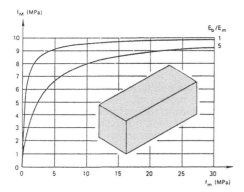

Fig. 2
F_M in dependance of E_b/E_m ratio
$f_b = 10 MPa$; $h = 19 cm$; $e = 1 cm$

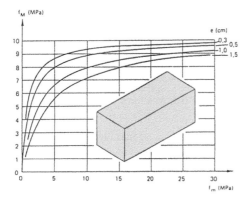

Fig. 3
f_M in dependance of f_m and mortar joint thickness
$f_b = 10 MPa$; $E_b/E_m = 5$

NOTE 1 : When subjected to vertical loads, masonry may be the seat of horizontal tensile stress in the unit near the mortar joint, which may lead to a premature splitting failure of the masonry (as is the case for instance with masonry made up of AAC units). These stresses results from hooping-counterhooping effects between unit and mortar.
Given ϵ_{mh} and ϵ_{bh} ultimate strains of mortar and unit in horizontally direction, the horizontal elongation ϵ_h if measured very closed by to the joint can be assessed as the mean value of ϵ_{mh} and ϵ_{bh}

$$\epsilon_h = \frac{\sigma_o}{2} \cdot \frac{\nu m}{Em} + \frac{\sigma_o}{2} \cdot \frac{\nu b}{Eb} \qquad (12)$$

If $\epsilon_{mh} > \epsilon_{bh}$ interference with strain in unit $[\epsilon_h - \epsilon_{bh}]$ will lead to a resultant tensile stress in unit ; this stress can be approximated by the following formula :

$$\sigma_{tension} = (\lambda + 2\mu) \cdot (\epsilon_h - \epsilon_{bh}) + \epsilon_b \qquad (13)$$

Note 2 : With the utmost rigour, the characteristics of the mortar joint are not homogeneous over the width of the masonry because of the differential desiccations between the centre and the external sides. This phenomenon, among others, as for instance the water absorption by succion from mortar into unit, may have caused the great differences noticed between the mortar characteristics, measured on the one hand by using a test specimen, on the other hand indirectly on the masonry. Thus in particular case of thin layer mortars, additives in mortars are capable to soften that difference

2.5- Specific formulation for AAC units and thin layer mortars

In accordance with l/h ratio and testing process to determine f_b, two modes of failure are possible:
1- tensile splitting failure (Poisson effect)
$$f_b = E_b \cdot \epsilon_{tu}/ b$$

2- crushing failure of the unit
$$f_b = E_b \cdot \epsilon_{bu}$$

If the joint is very thin and if the mortar is resistant, the behaviour of the masonry is ideally very similar to the behaviour of a monolithic wall. If the joint is thicker, it falls to ruin from outside to inside (see diagram below) as the load increases. Failure can occur either by crushing of the mortar in the case of a low-strength mortar and a thick joint, or more generally by vertical cracking of the units from both sides of the resistant core (see Fig. 5).

If the modes of failure observed on masonry and unit are similar, f_M can be assessed from the formula (11) which is to be adjusted by a reduction coefficient j so as to take into account the alteration of width of the joints under loading.
We may take for instance j = S_r/S_b = (l-e)/l;
$\epsilon_{equ} = \epsilon_{bu}$

If the two modes of failure (unit and masonry) are different, which is the general case, f_M cannot be directly deduced from f_b without any correction. the strength of the unit is to be adjusted as if the testing method did not induce hooping effects, it means to take

a) splitting failure   b) crushing failure

FiG. 4 units after testing

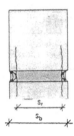

Fig. 5 Vertical cracking in units

$f_b = \min(E_b.\epsilon_{bu}, E_b.\epsilon_{tu}/ _b)$
instead of measured strength.
f_M can be assessed from the formula (11) adjusted by j, and by replacing ϵ_{bu} by $\epsilon_{equ} = \min(\epsilon_{bu}, \epsilon_{tu}/m_b)$.
in the two cases, we obtain:

$$f_M = f_b . \frac{1\text{-}e}{1} . \frac{1}{1 + \dfrac{e.f_b.\epsilon_{mu}}{h.f_m.\epsilon_{equ}}} \qquad (14)$$

In addition to this, with regard to masonry made up of units and mortar of different deformability, it is advisable to make sure that the quantity given in (13) does not exceed the maximum tensile stress which the constituent material of the unit is subjected to. This condition must be verified:

$$f_M < \frac{E_b.\epsilon_{tu}.}{(\lambda+2\mu).(\nu_m/E_m - \nu_b/E_b) + \lambda /E_b} \qquad (15)$$

Note 3: As for masonry for which the unit-mortar hooping-counterhooping effects are significant, confinement of the mortar results in an increase in the overall mechanical capacity of the masonry.
Formula (11) does not directly account for this phenomenon; still it shows that when e tends towards zero, f_M tends towards f_b, whatever the strength of the mortar. The masonry then behaves like a monolithic structure.

Note 4: Limit e -> 0 remains theoretical and the thickness of the joint shall in practice be higher than the size of the coarsest aggregate which makes up the joint mortar.

In the case of AAC masonry and units have generallly quite different modes of failure.
Masonry mode of failure is splitting perpendicular to the bed joints but, for shorter specimens i.e. units, it becomes diagonal shear, yielding higher apparent masonry strength. When the AAC unit is tested flat between the steel heads of a testing machine, its strength is due to a shear failure and consequently is higher than a tensile mode of failure would yield.
So we can propose the following formula :

$$f_M = Eb \quad . \frac{\epsilon t}{\nu} \quad . \frac{1\text{-}e}{1} \quad . \frac{1}{1+ \dfrac{e}{h} \dfrac{Eb}{Em}} \qquad (16)$$

$[Eb \dfrac{\epsilon t}{\nu} \text{ instead of } f_b]$

finally, because $\dfrac{e.Eb}{h.Em}$ ratio is very small compared to 1, whe have $f_M = \quad Eb \dfrac{\epsilon t}{\nu}$

(Experimentally, we generally obtain $f_M = 0.6 f_b$)

NOTE 5 : Moreover, the compressive elastic modulus of the mortar is probably about five that of the AAC unit. So, under compression the lateral extension of the mortar does not create a tensile strain in the unit

NOTE 6 : Even if the cube compressive strength of the mortar is exceeded, due to friction, the mortar placed between the mortar-unit interface remains confined within the joint : Paradoxically experiments on joints merely filled with sond may produce stronger masonry systems than those filled with mortar.

2.6 Wall Strength

2.6.1 General

The intended approach is to consider such a masonry wall like a monolithic construction because of :
- the very low thickness of joints
- the fact that joints don't constitute weak points in the masonry: higher characteristics mortar, adherence between units and mortar higher than tensile strength of AAC,...

However it is necessary to ensure adherence on the whole surface of unit (quality of mortar, care taken from bricklayer to their work,...)

2.6.2 Particular points

The calculation shall be organised
. to take in account the brittleness of AAC under concentrated loads (floor support for instance)
. to study influence of thermal and humidity criteria on overall behaviour of wall : initial bending for instance and thus an half heigth eccentricity different from zero in relation to ΔT, α and of course the thickness of wall
. to study influence of stiffening walls on mechanical slenderness.
. to study the consequence of stickness or not in vertical joints.

2.6.3 Buckling of walls - Principle of the calculation method

It is proposed to use CALCO model, briefly described hereafter, by means a graphic inter pretation of phenomena

All the theoretical calculation methods are based on the resolution of the following differential equation:

$$\frac{d^2e}{dx^2} = \frac{N.e}{E.I}$$

with e: eccentricity of load N in section of abscissa x
E: modulus of elasticity of the material
I: moment of inertia of the section

This equation cannot be integrated easily since EI product varies with x. Thus, I varies with x because of the cracking and E varies with x if the material is not both elastic and linear (see figure 6)

The principle of the calculation method consists in evading the difficulty by giving the differential equation its geometrical formulation as follows: (see figure 7)

$$\frac{d^2e}{dx^2} = \frac{\epsilon_2 - \epsilon_1}{h} = \frac{\Delta\epsilon}{h}$$

with: ϵ_1 and ϵ_2: unit strains of the extreme fibres
h: height of the section at abscissa x

To make things easier, let's consider for

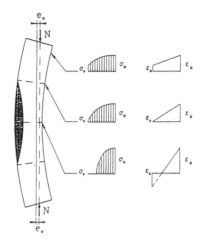

Fig. 6 - $\sigma-\epsilon$ diagrams at different levels of specimen

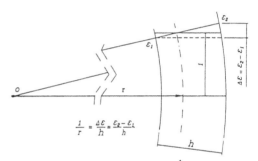

Fig. 7 - Relation ship between $\frac{1}{r}$ and ϵ_1 and ϵ_2

example a rectangular blank wall and a constituent material which is not resistant to tensile stress. The same reasoning applies to the other cases anyway.

Taking the law of behaviour $\sigma(e)$ of the material into account enables the variations of e/h to be represented on a diagram as a function of both the reduced eccentricity e/h and the mean stress applied. The following family of curves may then be obtained (see figure 8).

OBB'DD' envelope curve represents the set of points which correspond to the crushing failure of the material.
Curve ab (represented by a dotted line) is the locus of points corresponding to a zero strain on the opposite side of the loading, and thus corresponding to the probability of an incipient crack appearing in the section.
It goes without saying that the course of all

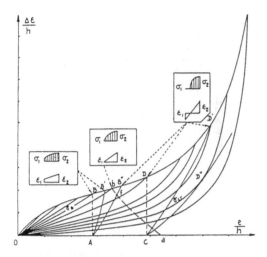

Fig. 8 - Relationship between $\frac{\Delta \epsilon}{h}$ and $\frac{e}{h}$ for a wall the material of which have tensile strength equal to O

these curves depends on both the geometry of the section considered and the law of behaviour of the constituent material.

The diagram makes it possible for us to understand the different probabilities of a wall failure, provided that the following six cases are examined:.

1- k equal to zero, e low (point A)
The representative point of the state of the median section moves along the vertical segment AB, since the increase of load does not lead to higher eccentricity in this section.

2- k rather high, e low (point A)

The representative point of the median section state describes the curve arc AB'. The increase of eccentricity is due to bending deflection. Failure occurs under a mean stress lower than in 1; the median section is still not cracked.

3- k very high, e low (point A)

Failure in B' will occur before cracking of the median section.

4- k equal to zero, e high (point C)

The representative point of the median section state describes the vertical segment of line CD. Cracking appears either right from the beginning of the loading, or during loading, depending on the position of C with regard to a.

5- k average, e high (point C)

The representative point of the median section state describes the path CD', and failure in D' occurs under a mean stress lower than in 4.

6- In the case of sufficient slenderness, the curve representing the evolution of the median section state may then be tangent to a curve $\sigma = \sigma_{cr}$ in D".
Any state that is located above point D" can in no way be taken into consideration since it would correspond to a stress lower than σ_{cr}. Failure occurs under instability under the critical stress σ_{cr}.

2.6.4 Calculation

The buckling function is to be determined by numerical simulations issued from probalistic data which are defined and justified from results of quality control, geometrical characteristics, behaviour laws, applied eccentricitries,...)

2.6.5 Validation

Calculation is to be valided by crosscheck testing : an experimental programm of work is scheduled for this purpose by testing wallets and storey heigth walls under concentric and excentric loadding.

3 CONCLUSIONS

Major development is given for design in case of vertically in plane loadbearing walls but others questions in relation to masonry design in case of horizontally loading in and out of plane for AAC masonry should also be finally taken up in wiew to contribute to an as complete as possible model code

NOTATIONS

f_M : strength of masonry;
f_b : strength of the unit;
f_m : strength of mortar;
E_M: modulus of elasticity of the masonry; (1)
E_b : modulus of elasticity of the unit;   (1)
E_m: modulus of elasticity of mortar; (1)
ν_b : Poisson ratio of the unit;
ν,μ : Lame coefficients of the unit;
ν_m : Poisson ratio of mortar;
ϵ_{Mu} : ultimate strain of masonry;
ϵ_{bu} : ultimate strain of the material of the unit;

ϵ_{bh} : ultimate strain of the unit in the horizontal direction;
ϵ_{tu} : ultimate strain of the unit (elongation);
ϵ_{mu} : ultimate strain of mortar;
ϵ_{equ}: equivalent ultimate strain;
$\sigma o'$: mean net stress;
ϵ_{bh} : ultimate strain of mortar in the horizontal direction
l : width of the unit;
h : height of the unit;
e : thickness of mortar joint;

REFERENCES

. DELMOTTE P., LUGEZ J.,.& POUSHANCHI M.,
    Flambement des murs en maçonnerie .Elaboration d'un modèle pour sections de forme quelconque. Cahiers du CSTB N° 2540 Liv. 325 dec. 1991

. DELMOTTE P., LUGEZ J., & J-D. MERLET
    Résistance des maçonneries sous charges verticales. Proposition d'un modèle simplifié de calcul. Cahiers du CSTB n° 2553 Liv. 326 janv.fev. 1992

(1) tangent modulus at zero

Advances in Autoclaved Aerated Concrete, Wittmann (ed.) © 1992 *Taylor & Francis. ISBN 90 5410 086 9*

Shear capacity of mortar joints in masonry with AAC-precision blocks

F. Fath
YTONG AG, R+D-Centre, Schrobenhausen, Germany

ABSTRACT: The mortar layer of a masonry brickwall is always an area of disturbance and has significant influence on the loadbearing capacity of a masonry structure. To achieve the optimum loadbearing performance the mortar has to match perfectly the properties of the blocks. The aim is to achieve a homogeneous and monolithic wall structure whose behaviour can be determined by the tools of static equilibrium calculations.

The bond shear capacity of the mortar layer in masonry under bending load (e.g. wind) or under shear loading (e.g. earthquake) is an important property. At the R+D Center of the YTONG AG tests have been carried out to adjust the mortar properties to the block. The test results showed that the thin layer mortar as defined by the German Standard DIN meets the requirements, i.e. its properties are very similar to the AAC blocks.

1. GENERAL PROBLEM

The loadbearing capacity of masonry is mainly determined by the properties of its components, i.e. blocks and mortar. The compressive strength of blocks as well as their deformation behaviour is well known and was the subject of many investigations. Compared to the blocks less is known about the mortar.

The aim of all investigations must be either to increase the loadbearing capacity or to improve the crack-resistance. In additiion to that masonry must become more predictible making the empirical formulas unnecesary. To achieve this aim, masonry must become more homogeneous and therfore almost monolithic. This can only happen when the properties of mortar and block are almost the same.

In the laboratory it is quite easy to make a mortar which matches perfectly a block, i.e. it acts like a block substitute. In practice this requirement poses many problemes. There is an enormous amount of diffrent materials, geometries and strengths on the market and it is impossible to make a special mortar for each case.

National standard have taken this into account and as a compromise ther are, for example, six different mortar classes known in the German Standard DIN. The range of compressive strengths spreads from 5 N/mm² to 20 N/mm². Each mortar, which is availible on the market must beclassified in these groups so that the masonry strength can be calculated from the stone strength and the mortar class.

In this discussion, precision blocks play a special part. Due to the very thin mortar layer (< 3 mm) and the large blocks the area of disturbance between the blocks is much less than in "normal" brickwork masonry. Reducing the amount of layers in a masonry wall increases the loadbearing capacity as proved in many tests. Consequently, the DIN Standard allows up to 40 % higher stresses in masonry made of AAC precision blocks an thin bed mortar compared to "normal" mortar.

This confirms the proposition that ideal masonry should be almost homogeneous and monolithic.

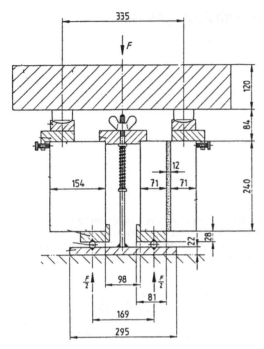

Fig. 1: Test apparatus for determining shear-bond-strength of mortar joints in masonry

Table 1: Test results of shear bond capacity of monolithic AAC-blocks when tested according to DIN 18555 part 5

Strengthclass G2				
Width	Length	Load ult.	fs.	Statistics
mm	mm	N	N/mm²	
116	242	30600	0,55	Mean Val
115	242	30400	0,55	0,623
117	243	40200	0,71	StdDev.
115	241	33800	0,61	0,082
116	243	40000	0,71	

Strengthclass G4				
Width	Length	Load ult.	fs	Statistics
mm	mm	N	N/mm²	
114	242	43800	0,79	Mean-Val.
115	242	49000	0,88	0,956
114	243	53000	0,96	StdDev.
115	243	58200	1,04	0,124
114	242	61000	1,11	

Strengthclass G6				
Width	Length	Load ult.	fs	Statistics
mm	mm	N	N/mm²	
116	242	76200	1,36	Mean-Val.
115	240	50200	0,91	1,196
115	243	71000	1,27	Std.Dev.
115	242	69400	1,25	0,197

Strengthclass G8				
Width	Length	Load ult.	fs	Statistics
mm	mm	N	N/mm²	
115	240	78000	1,41	Mean-Val.
115	242	69600	1,25	1,494
115	240	81000	1,47	Std.Dev.
114	242	92200	1,67	0,179
115	240	92000	1,67	

2. TESTCONCEPTION

There are not many results available regarding the shear strength of masonry and of the shear strength of single blocks. The basic idea was to test monolithic blocks and to compare the results with bond shear tests as are carried out for quality control purposes.

First of all the bond-shear-strength of monolitic blocks was tested according to DIN 18 555. Following this, tests on block-mortar-block sandwiches were analysed. If the bond-shear capacity of the sandwich is close to that of the monolithic block it can be assumed that the disturbance in a wall has reached a minimum which leads to better properties of the wall.

3. TESTING OF MONOLITHIC AAC BLOCKS

To define the requirements the question has to be answered "what must the shear-bond-strength of mortar be so that the behaviour of masonry is not unfavourably affected?"

The German Standard DIN 18555 Part 5 describes the testing method for determining shear-bond strength (f_s) of a block-mortar-block sandwich. The load acts diagonally at an angle of about 17° and the value of f_s is calculated by

$$f_s = F_u / 2*A \qquad (1)$$

where
$f_s =$ shear bond strength
$F_u =$ load
$A =$ cross-section of specimen

In practical application of AAC masonry

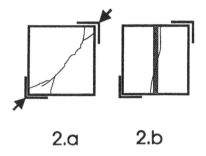

2.a   2.b

Figure 2: Cracking pattern of specimen when subjected to a shear bond testing according to DIN 18555. 2a: monolithic specimen; 2b: block-mortar-sandwich.

with compressive strength between 2 and 10 N/mm² only one type of mortar should be used. This avoids different types of mortar being mixed up on site.

Tests were carried out with the apparatus in. fig. 1 taking a monolithic AAC-block as a specimen instead of the normal sandwich. The test results are given in table 1.

It can be seen that the values depend on the strength class of the blocks as expected. The values increase with the compressive strength.

f_s = 0,62 N/mm² for G2
f_s = 0,96 N/mm² for G4
f_s = 1,20 N/mm² for G6
f_s = 1,49 N/mm² for G8

4.   EXPERIMENTAL VALUES OF BLOCK-MORTAR-SANDWICH

This test is part of the YTONG Quality controll in Germany and so it was not neccessary to perform extra test but only to analyze availible results. This analyses yielded mean-values as follows

G2 + thin layer mortar 0,6 N/mm²
G4 + thin layer mortar 1,0 N/mm²
G6 + thin layer mortar 1,2 N/mm²

Test results for G8 were not available, because this type is not yet on the market.

5. ANALYSIS AND CONCLUSIONS

Comparison of test results between monolithic AAC blocks and block-mortar sandwich specimen prove the similarity in shear bond capacity. This means that the mortar joint does not weaken the blockwork masonry.

However, taking a look at the crack pattern of the specimen it can be seen that there are differences in loadbearing behaviour. In the sandwich the crack is almost close to the joint (fig.2b) while in the monolithic block the crack follows the direction of the diagonally acting force (fig. 2a). This proves that the free deformation is prevented by the mortar layer wich ist much stiffer than the AAC block.

Further investigations have to be made to find out how far masonry can be improved when the mortar is better matched to the properties of the blocks. Tests are in preparation at the YTONG R+D Center concerning the earthquake resistance of masonry and thus also dealing with the shear strength of walls.

6. SUMMARY

Tests were carried out according to German Standard DIN 18555 Part 5 to determine the bond-shear strength of pure AAC and AACd blocks combined with mortar to simulate a masonry joint. The results show that regarding the loadbearing capacity the plain AAC specimen and the block-mortar sandwich are almost equal.

The different crack patterns, diagonal in case of monolithic AAC and parallel to the joint in the case of the sandwich, leave room for further investigations. From the crack pattern it is evident that there is still no continuous behaviour in the joint area.

At the YTONG R+D Center, tests are in preparation to investigate the earthquake resistance of AAC-masonry. These experiments will also contain some tests concerning shear transfer in mortar joints.

Advances in Autoclaved Aerated Concrete, Wittmann (ed.) © 1992   Taylor & Francis. ISBN 90 5410 086 9

The recent activities of CEN TC 125 – Masonry

Norman J. Bright
CEN TC 125, Kingsway Group plc, UK

ABSTRACT

CEN Technical Committee TC 125 is responsible for the preparation of harmonised European Standards for masonry products and the associated test methods. The Committee structure is explained. By delegating the technical drafting to Working Groups, which further subdivided the work which was then undertaken by Task Groups, good progress has been made. Concern is expressed about the length of time being taken by editing and translation.

INTRODUCTION

The Construction Products Directive (CPD) is a new approach directive issued by the European Commission. The Directive identifies six essential requirements for construction products and these are given concrete form in six Interpretative Documents which unfortunately have not yet been issued. The Interpretative Documents will create the necessary links between the essential requirements and the mandates for harmonised European Standards. The planning, drafting and adoption of Harmonised European Standards for construction products has been delegated to the European Committee for Standardisation (CEN). The procedures should respect the following principles:

- ° openness and transparency
- ° consensus
- ° National commitment
- ° technical coherence at the European and National level

When all of the documentation (particularly the Interpretative Documents) is issued it will be possible to mark construction products with the CE mark to show that they comply with the six Essential Requirements and are therefore suitable to be placed on the market.

CEN is the European Standards Organisation for both European Community Countries (EC) and European Free Trade Countries (EFTA). Normally the National Standards Organisation, e.g. DIN, BSI, UNI etc. is the member of CEN. Recently seven other Standards Organisations from outside EC and EFTA have joined as affiliated members. These countries are Poland, Hungary, Czechoslovakia, Cyprus, Turkey, Romania and Bulgaria. These countries now have the right to attend meetings but have no voting rights. Two of these countries, Poland and Hungary, have applied for Observer status with TC 125.

CEN TC 125

In mid 1988, CEN set up technical committee TC 125 to prepare harmonised European Standards for masonry units, ancillary components (e.g. lintels), mortars, renders and the necessary test methods for these products and for masonry assemblages. At its inaugural meeting in June 1988, CEN TC 125 created four working groups (WGs).

WG1	–	Masonry Units
WG2	–	Mortars and Renders
WG3	–	Ancillary Components
WG4	–	Test Methods.

Under each Working Group are a number of Task Groups (TG), each of which undertook the drafting of one or more Standards. Under WG1, for example, TG4 prepared the draft Standard for Autoclaved Aerated Concrete Masonry Units. In addition it was necessary for TC 125 to establish liaison with other relevant technical committees in CEN, e.g. those dealing with acoustics, energy conservation, aggregates for concrete etc. There is also liaison with other CEN TCs such as CEN TC 241 (responsible for gypsum and gypsum based products and plastering); CEN TC 250 SC6 - Structural Eurocode for Masonry (EC6) and with the International Standards Organisation Technical Committee ISO TC 179 - Masonry etc.

TC 125 also set up a Coordinating Group (CG) comprising the Working Group Convenors and one Representative of each country which did not hold a convenorship. The Coordinating Group subsequently formed two Task Groups, CG TG1 and CG TG2.

Various European Trade Associations have liaison with TC 125 and are entitled to have a representative at meetings of the Committee. One such association is the European Autoclaved Aerated Concrete Association (EAACA).

Position in 1992

The drafting of the Standards has taken longer than originally expected. The original timetable was unrealistic but very good work by a large number of people from most of the member countries, funded voluntarily by industry and other interests, enabled approximately sixty draft Standards to be approved by TC 125 at a meeting in Madrid in September 1990. They were approved to be issued as prENs (draft European Standards), subject to editing and translation into the three languages of CEN. [When a draft Standard is issued as a prEN it means it is issued to the National Standards Organisations for public enquiry.] This editing and translation process has taken until September 1992 to complete. It had been decided by TC 125 that all the available drafts should be issued as prENs concurrently because of the desirability of seeing the whole picture. Due to the length of time taken for the editing and translation process it has become necessary to sub-divide the drafts into groups and give priority as follows:

Group 1 : Masonry units and associated test methods
Group 2 : Ancillary components and associated test methods
Group 3 : Mortar and associated test methods
Group 4 : Masonry tests.

It was further decided that the lack of a small number of test method drafts should not be allowed to delay a whole group of prENs from going forward. Consequently the draft Standards for masonry units, and eighteen of their associated test methods, were issued in June 1992 and the comments should be received by the TC 125 Secretariat by December 1992. Three further test methods for masonry units, resulting from late requests to Working Group 4, are following after consideration by the members of TC 125 for approval to issue as prENs. The majority of the remainder of the drafts are expected to be issued as prENs at the end of September 1992 with comments due by the end of March 1993.

The Secretary of TC 125 has issued a letter to all the members of CEN with a pro-forma reply sheet to standardise the comment format. This will facilitate the huge task of collating the comments by the TC 125 Secretariat. It is also requested that replies clearly distinguish between major points of principle, normal points of technical correctness and editorial points.

At the end of the prEN period, after the comments have been collated, they will be considered by the relevant Working Group Convenors. They will, in conjunction with members of the working groups if necessary make proposals on how to modify the drafts to take account of the comments. It is hoped then that the drafts can be progressed to a meeting of TC 125 in May 1994 for approval to issue for voting. If this is successful the drafts should be issued for voting in Autumn 1994; ratified for publication by CEN in Spring 1995 and implemented towards the end of 1995.

Energy economy

In addition to the draft Standards referred to above a document "Masonry and masonry products - Methods for determining declared and design thermal values" is being drafted and is expected to be ready for issue as a prEN by about the end of 1992. Liaison with CEN TC 89

- Thermal Performance is the responsibility of the Convenor of CEN TC 125 CG TG1.

Structural Eurocode No.6 and other masonry application documents

CEN TC 125 is required to provide the test methods to support Eurocode No.6. These test methods include tests for the strength of masonry in compression, flexure, shear etc. Some of the test methods from RILEM TC 76-LUM have been used as a basis for these.

Liaison with CEN TC 250 SC6 is the responsibility of the Convenor of CEN TC 125 CG TG2. Together with his Task Group he will advise TC 250 SC6 on suitable application clauses relevant to structural masonry. In addition TC 125 CG TG2 will prepare under the guidance of the Coordinating Group the necessary TC 125 application documents which are not the responsibility of other Committees.

External rendering

Following a request from CEN and a meeting between the Chairmen of TC 125 and TC 241 a new Working Group 5 of TC 125 is being set up to draft a document on the application of external rendering. TC 241 is setting up a corresponding Working Group for plastering.

AAC masonry units and test methods

The prEN 771-4 - 'Autoclaved Aerated Concrete Masonry Units' has been drafted by experts from the member countries and should be satisfactory subject to a few detail corrections.

One of the problems at the outset was the question of the maximum size of a masonry unit. In the end it was decided to not give maximum dimensions for masonry units. However, storey height panels are excluded from TC 125 and are covered by TC 177. Although generally masonry units are simple regular parallelepipeds the Standard also allows tongue and grooves and depressions or indentations for grip holds etc. Another point to note is that the raw materials may not all comply with other CEN Standards where a suitable Standard does not exist.

The test methods are largely based on the test methods developed in RILEM AAC Committees and should not present any major problems.

A principle which it is hoped to maintain is that no country should be prevented from placing on the market a unit which is already in accordance with a National Standard. For all masonry units, probably the single most important property is the compressive strength and for materials such as AAC, which are also used for their thermal insulation properties, the density is also important. One aspect of compressive strength testing which has caused some difficulty in CEN TC 125 is the concept of "normalised" compressive strength. In addition some masonry units, including some AAC units are too large to test in a standard masonry unit compression test machine. The results are sensitive to the test method and it is important that the test report gives the actual test results, including if carried out on whole or part units, regardless of whether the report also includes any "normalisation" calculation.

The 'normalised' strength is obtained by applying a multiplier known as the shape factor to the air-dry compressive strength of masonry units.

The preparation and conditioning of the test specimens are recognised as being essential. The units may be accurately ground flat on the compression faces or thin cement mortar capping may be used. It has also been agreed for AAC that the units should be conditioned 'air dry' to a moisture content of $6\% \pm 2\%$ by weight. However, when testing capped samples it is difficult to check the moisture content of the sample. The draft as it stands allows a number of options for testing.

In addition to the compressive strength and density the dimensions of the units together with the dimensional tolerances will be included in the minimum descriptions of units. The allowable tolerances for dimensions depend on whether the units are intended for construction with normal thickness joints or thin layer joints.

Attestation of conformity

The Construction Products Directive sets out alternative levels of attestation of conformity with technical specifications such as a Harmonised European Standard. There has been considerable confusion within CEN technical committees on the subject of attestation of conformity and factory production control. Initially the CEN committees were asked to make proposals for levels of attestation for the products within their mandate. Within TC 125 there were very widely held views. Ultimately the CEN technical committees, including TC 125, were asked to do nothing on attestation of conformity until advice has been issued from CEN Central Secretariat. This advice is expected to be issued in September 1992 and it is hoped that the guidance will include assessment compliance with a Standard. Consequently, at the request of CEN Central Secretariat, there is nothing in the present TC 125 drafts on these subjects. TC 125 has been advised by CEN that the draft prENs should be issued without covering these subjects but they will be covered later.

Crucial to the assessment of compliance with the Standard is the reliability of the test methods. The reliability includes such matters as the accuracy with which the results can be reported and apart from the inherent accuracy of the measurement brings in the reproducibility of the methods. There is no proposal to systematically obtain precision data from the proposed test methods. While it is felt that this may not be important for the Structural Design of Masonry using EC6, it may be more important in case of any dispute concerning compliance with a Standard. For example, compressive strength can be tested in a number of ways. It is a fundamental concept that products being placed on the market will be tested by the same method so that, for example, the stated compressive strength will for a particular product be the same regardless of the country or laboratory where the test is undertaken.

COMMENT

From the foregoing summary of the current position in CEN TC 125 it can be seen that tremendous progress has been made. This has only been possible due to the good spirit of all of those participating. The main purpose of harmonised European Standards is to eliminate technical barriers to trade. When used together with application documents and Eurocode No.6 they will improve understanding between manufacturers, designers, constructors and regulators on the performance and use of masonry and thus encourage competence, growth and prosperity.

To enable this to happen some things

outside the scope of TC 125 are necessary to enable CE marks to be issued for masonry products produced in accordance with TC 125 Standards. Progress is being made on these concurrently with the writing of TC 125 Standards although they should have been in place earlier.

Reaction to fire and fire resistance testing are examples of problem areas in the preparation of CEN Standards and have delayed the publication of the Interpretative Document for Fire. As far as masonry is concerned, including autoclaved aerated concrete, this should not be allowed to cause any delay. Masonry units and mortars and renders are clearly non-combustible and this should be recognised. Until fire resistance grade test methods are published "default" grades tabulated in Eurocode No.6 can be used.

CONCLUSIONS

As this paper forms an interim summary of recent activities which are continuing, it is not sensible to try to draw firm conclusions. The guidance and 'mandates' from CEN do tend to change with time so it seems to be a mistake to take them too literally. A pragmatic approach is being adopted in CEN TC 125 and that has enabled the drafts to progress to their present stage. It is hoped that in future ways of speeding the 'non-technical' work of editing, translation, and distribution can be found to reduce the length of time it takes to issue a Standard.

REFERENCES

Official Journal of the European Communities No.L40/12 1989.
European Economic Community : Directive 89/106/EEC Construction Products (Construction Products Directive), Brussels.

European Committee for Standardisation, prEN 771-4 Specification for Masonry Units - Part 4 : Autoclaved Aerated Concrete Masonry Units, Brussels.

European Committee for Standardisation, prENV 1996-1-1 Eurocode-6 Design of Masonry Structures Part 1, Brussels.

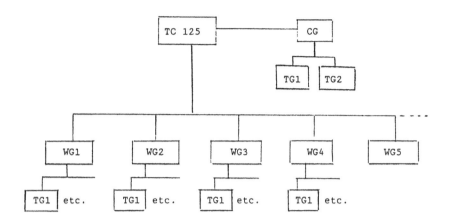

CEN TC 125 - MASONRY COMMITTEE STRUCTURE

7 Ecology and new developments

Advances in Autoclaved Aerated Concrete, Wittmann (ed.) © 1992   Taylor & Francis. ISBN 90 5410 086 9

Ecological aspects for the production and use of autoclaved aerated concrete

D. Hums
YTONG AG, R+D-Centre, Schrobenhausen, Germany

ABSTRACT: AAC is a building material which can be produced with a low consumption of raw materials and energy. Compared to other building materials, the use of AAC offers a remarkably high heat insulation and consequently a low energy consumption for the heating of buildings. AAC releases no materials, in case of proper application, which could be harmful to the human organism or the environment. The use of AAC demolition material is guaranteed by its reuse in the AAC production, its use as a road construction material or as fill material as well as by its removal to waste disposal sites without any difficulties.

1   INTRODUCTION: THE IMPORTANCE OF ECOLOGICAL ASPECTS IN THE CONSTRUCTION INDUSTRY

The construction of buildings has been considered traditionally from economical and architectural aspects. Recently, the construction industry has been faced with the ecological effects of construction. Ecology, as a science of the relationship between the living and an inanimate environment, is on the way to replace traditional considerations.

What are the reasons for this change?
We have to recognise the following essential developments:

- Due to an increasing health consciousness in nations with a very high standard of living, the question of the effect of building materials on the health of people comes up increasingly. Keywords such as asbestos, formaldehyde and pentachlorphenol serve as examples here.
- Initiated through the fierce discussion on the endangered environment, industry is asked in which way it harms the environment. Thus for the construction industry, the destruction of the countryside through the extraction of raw material or the burdening of the atmosphere by emissions have been discussed.
- The discussion about adverse effects on the environment, however, is also directed towards effects which originate from living itself, above all the heating of residential buildings and the emissions which are linked to this.

These developments have to be considered by the construction industry.

Consequently, for the AAC industry, the following questions have to be answered.

- What is the position of the AAC industry from an ecological point of view?
- Which weaknesses should be recognised? To which aspect must particular attention be paid in the future?

2   THE STATE OF NATURAL RESOURCES

The extraction of raw material such as non-metallic minerals from the earth's crust is becoming more difficult. On the one hand because certain raw materials have already been exploited intensively, - this is valid for certain fields, e.g. concrete gravel - on the other hand because it is getting increasingly difficult to obtain permission to exploit resources that are close to the surface. The non-metallic minerals industry with its high demand for raw materials is often confronted with the reproach that it affects nature permanently with its open-mining plants, e.g. by deforestation or impairment of ground water.

2.1 Availability of raw materials for the AAC industry

The main raw materials for the production of AAC are silicious materials such as natural sands and as well as lime and cement which are used as binders. Aluminium powder is employed as a gas-forming agent. Silicious sands are sufficiently available in most regions of the world. The quality of the sand does not have to meet high requirements, e.g. a silicic acid content of approximately 70 % is sufficient for the AAC industry. Sea sands have already been used recently by the AAC industry. The salt content has to be washed out with fresh water. It is especially important that AAC can also be produced from other silicious raw materials such as fly ash, oil shale ashes, steel mill sands and slags. This means that it is possible to gain a valuable raw material from wastes or byproducts, which would normally be dumped. In Great Britain for instance, 1.5 mill m³ of AAC are produced in 10 factories using fly ash. In Czechoslovakia, 6 plants produce 2.0 mill m³ of AAC. 70 % of the output are produced with fly ash.

Many countries in Central Europe e.g. Germany, do not use fly ash. This is only due to marketing considerations. The AAC industry does not want to be considered as a waste utilizer by the building owner.

Lime and cement are used as binders in the AAC industry. An unlimited supply of raw materials is available. The gas-forming agent aluminium powder is produced by milling aluminium. As a rule, waste from aluminium foils serves as the raw material.

2.2 Consumption of raw materials in the AAC industry

AAC is a building material with a low bulk density and a particularly favourable ratio between material strength and bulk density.

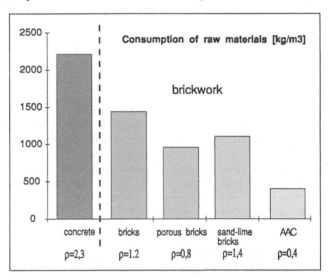

Table 1: Consumption of raw materials in the production of building materials

Understandably, the lower bulk density consequently causes a lower consumption of raw materials which is shown by way of comparison to other building materials in table 1.

3 ECOLOGICAL ASPECTS OF AAC PRODUCTION

3.1 The raw material-energy-valuable material cycle

Energy is needed to produce AAC. Most energy is used in the autoclaving of AAC, grinding of the in the factory prepared raw materials - so to say above all the sand and energy content of the raw materials. During the production process, substances are produced which were dumped in the past, discharged as used waste water or released into the air as emissions. A modern AAC plant produces no waste substances. All byproducts are recirculated and reutilized in the production process. Table 2 shows a typical circulation.

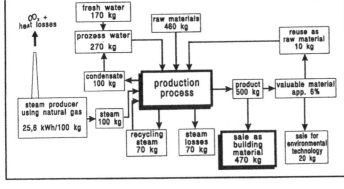

Table 2:   Simplified raw material and energy circulation in the production of AAC with dry-density 500 kg/m^3

Next to the circulation stages, the quantities that are consumed or reutilized are given as well.

3.2   Energy consumption

AAC characteristically shows a low energy consumption during its production, since the thermal process, which is typical for processes with non-metallic minerals, is effected at relatively low temperatures, namely at 200 øC. In the complete energy consideration, the energy contents of supplied raw materials have to be

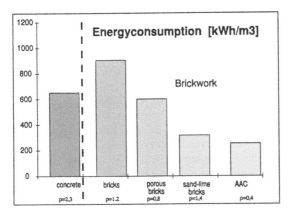

Table 3: Energie consumption in the production of building materials

included. It is therefore essential to consider all

energies which are used in this process, as well as regarding the efficiency when calculating the consumption of thermal and electric energy. In this manner the energy consumptions which are contained in table 3 are arrived at. The publication of Lutter (1) contains further details on this.

4   ECOLOGICAL ASPECTS DURING PROCESSING

4.1   Release of harmful materials when processing AAC

The AAC microstructure consists of calcium silicate hydrates, quartz and pores. To a low extent, depending on the raw material employed, it is also possible that there are calcites and dolomites as well as sulphate compounds such as anhydrite. The pores in AAC are filled with air and water vapour. AAC can therefore not release any gases other than water vapour during processing.

Even at high temperatures, e.g. in the extreme case of fires, no gas-producing substance forms in AAC.

AAC contains liquid water within a part of its pores. On the delivery the total water content ranges between 20 and 30 wt.% . In the case of high moisturization on the jobsite, the water content may increase, in an extreme case up to saturation of (ó 70 wt%). The water evaporates in the building, that is to say AAC drys out. No other volatile substances are contained in AAC.

When processing AAC, solid substances can be released in the form of dust.

Coarse dust particles, which come from sawing are harmless. Fine dusts, however, which contain silicic acids in high concentrations are dangerous. The silicosis risk for AAC dust has been investigated many times. The result however has been constantly the same; the

Energy consumption for the production and use of AAC and porous bricks

AAC 2 / λ = 0,12 W/mK - HLZ 6 / λ = 0,21 W/mK

	Energy consumption for heating resindtial buildings with a liftime of 50 years/ Wall thickness 0.3 m	Primary energy consumption for the production of the building material
AAC	5322 kWh/m³	279 kWh/m³
HLZ	8918 kWh/m³	604 kWh/m³

Table 4: Comparative ecological balance

quartz percentage is so small (quartz percentage in the fine dust (1 wt.%), that AAC is not considered as quartz containing on the basis of the MAK values (maximum workplace concentration) and according to TRGS 900 (Johannisburg convention).

Mineral building materials are currently dicussed, because of their content of fibre-shaped minerals. The asbestos problem as well as the problem of mesotheliom cancer which is caused by asbestos, are known in general. In the meantime, the fibrous glass and mineral wools which contain fibres fine enough to enter the lungs are being discussed.

For the carcinogenic effect of a fibre-shaped mineral, two conditions are necessary. First of all, the particle has to have a certain ratio between length and diameter and a given particle length. According to VDI guideline 3492.

for $L/D > 3$
$D < 3$ μm and $L = 5 - 100$ μM
according to Pott (3):
for $L/D > 5$
$D < 2$ μm and $L > 5$ μm
Secondly, the particles have to remain sufficiently long inside to body's tissue, i.e.

their solubility has to be low.

The minerals in AAC, which means above all tobermoritic crystalline structures, have the shape of bars or plates. Needle-shaped crystals in a mineralogical sense are hardly present. Furthermore, investigations on the solubility of these minerals have shown that they have a half-life period (of solubility) of 12 days in body liquids. Therefore they dissolve significantly faster than certain glass fibres (half-life period approximately 100 days, Bayer B1-Glass). These fibres have been categorized by Pott (3,4) as alsolutely harmless.

According to the present standard of knowledge it can be assumed that the minerals which are contained in AAC do not hold a carcinogenic potentials for people, because of their structure and their high solubility.

4.2   Possibilities to process AAC waste

AAC can be worked with simple tools such as saws and mills. Picking and beating of building components is therefore no longer necessary. Even the smallest fittings can be produced from AAC. It is therefore possible to reduce the waste of a AAC jobsite to a minimum by using off-cuts for fittings.

Processing AAC waste from the production, from the building site as well as from the demolition of an AAC house can be performed in 3 ways:
1.   Recycling in an AAC plant
2.   Production of new products wth applications which do not belong to the construction industry
3.   Dumping

All these 3 possiblilities, for which different requirements apply, are employed today. Lang-Beddoe (2) reports on them in detail.

5 ECOLOGICAL ASPECTS FOR APPLICATIONS IN BUILDING-ENERGY CONSUMPTION FOR HEATING

Among the already mentioned effect of the construction industry on the environment, the energy consumption for heating buildings is the decisive factor. In Germany, roughly 30 % of the entire energy consumption can be led back to the heating of residential buildings. The energy used to produce building material is low compared with the energy required for heating buildings. This is shown, by way of example, for AAC and brick for 1 m³ of building material in table 4. The energy consumption for heating was calculated for a lifetime of 50 years for a building. The stated consumptions were calculated for 1 m³ of building material with a wall thickness of 0.30 m, this a surface of 3.3 m².

It is apparent in how large the influence of the lambda value on the entire energy consumption is.

The characteristically low thermal conductivity of AAC contributes significantly to saving energy and thus to a reduction of emissions. This is why the AAC industry is involved in the task of further improving the lambda values.

The regulations on European Standards are currently being harmonized with respect to determination methods for heat insulation. The requirements will be stipulated for each nation. Germany strives for so-called low-energy houses, which have an energy consumption of 50 kW/m³ p.a. These requirements can nowadays be met with AAC single-shell construction methods.

REFERENCES

(1) Lutter, J.: 3rd RILEM International Symposium Zürich Oct. 14 - 16 1992
(2) Lang-Beddoe, I.: 3rd RILEM International Symposium Zürich Oct. 14 - 16 1992
(3) Pott, Schlipköter, Roller, Rippe, Germann, Mohr, Bellmann: Zbl. Hyg. 189 (1990) 563-566
(4) Bellman, B.; Muhle, H.; Pott, F.: Zbl. Hyg. 190, (1990), 310 - 314

Advances in Autoclaved Aerated Concrete, Wittmann (ed.)© 1992 Taylor & Francis. ISBN 90 5410 086 9

New research on the primary energy content of building materials

J. Lutter
YTONG AG, R+D-Centre, Schrobenhausen, Germany

ABSTRACT: The primary energy content (PEC) of porosed bricks, solid bricks, sand-lime bricks and AAC (= Aerated Autoclaved Concrete) has been determined. While doing so, the PECs of the combustibles, the raw materials and the fuels as well as of the energies required in the production process, have been considered. The investigastion shows that autoclaved building materials have a lower consumption of primary energy than fired bricks.

1 INTRODUCTION

In the course of the past years, the environmental consciousness of the public has increased constantly. In particular, the economical use of raw materials and energy directly employed by the consumer has been taken into consideration.

Beyond this immediate range of vision of the consumer, there is the use of energy for the production of goods used in everyday life. Therefore, the opinion of an environmentally-minded consumer should be more critical in such cases, so that he changes over to similiar products which serve the same purposes, but have a lower PEC.

This aspect, namely to preserve the energy resources and the environment, will be examined more closely by means of four frequently employed building materials.

To determine the PEC of a product, it is not sufficient just to determine the energies employed during the production process such as electric power as well as fuels and combustibles. Moreover in a correct energy account, it is unavoidable that the finished product is burdened with the energy required for the production of the raw and process materials.

2 PRODUCING THE ENERGY SOURCES

To generate electric energy, another energy source is employed whose energy content will be transformed. This process is linked to considerable losses and is defined by means of the power station efficiency. It has been assumed to be 42% according to Dubbel (1983) which are presently realizable in a gas/steam power station.

Combustibles and fuels are used in every industrial process. They are extracted from crude oil by means of a complicated procedure called cracking. During this procedure in the refinery, residues or exhaust gases from the process are used as combustibles. A basic refinery efficiency of 96 % has been adopted.

The energy consumption need to produce and transport the primary energy source will be neglected.

The calorific values L.C.V. of the produced combustibles or fuels are taken from Dubbel (1983). The PEC has been calculated from the calorific value while taking into consideration the refinery efficiency.

The following neglects the PECs of the employed machines, vehicles, packing materials as well as those of other process materials which are only needed to a minor extent.

2.1 Employed raw materials

The building material industries which are closer examined here are, as a rule, in the immediate vicinity of the raw material they employ the most.

The main raw material used for brick products is a suitable brick clay, or sand for the sand-lime brick and AAC industry. They are not subject to an industrial upgrading process. Therefore it is only necessary to set a PEC for its production and its transport to the plant premises.

This is not the case for the other raw materials employed such as polystyrene, lime, cement and aluminium powder. They are produced in an industrial preparation and upgrading processes. Consequently, they burden the finished product with their PEC. The consumption of grinding energy to produce pulverized aluminium powder amounts to approximately 3 kWh/kg. The resulting PEC for grinding has been included in table 1.

Industrial produced raw materials	kWh/t for raw material	Reference
Polystyrene	18250	Marmé (1982)
Lime	1055	Bacher
Cement	1530	Bacher, Scheuer (1992)
Aluminium powder	55000	Aluminium-taschenbuch (1954)

Table 1: PEC of industrial raw materials

3 PRODUCING BUILDING MATERIAL

3.1 Brick

Bricks are fired from brick clays, which are won from clay pits that belong to the plant's property. Conveying the clay is heavily impeded by weather factors and the viscosity of the material. A fuel consumption of one litre diesel/m³ for the finished product is assumed. The PEC of the clay includes the energies used for mining and processing. Calcination losses of the clay are assumed to be 10 wt.%.

The won clay is prepared, compressed and usually fired in tunnel kilns. The fired finished product is then packed and dispatched. A fuel consumption of 0.2 litre/m³ for the finished product is assumed for its dispatch.

Further energy quantities which have been employed were taken from West (1986).

The PEC used in the respective production phases are reproduced in figure 1. A product with a bulk density of 1.3 t/m³ consequently yields a PEC of 900 kWh/m³.

3.2 Porosed bricks

Porosed bricks are produced by a mixture of brick clays and organic combustible opening materials. To produce porosed bricks, a mixture of 99.5 percent by weight clay substance and 0.5 percent by weight polystyrene has been taken. The polystyrene is mixed in the brick clay and burns out during the calcination process.

To calculate the PEC of porosed bricks it is permissible, to take the calculated PEC of bricks from this work and that of polystyrene according to Marmé (1982) for the above mentioned mix proportions. The PEC of porosed bricks amounts to 604 kWh/m³ at a bulk density of 0,8 t/m³. The details are shown in figure 2.

3.3 Sand-lime brick

Sand-lime bricks are made from a mixture of sand, lime and water. This mixture is hardened in an autoclave. To calculate the PEC, the following mixture has been taken, which is based on the specifications of Grundlach (1980), to reach a stone bulk density of 1,4 t/m³:

lime	73 kg
sand	1286 kg
water	41 kg

For the PEC of quick lime it is necessary to set 1055 kWh/t for the production (Bacher). The consumption of electric energy per ton for the finished sand-lime brick can be taken from Fischer (1982). The specific consumption of combustibles/t for the finished product is given in Ebersbach (1982) with approximately 104 kWh/t for the finished product. The PEC of water for delivery and purification is 0.3 kWh/t of water according to the specifications given by the Stadtwerke Ingolstadt (water board for the area of Ingolstadt). This is so small that it can be neglected. The PEC used for the most important process phases is described in figure 3. On the basis of these data, a PEC of 341 kWh/m^3 results for a product with a bulk density of 1,4 t/m^3.

4 Autoclaved Aerated Concrete (AAC)

AAC is produced from a mixture of cement, lime, sand, water and aluminium powder. The finished mass is cut in the course of the further production process and hardened in an auto-clave.

The employed raw materials and the energies consumed during the production process are based on the specifications given in internal YTONG data (1989). The PEC has been determined for a mixture of a block of category G 2 according to the German Standard DIN 4165. The material used is shown in table 2.

	kg/m^3
Cement	65
Sand	290
Lime	45
Water	225
Aluminium Powder	0,53

Table 2 : Mix formula of an AAC product G 2

While taking into account the PEC within the process stages a PEC of 279 kWh/m^3 results for the product (see figure 4).

5 SUMMARY

Table 3 shows the calculated PEC of the considered products.

It is apparent that building materials which are not produced by a high-temperature process are favourable in primary energy consumption. Therefore the environmentally-minded consumer should use autoclaved building materials when-ever possible. The considerably higher PEC of brick products is above all due to the high-temperature process of calcination and the energy consuming polystyrene as a combustible opening material.

	Bulk density	PEC
	t/m^3	kWh/m^3
Brick	1.3	900
Porosed brick	0.8	604
Sand-lime brick	1.4	341
AAC	0.4	279

Table 3: Calculated PEC of different products

REFERENCES

Dubbel 1983. *Taschenbuch für den Maschinenbau.* 15. Auflage. Berlin: Springer.

Marmé, W., Seeberger, J. 1982. Der Primär-energieinhalt von Baustoffen. *Bauphysik* 4: 208-214.

Bacher. Persönliche Mitteilung von Herrn Bacher, Kalk- und Zementwerk A. Büechl. Regensburg.

Scheuer, A., Ellerbrock H.G. 1992. Möglich-keiten der Energieeinsparung bei der Zementherstellung. *ZKG* 45: 222-230.

Aluminiumtaschenbuch 1954. 10. Auflage. Düsseldorf: Aluminium-Verlag GmbH.

West, H. 1986. Energieverbrauch in der Zie-gelindustrie der EG. *ZI* 39: 4-7.

Gundlach, H. 1980. Kalksandstein-Technologie Teil V. *TIZ* 104: 98-104.

Fischer, A.M. 1982. Stromverbrauch und Leistungsbedarf in der Kalksandstein-Industrie. *TIZ* 106: 123-126.

Ebersbach, K.F. 1982. *Wirtschaftlichkeit des Dampfenergiebedarfs bei der Herstellung von Kalksandsteinen.* Hannover: Bundesver-band der Kalksandstein-Industrie e. V..

Internal YTONG Data (1989). Daten des internen Rechnungswesens über das Jahr 1989 von: YTONG Bayern GmbH, YTONG Nord GmbH, YTONG Wedel GmbH & Co KG, YTONG Südwest GmbH.

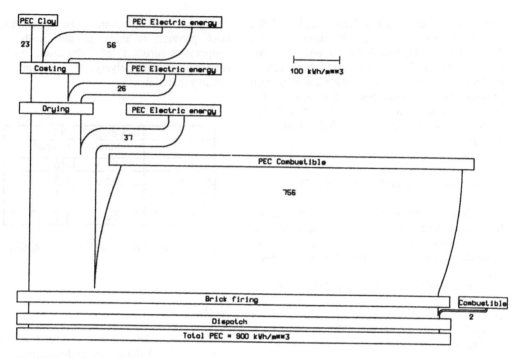

Fig. (1): PEC of a brick product

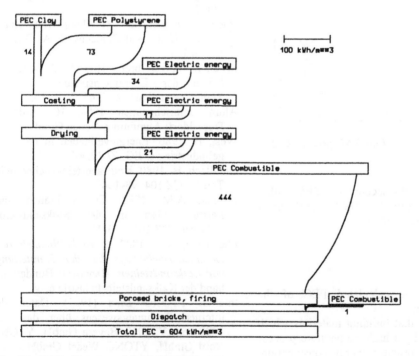

Fig. (2): PEC of a porosed brick product

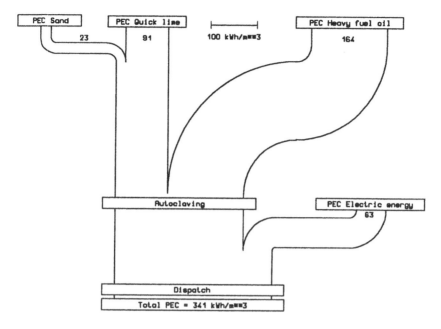

Fig. (3): PEC of sand lime brick

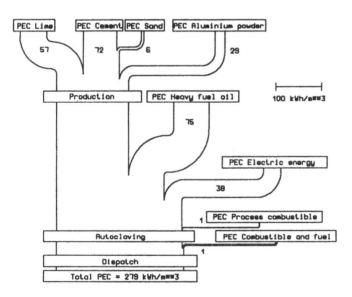

Fig. (4): PEC of an AAC product

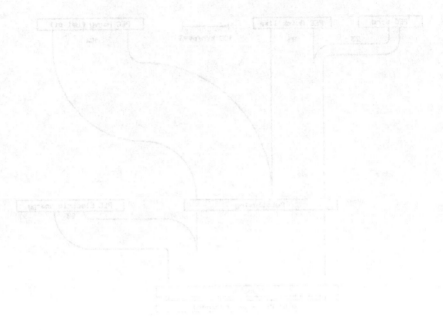

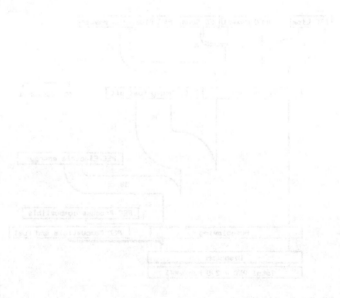

Fig. 12. Tree of an AAD product

Advances in Autoclaved Aerated Concrete, Wittmann (ed.) © 1992 Taylor & Francis. ISBN 90 5410 086 9

Possibilities to dispose of waste autoclaved aerated concrete (AAC)

I. Lang-Beddoe
YTONG AG, R+D-Centre, Schrobenhausen, Germany

ABSTRACT: In general there are three possibilities to dispose of waste AAC: dumping, recycling and its use in the construction of roads and embankments. Recycling of AAC will steadily gain in importance due to increasingly stricter laws for environmental protection, growing public awareness of environmental issues and lack of space for waste disposal sites. Currently, pureAAC-waste can be recycled without many difficulties because there are many applications for it and a large proportion of the pure material can be led back into production. The use of prepared AAC from demolition material is presently restricted to road construction, that is to say for noise barriers, subgrade and natural ground improvements. The applications in this field have to be investigated and expanded.

1 SURVEY OF OCCURRENCE AND DISPOSAL POSSIBILITIES OF RESIDUAL CONSTRUCTION MATERIAL IN GERMANY

The construction and mining industry are, with respect to the quantitative dimensions of the waste problem, the main producers of waste. The Federal Office for Statistics estimated the waste produced by the construction industry in 1989 at approx. 221 million tonnes p.a. /Willig 1991/. The individual percentual and quantitative shares of the different waste sources in the construction industry are shown in Fig.(1). It is apparent that demolition material and building site waste account with 32.6 mio. t/a for approximately 15 % of the entire waste produced by the construction industry.

The Federal Waste Law, which is in force since 27.8.1986,embodies the principle that "waste is fundamentally to be avoided and unavoidable waste is to be recycled". The German government therefore attempted to reach recycling quotas of 60 % for prepared building demolition debris, 40 % for building site waste and 90 % for scarified material. The actual recycling quotas achieved were much

Waste from the construction industry in the former federal states of Germany in 1989

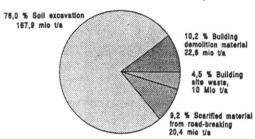

76,0 % Soil excavation
167,9 mio t/a

10,2 % Building demolition material 22,6 mio t/a

4,5 % Building site waste, 10 Mio t/a

9,2 % Scarified material from road-breaking 20,4 mio t/a

Fig. (1):   Share of the individual construction branches in the overall waste volume produced by the construction industry

lower. Less than 20 % for prepared building demolition debris, 0 % for building site waste and approximately 55 % for scarified material /Heckötter 1991, Freise 1990/.

At present the main proportion of waste produced by the construction industry is dumped since waste disposal costs are still relatively low, especially in rural areas. Consequently, the construction industry supplies the main share of material, 58 weight

Essential factors for recycling of construction material

1. Costs
   - Transport (Waste disposal site - Recycling plant)
   - Presorting of construction waste
   - Preparation (stationary, semi-stationary, mobile)
   - Quality control of recycled construction material through external and self-control
2. Returns through marketing of recycled construction material
3. Lower operating costs through the use of recycled construction material which is, in most cases, cheaper
4. Increasing costs of dumping

Fig. (2):   Essential factors for or against recycling of construction material

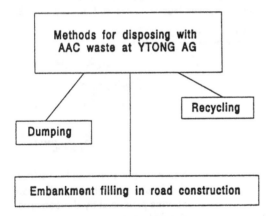

Methods for disposing with AAC waste at YTONG AG

Recycling

Dumping

Embankment filling in road construction

Fig. (3):   Methods for disposing with AAC waste at YTONG AG

percent, which is on waste disposal sites /Scheffold 1991/.

Nevertheless, essential factors of influence such as
- measures for environmental protection which will generally increase in strictness
- a limited number of suitable locations for waste disposal sites
- a low acceptance of new locations for waste disposal sites
- a waste disposal site volume which is limited
- increased costs of dumping
will lead to only non-recyclable residues being dumped /Cevrim 1991/.

Fig. (2) shows a general view of the most important decisive factors for or against recycling of construction material.

The most important criteria for recycling are the comparison of costs for dumping and recycling as well as the returns gained by replacing raw materials by recycled material and/or the marketing of secondary material /Labonte 1991/.

2 DISPOSAL OF WASTE AAC AT YTONG AG

Currently, there are many possibilities of dealing with AAC waste at the YTONG AG. Fig. (3) shows the most important methods used by YTONG AG.

YTONG AG is endeavouring to recycle the entire AAC waste material. Due to the increasingly stricter rules regarding dumping and environmental protection, the complete recycling of construction material sets a trend geared to the future.

3 DUMPING OF AAC WASTE

Dumping of AAC still presents few problems except for the increasing dumping fees. In most cases construction material waste is accepted by the owners of waste disposal sites for construction material without any further examination, provided that there is no serious pollution visible at first sight. Only in the federal state of North Rhine-Westphalia have guidelines been drawn up for concentration limits in waste eluates and the corresponding classification of different waste disposal sites.

Fig. (4) lists typical elution values of AAC from leaching tests according to German standard DIN 38414 part 4. For comparison, the permissible concentrations of pollutants in eluates and wastes for construction material waste disposal sites (Site category 2; Waste

Manual, vol. 3, ref. no. 4735, p. 12) are also listed.

It is plain to see that, during the elution test, the heavy metal contents fall below the concentration limits given for a category 2 waste disposal site. Thus, from a chemical point of view, AAC can be dumped without any problems.

4 RECYCLING OF AAC

The preparation and possible uses of recycled AAC waste require distinction between pure AAC rubble, i.e. hardened AAC from breakages and cutting off the surface layer, and AAC from demolition waste which is a mixture of various materials.

Unhardened waste which results for instance from sawing green, i.e. aerated concrete that has not been autoclaved, is directly returned into the production process.

4.1 Recycling of AAC rubble and/or surface waste material

Fig. (5) presents a general view over the possibilities for use of prepared, pure AAC rubble and waste material produced when the surface layer is removed. In a separate production process which comprises breaking, sieving, sizing and drying, granules with different size fractions are produced depending on the required application, and dusts with size fractions, which range roughly from 0 - 1 mm. A part of the dusts is returned into the production as a substitute for raw material.

The application of conditioned "pure" AAC are very different and cover a field which will certainly expand as time goes on. The AAC by-products cat litter and oil binder mentioned in Fig. (5) have already been marketed under well-known brand names for considerable time. AAC by-products are also used for sludge conditioning and as fill material and as an aggregate for plaster. There are, to some extent, industrial applications in the fields on flue and exhaust gas purification as well as sewage treatment which require further investigation.

Heavy Metal Contents Of AAC After 24 Hours Elution		
Parameter	Measured Values In The Eluate	Concentration Limit
pH - value	10,7 - 12,0	5,5 - 12
Cod mg O_2 /l	3,5 - 8,0	< 50
Conductivity mS/m	50 - 250	< 300
Fluorides mg/l	0,17 - 0,18	5,5
As mg/l	0,001 - 0,004	0,1
Fe mg/l	0,005 - 0,1	2,0
Mn mg/l	0,001 - 0,025	1,0
Cr mg/l	0,01 - 0,04	1,0
Cu mg/l	0,002 - 0,022	1,0
Ni mg/l	0,001 - 0,003	0,5
Zn mg/l	0,003 - 0,010	5,0
Pb mg/l	0,001 - 0,008	0,5
Cd mg/l	0,0001 - 0,0003	0,05
Hg mg/l	0,0003 - 0,0008	0,005
V mg/l	0,029 - 0,040	0,2
Tl mg/l	0,002 - 0,01	0,1

Fig. (4): Heavy metal contents of AAC after 24 hours of elution

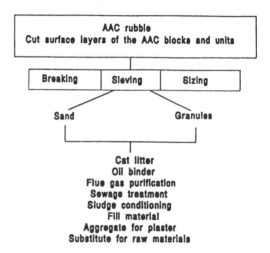

Fig. (5): Possible uses for prepared AAC

4.2 Recycling of AAC from demolition material

The preparation method for demolition material is more complex, but in principle the same as for pure AAC, i.e. breaking, sizing, sorting, storing and transporting. The problem with demolition material is the extreme diversity of the materials. In order to perform a selective recycling of construction material, it is necessary to separate the different types of material during demolition so they can be prepared separately and sorted. Nevertheless, the final recycled product is a more or less polluted material which only offers limited possibilities for application. Thus this recycled material is used especially in road construction.

5 THE USE OF AAC IN ROAD CONSTRUCTION

Another use of pure AAC rubble is for roads and paths as well as a fill in road construction. Pure AAC rubble has been used as a fill in road construction in Germany as well as in France.

Large quantities of AAC rubble are needed for such applications, approximately 35000 - 40000 m³. For such uses, material can be taken from stockpiles or waste disposal sites containing unprepared pure rubble. It is also possible to use recycling products made of AAC demolition material.

The suitability of pure AAC rubble or recycled demolition material for road construction and for fills has to be examined before its actual use. In the above mentioned cases, the necessary compaction and loadbearing values as well as the elution values had to be reached in accordance with ZTVE-StB (Additional technical regulations and guidelines for earthworks in road construction).

In order to promote the acceptance of recycled building material for road construction, the Association of German Recycling Companies was founded roughly one year ago. The purpose of this association is to introduce a quality control system for recycled building material used in road construction. This system is based on external and self-control and the award of a so-called RAL-quality symbol. Fig. (6) shows the most important tests that have to be performed for noise protection embankments, subgrades and natural ground improvements (construction material category III).

As can be seen from the list for testing regulations for construction material category III, very complex tests have to be performed. The quality of the recycled construction materials is therefore guaranteed for the customer.

In conclusion it can be stated that there are essentially three methods of disposing with AAC waste: dumping, recycling and employment in road construction. In the future, the development of methods for using AAC

Testing requirements for noise protection embankments, subgrade, natural ground improvements (category III) according to RAL-RG 501/1

1. Production, preparation, storage, material composition
2. Execution of self-control in accordance with the guidelines for quality control of mineral aggregates in road construction
3. Density
4. Sizing
5. Time-settlement behaviour
6. Elutriation
7. Grass-growing characteristics
8. Resistance against weathering, frost, freezing and thawing cycles etc.

Fig. (6): List of the most important tests for noise protection embankments, subgrade, natural ground improvements (category III) according to RAL-RG 501/1

waste will focus on the recycling of residual building material. To ensure an extensive utilization of occurring AAC waste, more possible applications for recycled AAC from demolition, i.e. polluted AAC, must be found.

REFERENCES

Willig, E. 1991. Politische und gesetzliche Vorgaben zum Bauschuttrecycling. Entsorgungs-Praxis Spezial 4: 2-4.

Heckötter, Ch. 1991. Einsatzmöglichkeiten für aufbereiteten Bauschutt. Entsorgungs-Praxis Spezial 4: 23-27.

Freise, H. 1990. Abfallrechtliche Probleme der Entsorgung von Erdaushub, Bauschutt und Straßenaufbruch. Entsorgungspraxis 11: 647-649.

Scheffold, K. 1991. Die Ablagerung von Abfällen - Notwendigkeit und Techniken. Entsorgungs-Praxis 4:150-159.

Cevrim, M. 1991. Abfallplanung - ein wesentlicher Bestandteil der Deponieplanung. Entsorgungs-Praxis 12: 750-760.

Labonte, D. 1991. Ökologischer und ökonomischer Nutzen von Bauschutt-recycling. Entsorgungs-Praxis Spezial 4: 7-11.

Recycling-Baustoffe für den Straßenbau, Güte-
und Prüfbestimmungen gemäß RAL-RG
501/1; RAL Deutsches Institut für
Gütesicherung und Kennzeichnung e.V.,
Bornheimer Straße 180, 5300 Bonn 1

Advances in Autoclaved Aerated Concrete, Wittmann (ed.) © 1992   Taylor & Francis. ISBN 90 5410 086 9

A study on using waste residue of ALC as a resource of building materials

N. Feng, X. Ji & J. Hu
Building Materials Laboratory, Department of Civil Engineering, Tsinghua University, Beijing, People's Republic of China

ABSTRACT: There is a great deal of waste residue in the production of ALC in China, eg. the waste residue of ALC (WRALC) remained in Beijing ALC Factory is up to 100 thousand tons every year. The WRALC not only pollutes the environment but also takes much land unless some measures are taken to make use of it. In this study, the WRALC is calcined at different temperatures and finely ground to replace partly the cement in concrete. The 3, 7, and 28 days strength of concrete (W/C = 0.3) with WRALC calcined at 800°C replacing 10% of cement is 9.0%, 27% and 15% higher than that of control concrete respectively. The 7, 28 days strength of concrete (W/C = 0.3) with strength enhancer (mixture of WRALC with natural zeolite at 20:80), which replaces 5% of cement, increases by 17% and 21% respectively. The strengthening mechanism of WRALC on cement concrete is described in this paper.

1 INTRODUCTION

China saw the ALC in 1964, at that time a production line was imported from Switzerland, the production capacity was about 150,000 m^3 per annum. In the period of 1984 to 1990, there were six more production lines imported from Poland, Romania and Germany; nowadays, there are 80 ALC factories in China, the total design production capacity is four million cubic meters per annum, the practical output is two million and one hundred thousand cubic meters, in which over 80 percent are ALC blocks.

There are mainly two series of raw materials for the production of ALC in China: (1) cement--lime--slag--sand; (2) cement--lime--fly ash--sand. The bulk density of the product is about 550 kg/m^3 with the compressive strength 3N/mm^2 ·

The reject rate of ALC is up to 10 percent in the production of ALC in China. According to statistics, the waste residue amounts to two hundred thousand m^3 in the production, and there is more and more

residue piled up mountain-high in many factories year after year, and it takes more and more land, in the windy weather, it usually flies about, which not only pollutes the environment but also affects the health of people.

In this study, the WRALC is dehydrated by calcination, and mixed up with the natural zeolite at a certain ratio, after they are finely ground together, the stength enhancer is obtained. The strength of cement concrete with this kind of strength enhancer replacing 5-20% of cement is 10-20% higher than normal concrete. In this test, when portland cement 450 kg/m^3 (strength grade #525 in accordance with Chinese Technical Standard), strength enhancer 50 kg/m^3, W/C = 0.30 and high-range water reducer 5 kg/m^3 are used, the slump of the concrete is 4-5 cm, and 28-day strength of the concrete is over 85 MPa, increased by 10-15% in comparison with control concrete without strength enhancer; in China, the cost of strength enhancer is only one third of that of cement, therefore, great benefits of

Table 1 Chemical compositions of NZ(%by mass)

Deposit	SiO$_2$	Al$_2$O$_3$	Fe$_2$O$_3$	CaO	MgO	Na$_2$O	K$_2$O	Loss of ignition
Zhang Jiakou	68.60	12.43	1.21	2.57	0.81	1.08	2.86	10.0

Table 2 Concrete compositions with WRALC calcined at different temperatures

No	Calcination temperature of WRALC	W/C	S (%)	Material(kg/m³ concrete)						Slump (cm)
				WRALC	Cement	Water	Sand	Gravel	UNF	
0	-----	0.3	35	--	500	150	630	1170	5	3.5
1	Uncalcined	0.3	35	50	450	150	630	1170	5	3.5
2	700ºC	0.3	35	50	450	150	630	1170	5	2.8
3	800ºC	0.3	35	50	450	150	630	1170	5	3.0
4	900ºC	0.3	35	50	450	150	630	1170	5	3.0

Table 3 Compressive strength of the concrete

No	Compressive strength(MPa)			Calcination temperature of WRALC
	3-day	7-day	28-day	
0	44.28 (100%)	55.39 (100%)	74.97 (100%)	-----
1	48.13 (108.7%)	60.60 (109.4%)	81.04 (108.1%)	Uncalcined
2	49.02 (110.7%)	70.01 (126.4%)	84.27 (112.4%)	700ºC
3	48.31 (109.1%)	70.40 (127.1%)	85.99 (114.7%)	800ºC
4	47.11 (106.4%)	63.48 (114.6%)	84.04 (112.1%)	900ºC

Table 4 The proportion of cement replaced by strength enhancer with different ratio of WRALC calcined at different temperatures to natural zeolite and test results

No	Calcination Temperature of WRALC	Strength Enhancer (WRALC:Zeolite)	Material(kg/m³)			Compressive strength(MPa)	
			WRALC	Zeolite	Cement	7-day	28-day
0	---	---	---	---	500	60.19(100%)	76.63(100%)
1	---	0:100	0	25	475	66.03(109.7%)	80.70(105.3%)
2	Uncalcined	20:80	5	20	475	67.89(112.8%)	87.40(114.1%)
3	700ºC	20:80	5	20	475	70.12(116.5%)	84.30(110.1%)
4	800ºC	20:80	5	20	475	70.66(117.4%)	92.65(120.9%)
5	900ºC	20:80	5	20	475	64.46(107.1%)	85.69(111.8%)
6	---	0:100	0	50	450	68.91(114.5%)	82.92(108.2%)
7	Uncalcined	35:65	17.5	32.5	450	68.86(114.4%)	83.52(109.0%)
8	700ºC	35:65	17.5	32.5	450	71.20(118.3%)	84.35(110.1%)
9	800ºC	35:65	17.5	32.5	450	65.66(109.1%)	87.44(114.1%)
10	900ºC	35:65	17.5	32.5	450	65.19(106.3%)	87.12(113.6%)

Table 5 Composition of concrete with different W/C ratios with and without the strength enhancer

No	W/C	With or Without Strength Enhancer	Materials(kg/m³)					
			Cement	Strength Enhancer	Water	Sand	Gravel	UNF
01	0.25	with	666	74	185	526	1119	7.4
		without	740	----	185	526	1119	7.4
02	0.30	with	555	61.7	185	565	1202	6.2
		without	617	----	185	565	1202	6.2
03	0.35	with	476	52.9	185	594	1262	---
		without	529	----	185	594	1262	---
04	040	with	416	46.3	185	615	1307	---
		without	462	----	185	615	1307	---

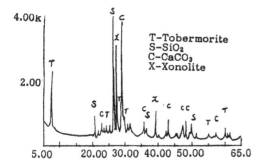

Fig.1 The XRD of WRALC

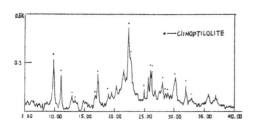

Fig.2 The XRD of NZ

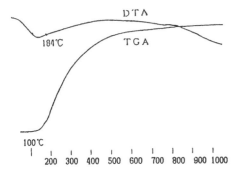

Fig.3 The TG-DTA of NZ

Fig.4 The SEM of NZ

technology, economics and environment will be obtained by using WRALC to produce strength enhancer.

2 MATERIALS USED IN THE TESTS:

1) Cement: Portland cement (strength grade #525);
2) Aggregate: river sand, Fm = 2.8 - 3.1; crushed gravel with the Dmax = 30 mm;
3) High-range water reducer: naphthalene UNF;
4) WRALC: The XRD of WRALC is shown in Fig.1. The main compositions of the WRALC are tobermorite, CSH(B), and a little of xonolite. The average diameter of fine particle of WRALC is 4.5 - 6 μm;
5) Natural Zeolite (NZ): The chemical compositions of NZ are shown in Table 1, the major mineral composition is the clinoptilolite, 65% by mass. The XRD, TG-DTA and SEM of NZ are shown in Fig. 2, 3 and 4, respectively.

3 TEST RESULTS AND DISCUSSIONS

1) Strengthening effectiveness of the WRALC calcined at different temperatures in cement concrete.

10% of cement in concrete is replaced by the WRALC calcined at 700°C, 800°C, and 900°C, respectively. The compositions of the concrete are shown in Table 2. Table 3 is the compressive strength of the concrete.

It is known from the Table 3 that when the WRALC calcined at different temperatures are used to replace 10% of the cement in concrete, the strength of the concrete at 3, 7 and 28 days is higher than that of concrete without WRALC, while the strength of concrete containing WRALC calcined at 800°C is the highest. The Table 3 also shows that the WRALC uncalcined concrete can also increase some of the strength, it may be due to the micro-crystallization effect of tobermorite in WRALC that the hydration of cement is accelerated.

2) Strengthening effectiveness of the strength in cement concrete

It is found by orthogonal test that there is an optimal ratio of the WRALC calcined at different calcination temperatures to the natural zeolite in order to obtain strengthening effect. The strength enhancer is the mixture of WRALC with natural zeolite finely ground together at different ratios. The tests in the Table 4 are

Table 6 Compressive strength and slump of the concrete in Table 5

No.	W/C	With or without strength enhancer	Compressive strength(MPa)			Slump
			3-day	7-day	28-day	(cm)
01	0.25	with	53.4(105%)	58.1(101%)	76.2(107%)	0.0
		without	51.0(100%)	57.4(100%)	71.2(100%)	0.0
02	0.30	with	59.5(115%)	69.9(117%)	80.8(115%)	2.5
		without	51.7(100%)	59.9(100%)	70.2(100%)	2.5
03	0.35	with	43.1(90%)	57.1(102%)	73.6(105%)	3.0
		without	47.7(100%)	56.2(100%)	70.2(100%)	3.0
04	0.40	with	34.6(85%)	48.3(89%)	66.4(99%)	4.0
		without	40.6(100%)	54.0(100%)	67.3(100%)	4.0

Table 7 Content of soluble SiO_2 and Al_2O_3 and strength of dehydrated WRALC mixed with CaO at 28 days

Calcination Temperature	700°C	800°C	900°C
Soluble SiO_2 (%)	23.42	23.31	22.33
Soluble Al_2O_3 (%)	5.89	6.24	5.62
Strength of dehydrated WRALC reacted with CaO(MPa)	7.3	7.05	3.44

Table 8 XRD Intensity of $Ca(OH)_2$ in hcp with and without WRALC dehydrated at 800°C

Position of peak	$2\theta=18.1°$		$2\theta=34.1°$	
Age	3-day	28-day	3-day	28-day
Intensity of $Ca(OH)_2$ in pure hcp	3.1	3.2	2.8	2.7
Intensity of $Ca(OH)_2$ in hcp with dehydrated WRALC	2.6	2.7	2.3	2.3

Table 9 Pore content(in $10^{-2}cm^3/g$) and accumulated pore content of the pure hcp and the hcp with strength enhancer

Pore diameter	7-day				28-day			
	Pore Content		Accum. pore content		Pore Content		Accum. pore content	
(A)	Pure*	Strength Enhancer*	Pure*	strength enhancer*	Pure*	strength enhancer*	Pure*	strength enhancer*
7500	0.285	0.186	0.285	0.186	0.045	0.139	0.045	0.140
1816	0	0.093	0.285	0.279	0.134	0.046	0.179	0.186
938	0.475	0.047	0.76	0.325	0.223	0.139	0.402	0.326
625	0.95	0.743	1.710	1.068	0.268	0.326	0.670	0.652
375	4.133	4.179	5.843	5.247	2.500	2.746	3.170	3.397
214	2.233	2.276	8.075	7.523	1.563	1.861	4.732	5.258
125	1.853	1.904	9.928	9.427	1.205	1.582	5.937	6.841
75	1.4251	1.254	11.353	10.681	1.652	1.163	7.589	8.004
54	0.523	0.836	11.875	11.516	0.833	0.698	8.482	8.702
42	0.143	1.300	12.018	12.817	0.714	0.326	9.163	9.028
38	0.143	0.325	12.16	13.142	0.223	0.233	9.420	9.260

* Pure=pure hcp, strength enhancer=hcp with the strength enhancer

repetition tests of the orthogonal test. The materials used in the test are the same as those in Table 2. The composition of control concrete without the strength enhancer (No. 0 in Table 4) is the same as that of the concrete No. 0 in Table 2. As for the concrete containing the strength enhancer (Table 4), the W/C ratio (W/C = 0.30), sand ratio (S = 35%) and the amounts of water, sand and gravel in the following concrete (Table 4) remain the same as those of the control concrete except the different proportion of the cement replaced by the strength enhancer and the different ratios of WRALC calcined at different temperatures to natural zeolite.

The test results show that the natural zeolite can increase the strength of concrete (No.1 and No. 6), while the strength enhancer can greatly increase the strength of the concrete. eg. when 10% of the cement is replaced by the strength enhancer (WRALC calcined at 700°C:zeolite = 35:65), the 7-day and 28-day strength of the concrete is increased by 9.1% and 14.1%, respectively. It is noticeable that the strength of the concrete containing the strength enhancer (WRALC calcined at 800°C:zeolite = 20:80), which replaces 5% of cement, is 93 MPa, increased by 21% as compared with the control concrete (No. 0). It can be used to make super-high strength concrete.

3) Strength effectiveness of strength enhancer in cement concrete with different W/C ratios.

Cement used in this test is Portland cement, (strength grade #425). Table 5 are compositions of the concrete with different W/C ratios with and without containing the strength enhancer. The strength enhancer is the mixture of WRALC calcined at 800°C with natural zeolite at the ratio of 40:60. The test results are presented in Table 6.

It is found from Table 6 that the strengthening effectiveness of the strength enhancer is affected by the W/C ratio , the larger the W/C ratio, the less the effectiveness of the strength enhancer. The strength of the concrete No. 01 (W/C = 0.25) is lower than that of concrete No. 02 (W/C = 0.30) possibly because the concrete is uncompactable caused by the too low W/C

ratio. It can be seen that the strength of the concrete will not decrease when using the strength enhancer to replace a part of cement if the W/C ratio does not exceed 0.4.

4 STRENGTHENING MECHANISM OF WRALC IN CEMENT CONCRETE

1) Structure changing process of the calcium silicate hydrate in dehydration.

The DTA curve of the WRALC and XRD

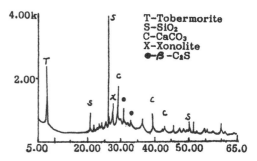

Fig.6 The XRD of WRALC calcined at 700°C

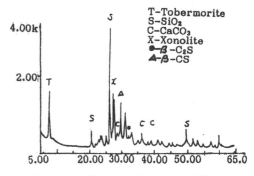

Fig.7 XRD of WRALC calcined at 800°C

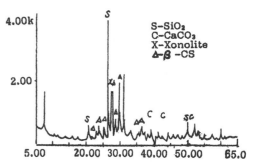

Fig.8 XRD of WRALC calcined at 900°C

Fig.5 The DTA of WRALC

Table 10 Micro-strength and CaO/SiO₂ on the surface of the interface zone

Micro-strength on surface of hcp(MPa)								CaO/SiO₂ (wt%)			
Distance to interface(um)	25	50	100	150	200	250	300	20	40	60	80
Hcp	46	36	73	95	117	112	117	2.3	1.2	3.6	1.9
Hcp+Strength Enhancer	73	97	109	116	120	125	123	1.7	1.9	2.3	0.1

patterns of the dehydrated phases at 700°C, 800°C, 900°C, are given in Fig.5, Fig.6, Fig.7, Fig.8, respectively.

The Fig. 5 shows that (1) there is an endothermic peak in the temperature range of 56°C - 202°C, the free water and inter-layer water of the calcium silicate hydrate in the WRALC are quickly dehydrated in this temperature range, but the structure of the calcium silicate hydrate is not affected;
(2) There is a wider endothermic peak in the temperature range of 258°C - 763°C, the O H⁻ions, which have different bonding force in the calcium silicate hydrate, are released gradually, the structure of calcium silicate hydrate is destroyed seriously, and at about 763°C the structure is completely destroyed; (3) At about 810°C there is an exothermic peak which corresponds to the formation of β-CS. It can be seen from Fig. 6 and Fig. 7 that there is β-C_2S in the WRALC dehydrated at 700°C, 800°C, it can also be seen from Fig. 8 that the tobermorite is turned into β-CS when the temperature is over 820°C. There is no new crystal phase formed in the dehydrated WRALC when the temperature does not exceed 810°C, and the WRALC, now with a high specific surface, is in the amorphous and substeady state and is of high reactive activity.

2) Activity of the dehydrated WRALC

It is due to the high reactivity of the dehydrated WRALC that there is strengthening effectiveness in cement concrete. It is found that there is high content of soluble SiO_2 and Al_2O_3 in the dehydrated WRALC (see Table 7).
The 28-day strength of the dehydrated WRALC mixed with water (water/solid = 0.45) is only 1.58 MPa, but when it is mixed with CaO and dehydrated WRALC:CaO:Water = 2:1:1.5, the strength is increased quite a lot (see Table 7). The dehydrated WRALC can react with $Ca(OH)_2$ produced in the hydration process of cement, and turns into CSH gel (as proven in Fig. 9), therefore the concrete is stengthened. The table also shows that the reactive activity is reduced when the temperature exceeds 820°C

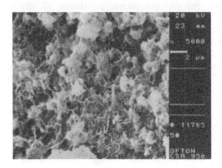

Fig.9 SEM of hydration products of WRALC dehydrated at 800°C reacted with CaO (autoclaved at 100°C for 16 hours)

because of the formation of new crystal phase in the dehydrated WRALC.

It is found by comparing tests between the pure hcp and the hcp containing the WRALC dehydrated at 800°C that the intensity of $Ca(OH)_2$ peaks in hcp with dehydrated WRALC replacing 10% of cement at 2θ = 18.1 and 2θ = 34.1 is weaker than that of the pure hcp (Table 8).

5 STRENGTHENING MECHANISM OF THE STRENGTH ENHANCER IN CEMENT CONCRETE

It is known from the above discussions that the dehydrated WRALC can react with $Ca(OH)_2$. The natural zeolite, which is of porous and micro-crystal structure, can also react with $Ca(OH)_2$ to produce CSH gel, on the other hand, the strength enhancer can also offer the outer space for the hydration products to fill in, hence improves the distribution of the hydration products.

1) The strength enhancer can improve the pore structure of the hardened cement paste

Table 9 is the pore content and accumulated pore content of the pure hcp and the hcp with the strength enhancer (WRALC dehydrated at 800°C:Zeolite = 20:80) which replaces 10% of cement at 7, 28 days. The water/solid = 0.30.
The Table 9 shows that the content of

harmful large pore (diameter > 938 Å) in hcp is greatly reduced by the strength enhancer, and the content of micro-pore (diameter <625 Å) in the hcp with strength enhancer is a bit higher. The strength enhancer can not only change the hydration products but also the pore structure.

2) The strength enhancer can improve the interface structure

The results of the micro-strength on surface of the interface zone are shown in Table 10. The computed CaO/SiO_2 analyzed by EDX (Energy Dispersion of X-ray) is also shown in Table 10.

Table 10 shows that the micro-strength of surface in interface zone of the hcp with strength enhancer is larger than that of pure hcp, and the CaO/SiO_2 in the interface zone is lower, because the active SiO_2 and Al_2O_3 absorb a great deal of the flake-like $Ca(OH)_2$ in the interface zone, the $Ca(OH)_2$ is harmful to the bond strength between the hcp and aggregate, so the interface structure is improved by the strength enhancer.

6 CONCLUSIONS

1. The WRALC dehydrated at 800°C is the most active, when 10% of cement is replaced by it the strength of concrete (W/C = 0.3) at 28 days is 15% higher than that of the control concrete.

2. The dehydrated WRALC can produce CSH gel in the presence of $Ca(OH)_2$, furthermore, it has the effect of micro-crystal and can accelerate the hydration process of the cement, hence increase the strength of the concrete.

3. The strength enhancer (mixture of dehydrated WRALC 800°C and zeolite finely ground together) can be used to make super-high strength concrete, when W/C = 0.3, 5% of cement is replaced by it, the strength of the concrete at 28 days is 93 MPa, 21% higher as compared with the control concrete.

4. The strength enhancer can improve the porosity and the interface zone structure, hence the strength of the concrete is increased.

REFERENCES

Pu Xincheng.1984. Science of Concrete. China Press of Architecture and Civil Engineering: 215–246.
Zhang Xiong.1990. Complex Cementitious Effect of Industial Waste Residue and its Computerized Database Model. Treatise of Tongji University: 15–18
Yuan Runzhang.1988. Structure Changing Process of Calcium Silicate Hydrate in Dehydration. Journal of Wuhan Technology University No.3: 8–13.
Gao Qiongying et al.1990. Study on the Rehydration of the Dehydrated Phases of CSH. Journal of Wuhan Technology University No.1: 25–33
Feng Naiqian.1988. The Strength Effect of the F-Mineral Admixture on Cement Concrete. Journal of Wuhan Technology University No.4: 43–47

Advances in Autoclaved Aerated Concrete, Wittmann (ed.)© 1992 Taylor & Francis. ISBN 90 5410 086 9

Autoclaved cementitious products using pulverised fuel ash

R.A.Carroll
Marley Building Materials Limited, Birmingham, West Midlands, UK

J.C.Payne
Marley plc, Sevenoaks, Kent, UK

ABSTRACT: Hydrothermal reactions of pulverised fuel ash (PFA) and lime at 184°C have been investigated using an experimental autoclaving facility. The process is characterised by the production of calcium silicate hydrates, which crystallise to tobermorite, and the formation of a hydrogarnet phase. Experiments using three PFA samples show that significantly different compressive strengths are achieved in autoclaved aerated concrete (AAC). The performance of a PFA sample in autoclaved products can not be predicted from either its oxide composition or mineralogy.

1 INTRODUCTION

This paper considers the reactions of PFA at autoclaving conditions typically used in the manufacture of AAC. New research facilities are reviewed, a description of the hydrothermal behaviour of PFA is given and some experimental results are discussed. Cement chemist's notation for compounds is used where appropriate in the text.

AAC has an intercellular matrix bound together by calcium silicate hydrates (CSH phases). The main raw materials required for manufacture are a lime source, to provide calcium ions, and a reactive silica. Quartz, if finely ground, is extensively solubilised at typical autoclaving temperatures (170 to 200°C). Ground, high purity quartz sand is therefore the most common siliceous material used for AAC production.

A widely used alternative to ground sand is PFA. In the UK, about 0.8M tonnes per annum of PFA is used in the manufacture of AAC.

This more complex raw material has been less fully studied for this particular application than ground sand. Few descriptions of the hydrothermal reactions between PFA and lime or cement have been published. There are no tests which replicate conditions during autoclaving. The performance of PFA within conventional concrete has been extensively researched and numerous test methods have been devised. Physical properties have been stressed rather than measurements of pozzolanic reactivity. The relevance of such tests for autoclaved products has been poorly covered in the literature.

2 RESEARCH FACILITIES

An adaptable experimental autoclaving facility has been built to investigate materials and processes at conditions comparable to industrial practice. Five research autoclaves, each of 0.58 m³ internal volume and supplied by an external steam source, have been installed at the Birmingham Laboratory of Marley Building Materials Limited (Figure 1).

Figure 1   Research autoclaves

Microprocessors operating the inlet, vent, and condensate valves ensure independent control of each vessel. Four thermocouples of 1.5 mm diameter, fitted within each vessel, enable the internal temperature of specimens to be measured during autoclaving. All stages of steam curing can be programmed and monitored.

The research autoclaves provide controlled curing for a range of specimens, including :

* pastes of PFA/calcium hydroxide or PFA/cement
* small scale AAC mixes
* uncured blocks from an AAC factory

Improved laboratory methods to characterise the behaviour of PFA are being sought. Simple ash/calcium hydroxide pastes cured at 184ºC are useful in providing basic data regarding hydrothermal performance.

A small scale mixing and casting facility at the Birmingham Laboratory can produce AAC under controlled conditions.   Mix

volumes ranging from 0.014 to 0.018 m^3 are produced, ensuring that a suitable density range is achieved. Normal experimental practice is to determine the compressive strength/density relationship for a given formulation using a standard autoclaving cycle.

AAC produced in this manner appears to be a satisfactory model of full scale manufacture. Investigations of raw materials, mix formulation and autoclaving practice have been successfully undertaken.

3   A DESCRIPTION OF HYDROTHERMAL
       REACTIONS

3.1 *Quartz products*

AAC produced using quartz, has calcium silicate hydrates as the main reaction products. The rates of formation, thermodynamic stabilities and the crystalline morphologies of the phases determine the structure and binding within the matrix. Numerous calcium silicate hydrates have been synthesised and these were reviewed by Taylor(1965). These phases have different mechanical strengths (Butt et al 1961). The strength of semi-crystalline C-S-H(I) was notable but high-lime phases such as α−dicalcium silicate hydrate ($\alpha-C_2SH$) had low strength. Ludwig and Pohlmann (1983) considered high strength in autoclaved products was due to the low density, high surface area of certain CSH phases. The high specific volume and fine crystal structure of such material provide numerous points of contact which act as bonding sites throughout the matrix. Beaudoin and Feldman (1979) regarded a mixture of poorly crystallised C-S-H(I) and tobermorite($C_5S_6H_5$) as responsible for maximum strength in autoclaved products.

Experimental results, such as those of Kalousek (1955) indicate that a reduction in the molar lime/silica ratio (C/S) of CSH phases occurs during autoclaving with an associated increase in crystallinity. An initial reaction product is often CSH gel, which converts into semi-crystalline C-S-H(II) or C-S-H(I), which in turn crystallises into tobermorite. A common reaction pathway was proposed for quartz or amorphous silica by Kalousek. Significant differences in formation rates of the phases are observed depending on the type of silica used.

3.2 PFA products

PFA is a by-product of the combustion of finely ground coal. Each ash particle cools from a slag droplet formed from a discrete coal fragment. Since adjacent coal fragments may differ in mineralogy the resulting PFA is a complex heterogeneous raw material. Most particles are spherical and have a range of silica, alumina and iron contents. The crystalline phases present are quartz, mullite, magnetite and haematite but amorphous alumino-silicate glass is the major constituent. The heterogeneous nature of PFA was stressed by Diamond (1986) who noted that composite structures were common. For example, mullite or magnetite crystals have been observed embedded in a spherical glass particle. Similar particles can have different internal structures and may exhibit dissimilar reactivities during autoclaving.

Results from experiments carried out at the Birmingham Laboratory of Marley Building Materials may be used to discuss aspects of the hydrothermal reactions of PFA.

Calcium hydroxide reacts rapidly with PFA at typical autoclaving temperatures. Figure 2 shows the consumption of calcium hydroxide in a paste containing a British PFA sample (RC1). An extraction method originally developed by Franke(1941) was modified to measure the free lime during autoclaving. After 6 hours, all calcium hydroxide (portlandite) has reacted. XRD results also

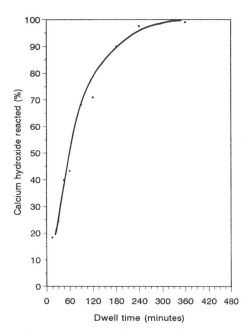

Figure 2   Reaction of lime with PFA at 184°C

confirm that portlandite is not present. This may be explained by the rapid reaction of lime with silica solubilised from amorphous glass within the PFA. This agrees with Kalousek (1955), who noted that amorphous silica is more reactive in the early stages of autoclaving than quartz. There is little evidence for the formation of high-lime reaction products, such as α-C_2SH. Differences in the reaction rates of PFA samples have been observed.

PFA based products show a steady formation of a CSH phase during the early stages of autoclaving. This may be identified by the growth of a strong XRD peak in the region 0.307 to 0.309 nm. A peak at 0.304 nm, due to C-S-H(I) or CSH gel, is often observed. The CSH phase continues to form with autoclaving. After approximately 2 to 3 hours, crystalline tobermorite appears as a reaction product. This is evident by a peak near 1.14 nm due to the basal reflectance of the phase. These observations are consistent with the formation of a poorly

299

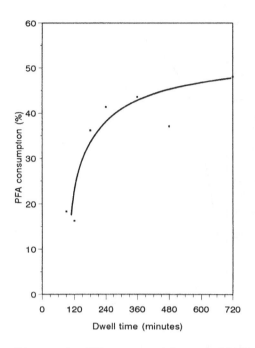

Figure 3   PFA consumption at 184°C

crystalline calcium silicate hydrate such as CSH gel and its crystallisation into tobermorite, possibly via C-S-H(I) as an intermediate. Formation of the initial CSH phase levels off between 4 and 6 hours but transformation into tobermorite continues.

PFA contains a significant amount of aluminium which is liberated along with silicate ions during autoclaving and this influences the hydrothermal reactions. Aluminium-substitution of the semi-crystalline CSH phases occurs which increases the rate of crystallisation to tobermorite. Kalousek (1957) noted that up to 5% Al_2O_3 could be accommodated within a calcium silicate hydrate. A more important effect with PFA is the formation of a silica-substituted hydrogarnet. Indeed, the presence of such a phase may be of equal significance to the growth of calcium silicate hydrates. A unit cell dimension between 1.225 and 1.235 nm has been noted for the hydrogarnet in PFA based materials tested at the Birmingham Laboratory. This

suggests a phase similar to katoite (C_3ASH_4) reported by Passaglia and Rinaldi (1984).

Hydrogarnets have dense, octahedral crystals which provide few points of contact and therefore contribute little to the strength of the matrix. Results on specific phases are not available, but Jambor (1972) noted that materials containing hydrogarnet had a coarse microstructure of low strength.

The iron within PFA may modify the reaction since multivalent ions such as Fe^{3+} can substitute for aluminium ions within hydrogarnets, resulting in an increase in the unit cell dimension. Zur Strassen (1958) and also Sauman (1973), confirmed that high iron contents favoured the formation of hydrogarnet. Sauman et al (1975) considered the hydrogarnet composition was given by the formula $C_3A_{1-y}F_yS_xH_{6-2X}$.

A gravimetric method has been devised to estimate the ash reaction in PFA/calcium hydroxide pastes. The insoluble residue after treatment with hydrochloric acid is taken as a measure of unreacted PFA. Uncombined calcium

After 12 hours autoclaving at 184°C, PFA consumption begins to level off at about 50%, despite a glass content over 80%. This suggests that a significant amount of glass within PFA remains unreacted under industrial autoclaving conditions. The precipitation of insoluble reaction products may inhibit silica solubilisation. Such an effect in autoclaved specimens containing amorphous silica was noted by Kondo(1965). Alternatively, a gradation in the reactivity of the glass within PFA may exist. Diamond (1986) judged that PFA was composed of several glasses which underwent pozzolanic reactions at different rates.

4 SMALL SCALE AAC MIXES

Factory experience at Marley Building Materials has shown that the compressive strength of AAC is greatly influenced by the PFA source. An experiment to study this effect, using the research facilities at the Birmingham Laboratory, is discussed below.

4.1 Experimental

Three PFA samples were evaluated within small scale AAC mixes using a common autoclaving cycle. Two of the ashes tested, reference RC1 and DD2, were collected from UK power stations. The ash reference GW3 was obtained from a USA power station. The experimental mix used is given in Table 1.

Analyses of the PFA samples were obtained. Oxide compositions were determined by x-ray fluorescence (XRF) and mineralogy was assessed by quantitative x-ray diffraction (XRD). The 45 micron residue was determined for each ash in accordance with BS 3892: 1982.

For each PFA sample, a number of small scale AAC mixes were produced of different heights by varying the aluminium powder addition. After rising and stiffening, the test mixes were steam purged at atmospheric pressure for 30 minutes prior to autoclaving. The autoclave cycle involved 120 minutes linear pressurisation to 10 bar (184°C), a dwell period of 360 minutes, followed by 120 minutes linear depressurisation.

Specimen 200 mm cubes were cut from the autoclaved mixes using a diamond saw. Two opposite faces, perpendicular to the direction of the rise, were milled parallel to each other. The cubes were saturated in water for 24 hours and tested for compressive strength with the milled faces in contact with the platens of the compression test machine. The cubes remained intact after failure and were dried at 105°C for 7 days. The cubes were weighed and oven dry densities calculated using the original dimensions.

Sub-samples of each type of AAC were analysed by XRD using CuKα

Table 1. Experimental AAC mix

Raw material	Weight (g)
PFA	4420
cement (OPC)	2084
quicklime (CaO)	503
anhydrite	180
mix water	3400
aluminium	4 to 8

Table 2

PFA Analysis

	RC1 (%)	DD2 (%)	GW3 (%)
oxides			
SiO_2	50.7	51.8	55.4
Al_2O_3	26.1	25.3	33.0
Fe_2O_3	8.9	11.4	3.6
CaO	1.7	1.5	0.9
MgO	1.7	1.0	1.2
Na_2O	1.5	1.1	0.3
K_2O	3.4	1.5	2.6
TiO_2	1.0	1.0	1.8
SO_3	0.7	0.5	0.2
LOI	4.3	4.9	1.0
mineralogy			
quartz	3.9	2.6	5.4
mullite	5.8	3.6	14.2
magnetite	0.8	1.1	0.2
haematite	0.5	0.4	0.4
glass	84.7	87.4	78.8
carbon	4.3	4.9	1.0
45 μm residue	13.5	20.1	14.7

Table 3   XRD results for AAC mixes

	Relative peak intensity (%)		
d-spacing (nm)	RC1	DD2	GW3
CSH phases			
1.140–1.180	56	56	55
0.308–0.309	58	59	53
0.304	33	55	33
0.298	34	--	35
0.280	--	32	30
0.182	14	15	17
hydrogarnet			
0.276–0.277	47	48	40
0.226	26	24	22
0.200	26	25	23

radiation. Rutile was interground at 10% addition level, as an internal standard. Scans from 5 to 60 degrees 2θ was carried out.

DTA/TGA analysis up to 1000°C in a nitrogen atmosphere was also undertaken.

4.2 Results

Analyses of the three ashes are summarised in Table 2.

DTA traces obtained from AAC made with each of the ashes were relatively featureless but proved useful in discounting certain phases. No peak at 400 to 550 °C was evident, suggesting that any portlandite ($Ca(OH)_2$), produced from either the cement or quicklime had been consumed during autoclaving. The absence of a peak at 460 to 480°C showed that $\alpha-C_2SH$ was not present in any of the AAC specimens. A shallow endotherm at 780 to 880°C was apparent in AAC made with GW3 or DD2. The presence of calcite ($CaCO_3$) was discounted since effervescence did not occur when specimens were treated with dilute acid.

Figure 4 shows the compressive strength/density relationships determined for AAC cubes.

The XRD traces obtained were complex but Table 3 summarises the main peaks attributed to CSH phases and hydrogarnet. Peak intensities quoted are relative to the main rutile peak at 0.325 nm.

4.3 Discussion

Similar silica contents are evident in ashes RC1 and DD2. A marginally higher silica content occurs in GW3. The high aluminium and low iron contents of GW3 is notable. There are distinct differences in the mineralogy of the ashes. DD2 has the highest glass content but the lowest quartz and mullite levels. Conversely, GW3 is relatively low in glass but has the highest

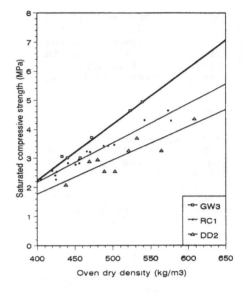

Figure 4 Effect of PFA source on the strength of AAC

quartz and mullite levels. The ash may be derived from a low-iron coal which has a poor fluxing action during combustion. The low loss on ignition figure of GW3 suggests efficient combustion.

Crystalline tobermorite is clearly indentified in the three AAC specimens by the basal reflectance at 1.14 to 1.18nm. Most peaks which can be attributed to calcium silicate hydrates have similar intensities whichever ash is used. However, AAC made with DD2, has a large peak at 0.304 nm, which suggests that a significant amount of semi-crystalline CSH gel or C-S-H(I) is present.

The position of the basal peak for tobermorite in AAC, appears to be related to the ash used. This suggests differences in the aluminium-substitution within the phase. Kalousek (1957) and also Diamond et al (1966) showed aluminium-substitution increased the basal spacing of tobermorite. The peak positions noted with GW3 and RC1 are 1.14 and 1.15 nm respectively, indicating aluminium contents of 10 to 15%. With DD2 a peak occurs at 1.18 nm, suggesting high levels of aluminium within the tobermorite.

Hydrogarnet is evident in the XRD traces obtained with all three sets of AAC. This is most obvious by a strong peak in the region 0.276 to 0.277 nm. Overlap with peaks due to anhydrous calcium silicates from cement is possible. There is however, a peak at 0.20 nm which is characteristic of hydrogarnet. AAC made with GW3 has less intense hydrogarnet peaks than in specimens made with either RC1 or DD2. This suggests lower levels of the phase associated with the use of GW3.

Semi-crystalline CSH phases such as C-S-H(I) are reported to have a DTA exotherm in the region 830 to 950°C. This is absent in all three AAC specimens. Kalousek and Prebus (1958) noted that with well crystallised tobermorite the endotherm is not observed.

Significant differences are apparent in the strength of AAC made using the three ashes. At a particular density, GW3 gives a higher strength than that achieved with DD2. Intermediate strength material is obtained with RC1. As suggested by XRD results, this may be due to the relative amounts of hydrogarnet formed. The position of the basal peak of tobermorite may indicate that aluminium ions are liberated readily during autoclaving from DD2. The ash also has a high iron content. Both these factors will assist hydrogarnet formation. Conversely, much of the aluminium in GW3 is combined in mullite, making it less available during autoclaving, and the ash has a low iron content.

It is interesting to compare the strength results from the experiment to studies of ashes used in conventional concrete. Helmuth (1987), during a review of pozzolanic reactions considered that the important PFA properties were fineness, glass content and silica or silica plus alumina. Watt and Thorne (1965), noted the importance of glass content but also cited a relationship between the long term strength of mortar and the silica plus alumina content of PFA. At normal

temperatures it is generally assumed that the glass alone in PFA undergoes pozzolanic reactions. The other siliceous components such as quartz and mullite are assumed to be unreactive. Neither the oxide analysis or mineralogy of a PFA sample, is adequate to predict performance within AAC. It should be noted that GW3, the ash with the least glass, containing the most mullite and quartz, gives AAC with the highest compressive strength.

The techniques described and others outlined by Payne and Carroll (1991) may prove the basis for test methods to evaluate PFA used in autoclaved products.

5 CONCLUSIONS

During the autoclaving of PFA based products, parallel to the formation and crystallisation of calcium silicate hydrates, is the rapid synthesis of a stable hydrogarnet phase. These two processes may be regarded as competitive since the strength of the autoclaved matrix will be determined by the relative amounts of each type of phase formed.

Differences in the compressive strength of AAC can be attributed to the PFA sample used. It has not been possible, however, to predict the performance of an ash from either its oxide composition or mineralogy.

Test methods to assess PFA for use in conventional concrete are inadequate for autoclaved products such as AAC. Techniques which model the hydrothermal curing of current manufacture are required.

6 REFERENCES

Beaudoin, J.J. and Feldman, R. F. 1979. Journal of Material Science, vol.14, pp.1681-1693.
Butt, Y.M.;Rashkovich, L.N.; Kheiker, D.M.; Maier, A.A. and Gracehva, O.I. 1961. Silikat-Technik, vol.12, part 6, pp.281-287.

Diamond, S. 1986. Cement and
  Concrete Research, vol.16,
  pp.569-579.
Diamond, S.; White, J.L. and
  Dolch, W.L. 1966. Amer.
  Mineral., vol. 51, pp.388-401.
Franke, B. 1941. Z. anorg. allgem.
  Chem.,vol. 249, p.180
Helmuth, R. 1987.PFA in Cement
  and Concrete. PCA Publication
  Skokie, Illinois. p.169.
Jambor, J. 1972. Proceed. Sec. Int
  Symp. Sci. Res. Silic. Chem. I
  48, Brno.
Kalousek, G.L. 1957. Journal of
  the American Ceramic Society,
  vol.40, No.3, pp.74-80.
Kalousek, G.L. 1955. Journal of
  the American Concrete
  Institution, pp.989-1011.
Kalousek, G.L. and Prebus, A.F.
  1958. J. Amer. ceram. Soc., vol.
  41, p.124
Kondo, R. 1965. Symposium on
  Autoclaved Calcium Silicate
  Building Products. Society of
  Chemical Industry.London.pp. 92-
  97.
Ludwig, U. and Pohlmann, R. 1983.
  Tiz-Fachberichte. 107, (11)
  pp.826-833.
Passaglia, E. and Rinaldi, R. 1984
  Bull. Miner., vol. 107, pp. 605-
  618
Payne, J.C. and Carroll, R.A. 1991
  Proceedings: 9th International
  Ash Use Symposium. American Coal
  Ash Association, Washington,D.C.
  vol. 1, 26-1.
Sauman, Z. 1973.Proceed. XI th
  Siliconf. Budapest, pp.461-473
Sauman, Z; Cerna, J; Lach, C and
  Zdenek, V. 1975. Silikaty, vol.2
  pp.105-115.
Taylor, H.F.W. 1965. Symposium on
  Autoclaved Calcium Silicate
  Building Products. Society of
  Chemical Industry. London.
  pp.195-204.
Watt, J. & Thorne, D.J. 1965.
  Journal of Applied Chemistry.
   vol.15, pp.584-594.
Zur-Strassen, H. 1958. Zem-Kalk-
  Gips., vol.11, pp 137-143.

Advances in Autoclaved Aerated Concrete, Wittmann (ed.) © 1992   Taylor & Francis. ISBN 90 5410 086 9

Ashes from fluidized bed combustion power plants as a potential raw material for the production of autoclaved aerated concrete

M. Jakob & H. Mörtel
Institut für Werkstoffwissenschaften III, Erlangen, Germany

ABSTRACT :   In the course of development of materials, that include by-products, the attempt was made to substitute sand in autoclaved aerated concrete by using ashes from fluidized bed power plants. These ashes were characterized, for containing other phases in comparison to well known ashes from "ordinary" pulverized coal combustion power plants. With lime, portland cement, anhydride, sand and the ashes aerated concrete was produced and steam cured under 190°C and 12 bar for 650 minutes. Physical properties obtained are as follows:

Flexural strength	0.6-1.9	MPa
Compressive strenght	1.3-5.2	MPa
Bulk density	0.43-0.57	g/cm^3

1. INTRODUCTION

Ashes result in huge amount by the combustion of coal in power plants. In the case of pulverized mineral coal combustion power plants about 90 % of the 7 million t. of the ashes are utilized in Germany.

At power plants working with fluidized bed combustion methods only some percent of the resulting ashes are utilized. To remedy this unsatisfactory state the attempt was made to use the fludized bed combustion ashes in autoclaved aerated concrete.

The goal is to save room for deposition of the ashes and raw materials for the production of aerated concrete.

In fluidized bed combustion power plants normally two different types of ashes occure, the coarse bed ashes and the fine fly ashes. There are different possibilities to use the ashes separate, but in general they are mixed. Unfortunately in many power plants it is not possible to get the separate ashes. Especially for aerated concrete only the fine fly ashe can be used, so that the entire ash has to be separated or to be grinded. So such an ash was grinded with a

Tab.1 : chemical compound of ashes; liw : loss in weight;

Ash	SiO_2	Al_2O_3	Fe_2O_3	CaO	MgO	K_2O	SO_3	liw.
#1	39.7	18.9	4.9	22.4	1.81	3.18	6.49	2.33
#2	33.3	18.6	5.37	23.5	2.08	2.82	8.99	4.19
#3	25.9	14.5	5.58	24.8	1.19	1.11	3.36	24.6
#4	15.5	9.27	2.87	49.4	0.93	2.54	8.55	7.7
#5	43.4	29.3	5.6	3.4	3.21	1.22	1.22	9.5

(in weight percent)

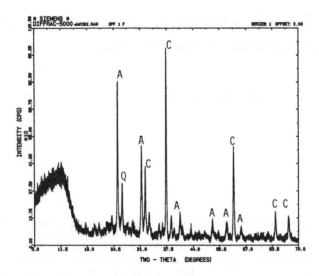

Figure 1 : XRD of fly ash (Q) quartz, (C) lime, (A) anhydride.

Table 2 : physical properties of fly ashes

Ash	specific surface [cm^2/g]	Density	d_{50}
#1	4500	2.83	23μm
#2	5700	2.76	19μm
#3	6500	2.85	18μm

(specific surface : Blaine)

ball mill and seperated with an air classifier.

2. ANALYSIS OF THE ASHES

The chemical analysis of the fly and bed ashes is shown in Table 1. For comparison the data of a typical fly ash coming from pulverized coal combustion is given in number five.

It is striking, that the ashes #1 to #4 beeing fly ashes from different fluidized com-bustion power plants and one bed ash (#4) include more SO_3 than #5. This is because of the addition of limestone together with the coal into the combustion room. This addition and the formation of $CaSO_4$ is responsible for the high capacity desulphuration in flui-dized combustion power plants. The loss of weight is mainly dependend on the burning out of the rest coal C.

The phase composition of the ashes was measured with XRD. As seen in Fig.1 the main parts are quartz, lime and anhydrite. Some not definitly identified peaks should be clayey minerals as illite or muscovite. The main difference to ordinary fly ash is the missing of glassy parts and of mullite.

The physical characteristics of the fly ashes is shown in Table 2. #3 is the not treated one, whereas #1 and #2 have been grinded and air classified.

The morphology of the fly ash was observed with SEM. The essential difference, confirming the XRD results, to ordinary ashes is the lack of the glassy, spherical particles. The observed particles are much more irregular. This is because of the lower combustion tem-

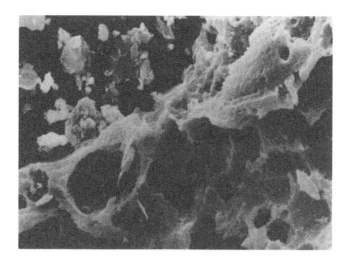

Figure 2: SEM photo of a typical fluidized bed combustion fly ash

peratures in fluidized bed combustion plants. The temperatures are in general not higher than 930 °C, while in pulverized coal combustion plants they can reach up to 1500 °C.

3. PRODUCTION OF AUTOCLAVED AERATED CONCRETE

The raw materials for the production of autoclaved aerated concrete are in general lime, sand, portland cement, an-hydride and aluminium powder as foaming agent. Aerated concrete with a different C/S (CaO/SiO_2) ratio was produced by substituting sand by fly ashes of different power plants or by the prepared fly ashes.

The C/S ratio is inter alia responsible for the developing CSH - phases. Therefore it has been varied from 0.3 to 1.2 to get the maximal strength of the aerated concrete. Cause the ashes containing lime, the standard share of binding agents in the recepture has been reduced from 26 % to 22 %.

One difficulty during the production of the areated concrete was the high reactivity of the ashes, which in the beginning caused damage by developing too high temperatures in the fermenting process. This could be solved by reducing both, the water/solid ratio and the water temperature. The temperature development can be explained with the free CaO and the high specific surfaces of the ashes.

After drying the product was steam cured under the ordinary production conditions in an industrial used autoclave (at YTONG laboratories in Schrobenhausen Germany). The temperature was 190°C and the autoclaving time 650 minutes.

4. TESTING OF AUTOCLAVED AERATED CONCRETE

The autoclaved material was cut into 4 x 4 x 16 cm prisms. The flexural strength was tested by a 3 point bending test method. The span length was 12 cm. Out of the remainding rest 4 cm cubes were cut and subjected to compressiv strength tests. To compare the different densities the so called A - parameter was calculated, in which the sample density is taken in consideration.

In the following equation the symbols have the meaning :

A : A-parameter/value
σ : flexure strength
δ : density

$$A = \frac{\sigma}{0.016 * \delta^2} \qquad (1)$$

It is known, that using ordinary fly ash in aerated concrete the curve in figure 3 has a maximum. As seen in figure 3, in which fly ash from fluidized combustion

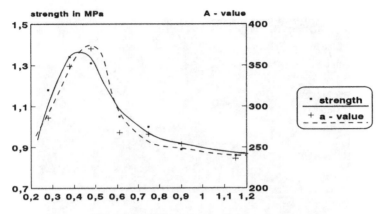

Figure 3: C/S ratio vs A-parameter

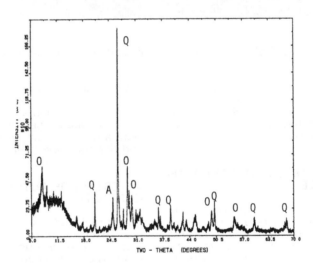

Figure 4 : typical XRD of autoclaved areated concrete using fluidized combustion fly ash. (O) : oyelite, (Q) quartz, (A) anhydride

plants is used, it is possible to find batches with a maximum strength. The flexure strength, determined by three point bending tests is much more sensitive than the compressiv strength, therefore here only the flexure strenght is figured. The value of the compressive tests are about 2.5 times higher.

Samples of the autoclaved areated concrete were analysed with XRD. The detected phases were oyelite, anhydride, quartz and some other CSH (tobermorite) phases. Samples with high C/S ratio showed also C3A2SH4 - hydrogarnet phases.

The shrinkage of the ashes was measured according the method refered in (Xu, 1991). This method was used till the end

of the shrinkage. The temperature was about 25 °C and the air moisture about 55 %. The aerated concrete had decreases in lenghth about 0.20 to 0.35 mm/m. These values are relatively low and can be accepted.

5. CONCLUSION

Autoclaved aerated concrete, with use of ashes from fludized coal combustion power plants can be prepared. It is necessary to harmonize the recipe of the areated concrete, to get high strength. A behaviour like figure 3 could be detected with different original fly ashes and with the prepared (grinded, air classified) ones. The resulting values of strength and shrinkage can be accepted.

From the business point of view only the seperate fly ashes may be acceptable, because additional benefication (grinding, air classifying) is too costly. Another problem is the availability of ashes with constant compounds.

The durability of autoclaved aerated concrete containing ashes from fludized combustion power plants should be investigated in the future.

6. REFERENCES

[1] Xu, X. Mörtel, H. Dauerhaftigkeit des Flugasche enthaltenden Gasbetons. Diplomarbeit Universität Erlangen-Nürnberg 1991

[2] Mörtel, H. Flugasche als Rohstoff für die Gasbetonherstellung. Rezepturoptimierung AIF Forschungsvorhaben 562/84; 1986

[3] Kautz,K. Untersuchungen an Aschen aus Wirbelschichtfeuerungen; VGB technisch wissenschaftliche Berichte "Wärmekraftwerke" VGB TW 667/71

[4] Mörtel,H. Verwertungsmögichkeiten von Flugaschen in der Silikatbetonindustrie, VGB Kraftwerkstechnik, Essen (1988) Band2, S C15-C19

Advances in Autoclaved Aerated Concrete, Wittmann (ed.) © 1992   Taylor & Francis. ISBN 90 5410 086 9

Extension of the range of primary materials for the production of AAC

I.Weretevskaja & N.Sudelainen
Silicate Concrete Institute, Tallinn, Estonia

ABSTRACT: The authors of this paper will present here zones of composition of industrial secondary products (fuel, ashes, slags) and their suitability as raw materials for the production of autoclaved cellular concrete.

The concept of raw material base extention and increase of cellular concrete production by the use of bulky secondary products is found on the utilization of local materials, which allows to reduce production costs and energy consumption as well as to improve the ecological situation. Autoclave technology of cellular concretes based on synthesis of new minearls at increased temperature and pressure makes it possible to utilize many industrial by-products as raw materials.

During 30 years we have accumulated a great deal of experience in research on the composition and properties of different industrial waste: fuel ashes, metallurgical slags, ore concentration waste as well as experience in their utilization for production of cellular autoclaved concretes. Mineral waste of metallurgic, mining and power industry are mainly silicates and aluminium silicates, calcium, magnesium and silicon oxide. Due to this, under autoclave treatment such products possess hydraulical properties of cementious binding of concrete.

To estimate the use of industrial waste in building materials, there are classifications of products based on their aggregate conditions (Volzhensky, 1969) and chemical activity (Bozhenov, 1978). Best suited for building material industry are hard powdery waste products. The chemical activity of products depends upon their chemical and mineralogical composition. Special attention in our research was paid to metallurgical slags and fuel ashes, as most bulky industrial waste. Fig. 1 shows zones of composition of the products on the basis of highly durable autoclaved cellular concretes which were produced.

According to the correlation of the ciesomposition of the main components, i.e. calcium oxide, silicon oxide, the sum of aluminium and ferrous oxides, the secondary products are divided into two main grops. The criterion of such a division is their content of calcium oxide. Products with high content of calcium contain between 25% and 56% CaO, acid products up to 10% CaO. Both slags and ashes have a glassy phase with higher reactivity in common. Slags and ashes with high calcium content possess their own binding ability at autoclave hardening.

The chemical activity of secondary products under the conditions of autoclave hardening is determined by the coefficient of basicity as suggested by Bozhenov (1978):

$$K_{bac} = \{(CaO + 1.03MgO + 0.6R2O) -$$

$$(0.55Al_2O_3 + 0.35Fe_2O_3 + 0.7SO_3)\}/0.93SiO_2$$

This coefficient takes into account the sequence of interreactions between the main oxides of the mixture during the formation of low-base calcium hydrosilicates in the final product.

While selecting the composition of a aerated concrete mixture, products having a coefficient of basicity $K \geq 1$ completely or partially substituted the binding agents - lime and portland cement. With coefficient $K < 1$ and not less than 50% of silicon oxide, the product was used as siliceous component in the mixture. With small content of silicon oxide quartz sand was added to the raw material mixtures.

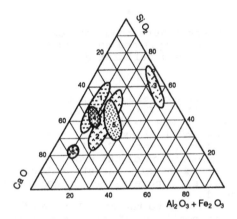

Fig. 1 Zones of composition of secondary products

1-shale ash, $K_{bac} = 0.7 - 1.25$

2-acid ashes of Kansko-Achinsk coal, $K_{bac} = 0.2 - 1.4$

3-acid ashes of black and brown coals, $K_{bac} = -0.3 - 0.2$

4-blast-furnace granulated slag, $K_{bac} = 0.8 - 1.26$

5-steel-smelting slags, $K_{bac} = 0.4 - 1.15$

6-Portland cement $K_{bac} = 2.43$

Different technological processes, as a result of which industrial waste products are formed, are the cause of chemical, mineralogical and physical heterogeneity of such waste products. Variations of the composition are often observed with the waste of one and the same industry. More constant with regard to their properties are waste products formed at strictly regulated technological peocesses of the main product, for example, metallurgic slags,

especially granulated ones. Among fuel ashes, acid ashes are more homogeneous. Therefore both dry ashes and ashslag mixtures from ash dumps are suitable for autoclave concretes. High calcium ashes contain hydraulically active minerals and they ought to be used in dry condition. The composition of high calcium ashes varies substantially depending on the fuel composition, the conditions of burning and the removal of the ash. High calcium ashes have many components and are polydisperse. While fly ash is removed differenciation of seperate fractions takes place with respect to their dispersity and composition. In this case one should choose the ash fraction that suits the conditions of aerated concrete techology. Thus, for example, to manufacture aerated concrete on the basis of ashes from Estonian shale deposits, ash collected in cyclones is chosen. Shale cyclone ash contains 40-53% CaO, including 15-28% disconnected (free) CaO, which makes it possible to use it for the production of aerated concrete as mineral binder (Goryainov et al. 1987).

The greatest economic efficiency in the production of areated concrete is acheived, if fly ash of electric power stations is used as raw materials, due to the reduction of price and energy needed for grinding. For the production of autoclaved cellular concrete there is no necessity to radically change the usual technological set-up of production. In some cases, however, it becomes necessary to introduce corrections in the technology to ensure optimum conditions for the synthesis of new formations and required properties of the final product. Therefore, in every specific case while organizing production of cellular concrete, it is necessary to have complex technological tests of raw materials to choose optimum composition and technological parameters, taking physical and chemical particularities of the product into consideration. The selection depends essentially on ash properties, fineness of grinding of raw materials, mix composition, conditions of forming and thermal treatment.

In technological research for production of aerated concrete, a lime/cement ratio of 0.8-1.0 was used, The only exception is aerated concrete on the basis of slag ash, used as binder without utilization of lime and cement.

Parameters such as compressive strength, slump and frost resistance of aerated concrete and the duration of raw mix maturing before cutting serve as criteria for optimisation of the technical parameters. Tradi-tional moulding method and impact method with vertically directed impulses were used. The impact method of forming cellular cement allows us to decrease the water content of the mixture by 15-25%, which is especially effective while using

Table 1: Characteristics of aerated concrete on the basis of secondary products

Fuel ash, electric power station	Ash (slag) content kg/m^3	Method of for-ming	Properties of aerated concrete		Shrinkage mm/m	Coefficient of frost resistance at 35 cycles
			Density (Dry) kg/m3	Com-pressive strength MPa		
Karaganda, Frunze termal power station	350	impact	599	3.84	0.6	0.95
Kuznetsk, Novgorod therm.el.cent.	280	impact	632	4.11	0.57	0.87
Azeisk, Novoziminsk therm.el.station	325	impact	585	4.20	0.33	1.07
Intinsk, Kondopoga therm. el. cent.	350	mould.	566	3.51	0.5	1.0
Ekibastuz, Krasnogorsk therm.el.station	350	impact	537	4.6	0.7	1.0
Podmoskov, Stupinsk therm.el.station	440	mould.	645	4.51	0.56	0.85
Novo-Achinsk, Krasnoyarsk Hydropower station	380	mould.	592	4.1	0.51	1.0
Baltic shale Baltic Hydro-electric station	345	mould.	600	4.0	0.5	0.90
Blast furnace granulated slag	125	mould.	598	4.78	0.45	1.02
Blast furnace slag from dump	175	impact	597	3.8	0.43	0.98

highly dispersed porous acid ashes. Impacts ensure dilution of viscous thick mixtures and normal dispersion of the mass. When impacts are over, the mixture regains its original strength very quickly, which considerably shortens the exposure time of the massive blocks before cutting into elements (Goryainov et al. 1987). Autoclave treatment is made at steam pressure of 0.8-1.2 MPa.

In some cases, as for example, for the manufacturing of aerated concrete on the basis of high calcium slag ash, it is necessary to have a two-stage thermal treatment.

Thorough investigations of the technological process under laboratory and industrial conditions showed that cellular concretes produced on the basis of ashes and slags are not inferior to concretes made of traditional raw materials. The most typical properties of them are given in Table 1.

The addition of ash and slag for the production of cellular concrete of medium density 600 kg/m^3 is 280-440 kg/m^3 and 115-175 kg/m^3 respectively. The compressive strength of cellular concrete on the basis of chemically active secondary products corresponds to class B 2.5-3.5, which corresponds to 3.5-5.0 MPa. Cellular concrete endures not less than 35 cycles of tests for frost-proof, shrinkage at drying does not exceed 0.7 mm/m.

To utilize industrial waste for the manufacturing of cellular concrete the Silicate Concrete Institute (the company 'YAICO') possesses elaborations of technology, equipments and technological line complexes of different output(20-220 thousands m^3 per year) of nonreinforced and reinforced elements of various dimensions.

REFERENCES

Bozhenov E. 1978. Technology of autoclave materials, Leningrad
Galibina E. & Weretevskaya I. 1973 Compound, processes ensuring strength properties of slag ashes and their utilization in Building, industry. Reports Ibausil 5
Goryainov K., Povel E., Weretevskaya I., Dombrovsky A. & Sazhnev N. 1987. Investigations on selection of technological parameters and conditions for manufacturing cellular concrete on the basis of ashes of electric power station of Middlo Urals, Tallinn Collection of works of NIPISilicatbeton
Volzhensky A., Burov Y., Vinogradov B. & Gladkih K 1969. Concretes and particles made of slag and ash materials. Moskow

Advances in Autoclaved Aerated Concrete, Wittmann (ed.) © 1992   Taylor & Francis. ISBN 90 5410 086 9

Sintered foamed fly ash and acid compositon and fly ash floater and acid material as alternatives to cellular concrete

Carolyn M. Dry

University of Illinois, Champaign, Ill., USA

ABSTRACT: A material which uses energy waste products has been developed for uses similar to cellular concrete. It is also foamed used alumina. The foamed sintered (600^0 C) fly ash/alumina/acid composition is lightweight, 800 kg/m^3, thermal conductivity 0.0062 W/mK, and of moderate compressive strength 5 000 - 7 000 kPa. This material is used as the center of structural sandwich panels. A fly ash material made of sphere was then substituted to improve insulative/weight properties. This material's improved performance has a lower density and a higher insulative value and higher compressive strength. It has a microstructure of large discontinuous spherical pores, while the other material has a large number of smaller pores uniformly distributed. Studies by other researchers on cement verify that the improved thermal and lightweight performance as well as reduced compressive strength can be expected as a result of these changes in microstructure. A densified fly ash with large tubular cells is being researched as a way of further increasing compressive strength while reducing the density and increasing thermal value.

1 FOAMED FLY ASH AS RELATED TO CELLULAR CONCRETE

The foamed fly ash/phosphoric acid alumina material is quite similar to cellular concrete in that the origin of the air voids is a reaction aluminum which produces hydrogen gas. The "typical properties of insulating cellular concrete are density 300 to 1100 Kg/m^3; compressive strength 300 - 7 000 kPa, thermal conductivity 1 to 0.3 W/mK (Mindess & Young, 1981).

Figure 1. Sandwich panel

The void structure is discreet, nearly spherical 1-1 mm. bubbles" (Mindess and Young, 1981). It is a low-density, low-strength material used for nonstructural insulating purposes, usually. The foamed fly ash material is an inexpensive, medium density 800 Kg/m^3 medium compressive strength 5 000 - 7 000 kPa and of good insulative value material, thermal conductivity 0.0062 W/mK. This material is designed to be used in the center of sandwich panels, exterior applied insulation, or for fire resistance or moderate strength masonry units. The fly ash material is also similar to PALC, a precast auto- claved, lightweight ceramic used in buildings by Misawa Homes of Japan. A silica rich waste material, it produces a nonstruc- tural panel and provides good sound insulation, fire resist- ance, and is durable. It can be made into wall sections and then hardened in an autoclave with high heat and pressure. Bubbles are trapped during autoclaving. It has good insulative value, is lightweight, and used for curtain wall-like construction, with separate structural members for support. The replacement for imported wood and the proximity to large markets made it very cost-effective. (Sackett, 1986).

2 RESEARCH ON FLY ASH MATERIAL

The goal of this research done under the Advanced Construction Technology Center at the University of Illinois sponsored by the U.S. Army Research Office was to produce a material like PALC, made from waste, but which would be structural, as well as low-cost, lightweight, and insulative. It would be used for building components much like cellular concrete is used.

We chose to use energy consumption materials for production of the sandwich panels, blocks, exterior insulating, and other components. Ashes gave the best material properties. For sandwich panels the parameters were: 1.) moderate structural capacity, 2.) high strength-to-weight ratio, 3.) inexpensiveness, 4.) casting of chases into the material, 5.) low temperature sintering, 6.) good insulative value, 7.) flexural strength, 8.) durability.

We chose ash materials, bottom ash and fly ash, which give the above parameters and also 1.) use common chemicals (phosphoric acid and ashes) which can bond the two materials together for a sandwich panel, and 2.) use low temperature sintering for chase in-casting. (The goal was to fire below 600^0 C so that aluminum chases could be used.) Compositions were optimized, then the two materials were combined as a sandwich panel. A mixture of fly ash, alumina, and phosphoric acid was developed and evaluated as a lightweight foam sandwich panel interior material with low firing temperature. Bottom ash and phosphoric acid were developed and evaluated as a strong exterior material with a low firing temperature (Dry 1991).

Most sandwich panels produced do not bond well between layers, and the safety of the foamed insulation (interior) material is in dispute. We developed sandwich panels which are lightweight, made of a thin tough structural material on the outside and lightweight insulative material on the inside (Figure 1). They are bonded well, transfer shear, and use safe materials.

Table 1. Chemical analysis

	Raw Fly Ash	Calcined Fly Ash (900 C)	Ammonia Treated Fly Ash	Fly Ash & H_3PO_4
SiO2	39.58	39.51	38.00	28.55
Al2O3	19.49	19.41	18.58	13.11
Fe2O3	6.50	6.56	6.18	4.89
CaO	21.57	21.59	20.88	14.89
MgO	5.05	5.04	4.74	3.25
K2O	0.64	0.65	0.61	0.45
Na2O	1.83	1.83	1.69	1.21
TiO2	1.43	1.43	1.33	0.97
P2O5	1.14	1.14	1.06	16.98
MnO	0.037	0.038	0.03	0.015
MnO2				

Table 2. X-ray analysis: raw fly ash, calcined alumina, phosphoric acid, unfired.

Compound	Powder File	% of Diffraction
- Quartz SiO_2	(5-490)	8%
- Al_2O_3 Corundum	(10-173)	33%
$Ca(H_2PO_4)_2$ H2O	(9-347)	52%
Unexplained lines		7%

Raw fly ash, calcined alumina, and phosphoric acid at 200^0 C.

Compound	Powder File	% of Diffraction
SiO2 @-quartz	(5-490)	4%
Al2O3 Corundum	(10-173)	21%
My3Si2O5(OH)4 Chlorite	(22-1155)	3%
Ca(H2PO4)2·1H2O	(9-347)	37%
Zeolite 5A	(19-1183)	10%

Raw fly ash, calcined alumina, and phosphoric acid, fired at 250^0 C.

Compound	Powder File	% of Diffraction
Quartz SiO_2	(5-490)	13%
Al_2O_3 Corundum	(10-173)	57%
$Ca(H_2PO_4)_2$ H_2O	(9-347)	4%
$CaHPO_4$	(9-80)	14%
Unexplained lines		12%

Fly Ash Material

The fly ash used in this research is a fine powder from the stacks of coal-fired furnaces in power plants in Arkansas (Table 1 and Table 2).

The x-ray analysis shows the primary bond being formed is a calcia-phosphate bond at 200^0 C.

After firing at 250^0 C, the calcia-phosphate bond has changed as evidenced by the x-ray analysis. (Dry, 9th International Ash Use Symposium Proceedings, 1991)

Early research conclusions on fly ash are:

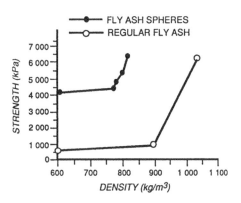

Fig. 4 Compressive strength of samples made of fly ash spheres vs. samples made of regular fly ash

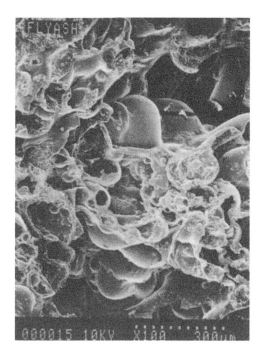

Fig. 2. SEM of foamed fly ash/acid material

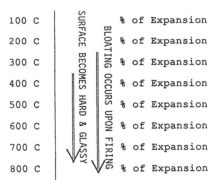

Fig. 3. Surface condition as related to temperature:fly ash, alumina, and phosphoric acid

1. Fly ash, alumina, and phosphoric acid undergo an exothermic reaction to produce a lightweight material.
2. A calcia-phosphate bond is formed but is replaced by corundum above 350^0 C. sintering. The surface hardens and the material bloats further (SEM Figure 2 and Figure 3).

Hollow fly ash spheres are the part of the fly ash which floats above 40 percent of the

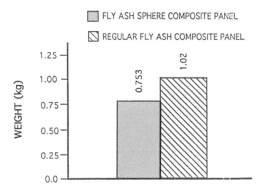

Fig. 5. Weight of composite panel made of (5.08 cm x 10.16 cm x 15.24 cm) fly ash spheres vs. composite panel made of regular fly ash

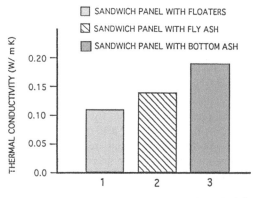

Fig. 6. Relative thermal values of sandwich panels made of floaters, fly ash, and bottom ash material

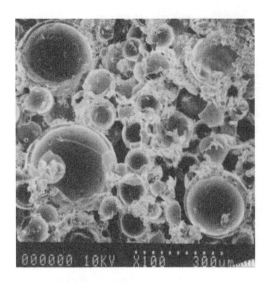

Fig. 7.   SEM of fly ash containing floaters phosphoric acid

Fig. 8.   Photo of exterior applied insulation samples made from floater fly ash material

fly ash.   Testing was done using this separated fly ash product to replace the fly ash.   Like the fly ash from which it is separated, these spheres are higher in calcium content than most fly ashes.   The research conclusions are: 1.) the fly ash spheres mixed with phosphoric acid give density range of 800 kg/m³ and related higher compressive strengths than the foamed fly ash (Figure 4), and 2.) the baseline insulation value using spheres is increased.   The use of hollow fly ash spheres reduces the weight (Figure 5) and increases the insulative value

Figure 9.   Photo of sample masonry unit made of foamed fly ash composition

of the sandwich panel Figure 6).

Cell size and shape as related to surrounding material were found to greatly influence microstructural factors, and therefore strength, weight, and insulation value in the two fly ash sintered materials studied.   In the first case, ash mixed with a water/phosphoric acid solution and alumina foamed and produced a microstructure consisting of a large number of small open cells dispersed relatively uniformly throughout.   The material has a density of 800 kg/m³ compressive strength of 5 000 - 7 000 kPa, but in some cases, as high as 14 000 kPa, and thermal conductivity 6.27×10^{-3} W/mK.   In the second material made of fly ash closed spheres (floaters) with acid and water, the microstructure of the sintered material had large closed spherical cells which were discontinuous (Figure 7).   This material had greater compressive strength.   This is to be expected if one considers the open cells as made of edge/stuts while closed cells have the added strength of the cell wall itself, and any enclosed gas pressure (Gibson 1988).   This material was lighter in weight, 590 kg/m³ and of higher thermal value of 0.0041 W/mK thermal conductivity.   Closed cells have less heat loss due to conduction through gas, the dominant mode of loss (Gibson 1988).   This material was made into exterior applied

318

Fig. 10. Photo of samples of densified fly ash with void s

insulation products and masonry units (Figure 8 and Figure 9).

Densified Fly Ash with Tubular Cells

Densified fly ash with tubular cells was produced with long tubular voids (Figure 10). The fly ash was densified by use of higher temperature and/or more acid (Figure 11). Straws, which burned out, produced the tubular voids. The goal is to move to higher compressive strength while maintaining lightweightness and high thermal value. Preliminary studies are inconclusive.

The linear cells made from straws can be thought of as a honeycomb structure. The analysis of compressive fracture "in place" (onto the honeycomb tube ends) for this brittle solid reveals that the cell wall fractures after first bending, then thickening, or densifying (Gibson 1988).

"When loaded in-plane the cell walls at first bend, giving linear elasticity which can persist to as large as straws, 10%. This is possible because the honeycomb is like a spring. The geometry allows a large distortion of the structure with only small strain in its members.

In a brittle material, strength determines the maximum tolerable gradient. If the honeycomb contains a defect, it may fail at a lower stress (than otherwise possible)" (Gibson 1988).

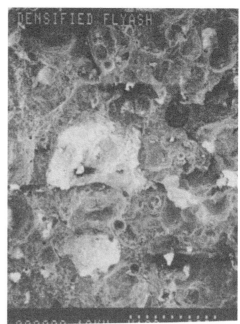

Fig. 11. SEM of densified fly ash composition

This explains the theoretical basis for going to a honeycomb-like linear cellular structure using densified, less flawed brittle material for the cell walls.

CONCLUSION

Development of a densified honey-comb cellular fly ash material is being researched. Theory shows that this is the way to increase compressive strength, yet maintain lightweightness and insulative properties.

Two structural materials made of fly ash and phosphoric acid showed that compressive strength can only be moderately increased if lighter density and the good thermal properties are maintained as in the floater fly ash material.

REFERENCES

Mindess, S. and Young, F.J. 1981. Concrete, Prentice-Hall, Publisher, Englewood Cliffs, New Jersey, p. 595.

Mindess, S. and Young, F.J. <u>Ibid.</u>

Sackett, J.G., May 1986. "Japan's manufactured housing capability," pp. 51-58.

Dry, C.M., January 1991. "Durability of sandwich panels made from phosphate bonded/sintered fly ash and bottom ash, 9th International Ash Use Symposium Proceedings, Electric Power Research Institute, pp. 77-1 to 77-21.

Dry, C.M., Ibid., pp. 77-3, 77-4.

Dry, C.M., Ibid., p. 77-13.

Gibson, L.J. and Ashby, M.F., 1988. Cellular solids, structure, and properties, Pergammon Press, Oxford, England.

Gibson, L.J. and Ashby, M.F., Ibid., p. 205.

Gibson, L.J. and Ashby, M.F., Ibid., p. 71.

Gibson, L.J. and Ashby, M.F., Ibid., p.113.

Gibson, L.J. and Ashby, M.F., Ibid., pp. 112-113.

Advances in Autoclaved Aerated Concrete, Wittmann (ed.) © 1992   Taylor & Francis. ISBN 90 5410 086 9

Autoclaved coal ash lime bricks

G.Usai
Università di Cagliari, Italy

ABSTRACT: Re-using of industrial by-products- as Coal Ashes of Thermal Power Plants - as raw-materials in the production of building materials, represents a valid solution of the environmental problem of disposal. In this research the production of masonry bricks by the autoclave hydration of compacted ash-limes pellets, with other activators, has been carried out. Furthermore the porous structure of the hydration products has been related to their mechanical properties. These materials possess good technical quality.

1. INTRODUCTION

The unfavoureable result of the popular consultation on "Nuclear Energy" (8th, november 1987), has compelled the Italian Government to revise his whole energetical Policy. The national Electrical Power Organization, ENEL, has particularly turned again his care to traditional fuels as Coal of foreign or national production, Enel (1989).

Two problems regard this solid fuel, mainly: desulfuration and the re-use of the great mass of combustion Ashes, until 20% of the mass of burned fuel,Enel (1982). Fly Ash is an annoying and harmful waste that has widely employed in Cement and concrete manufacturing, owing to his high pozzolanic activity, Massazza (1962); it is necessary, on the other hand, to study alternate re-uses of the great mass of Ashes difficult to dispose off. In this paper some researches are referred, on the possibility of produce, by autoclave hydration under high pressure, bricks made by mixes of Fly-Ashes and Furnace Slags (heavy ashes), and other activators.

2. EXPERIMENTS

2.1 Materials

Fly-ashes (CPV), supplied by ENEL Power Plant of Porto-Vesme, Cagliari,Italy; Furnace-Slags (CCS) supplied by Ferrovie Meridionali Sarde, Carbonia, Italy; Portland Cement Clinker, supplied by Italcementi, Samatzai, Cagliari, Italy; Pure $Ca(OH)_2$, Lime, Carlo Erba, Milano, Italy;

Chemical compositions are shown in table 1.

2.2 Methods

Cylindrical samples, height 2 cm, diameter 4 cm, were prepared by an hydraulic press, moulding pressure 20 N/mm^2. The mixes containing Fly Ashes,Furnace Slags,clinker or lime were prepared, moulded thed autoclaved at 5,10 or 20 bar, see Fig. 1, at various hydration periods. The mixes were of following proportions, mass: CCS with 10% lime, CCS with 0,5% clinker, CPV 1 part, lime 2 parts, CPV 1 part, Clinker 0,5 parts.

2.3 Testing

Compressive strenghts were measured by an Comazzi Dynamometer.

Porosities were obtained by means of a Carlo Erba 200 Porosimeter; the X-ray diffraction Patterns were obtained by a G.E. XRD-5 Diffractometer using the Cu-kalpha-1 radiation.

The results are shown in Figures and tables.

Other tests were carried out according Amandola, (1980).

Table 1:Chemical Compositions of Materials.

	CCS	CPV	P.Clinker
Al_2O_3	16%	7.11	5.10%
Fe_2O_3	31		4.80
SiO_2	31.50	67.80	25.30
TiO_2	1.5		
CaO	15	6	60.20
n.d.	4.4	1.17	0.10
l.o.i.	0.5	11.1	4.50
MgO		3.6	
inc.		1.0	
SO_3		2.22	

Bulk density: CCS 1.25 Kg/liter
             CPV 0.85 Kg/liter
Granulometry (both); passing the sieve of
             4500 m/sq.cm (DIN
             Normen 1170).

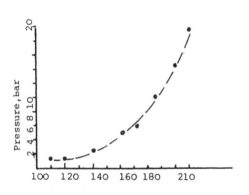

Temperature °C

Fig,1;Pressure-temperature trend,Auto-
clave "RMU-Giazzi",Italy.

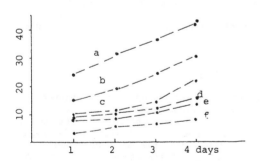

Fig.2:Compressive Strenght,N/mm^2,ver-
tical,v/s curing time.a:CCS-lime 20 bar;
b:CCS-lime 10 bar;c:CCS-clinker 20 bar;
d:CCS-lime 5 bar;e:CCS-clinker 10 bar;f:
CCS-clinker 5 bar.

3. RESULTS AND DISCUSSION

3.1 Compressive Strenghts

Many experimental attempts have been made
in order to ascertain the optimal mass
proportions between the two ashes and the
activators, lime or Portland Cement
clinker, this one chosen for confrontation
    The Fly Ashes CPV, a material with high
pozzolanic activity, Mariani (1980), needs
a higher proportion of activator than the
Furnace Slag, an ash of a sardinian
pitch-lignite, Giarratano (1935), which is
a low energy pozzolan, Amat (1980). In
Figures 2,3 and 4 the growth of the
compressive strenghts of bricks, v/s the
curing pressure or the curing time, is
shown. In the same conditions, i.e. with
the same mass of activator, the reaction
of the two ashes with the clinker is more
complete, probably because this material
gives, by hydration, a greater mass of
amorphous or microcrystalline resistant
phases, than the lime, Colleopardi (1980).
    Furthermore the hydration at 5 or 10
bar, even put off until four days, not
gives the strenghts obtained at 20 bar,
at shorter curing times. This fact shows
that the curing pressure (or curing
temperature, see Fig.1) is the preminent
factor in hydration reaction; probably,
with greater mass of clinker, the furnace
slags CCS can achieve the strenghts of
the Ash-lime bricks.

3.2 Porosities

The porosities of the samples can
obsviously correlate with the strenghts,
see Figure 5. Some differences, between
the Fly Ash bricks and Furnace Slag bricks,
because the clinker activated Bricks-with
the same mass of activator- are less porous:
the silica-lime reaction is here more
complete, Massazza (1979), and gives more
compacted structures.
    It is probable, furthermore, that the
reaction products of the clinker
precipitation into the greater pores of
the structure, have reduced the total
porosity, making the materials more
compacted with many "little" voids, Joisel
(1973).

3.3 X-rays diffraction patterns

Figures 6,7 and 8 show diffractometries of
"green" mixes and hydrated materials.
    For Furnace Slag-clinker mixes, we can
note the attenuation and disappearance

of Portland Cement compounds peaks, and
the formation of an extended amorphous
phase structure, due to clinker hydration
and to ash-lime reaction. Some peaks (f),
related to the low reactive Ferro-oxide
phases, Michel (1964), are even evident
after the hydration. For Furnace Slag-lime
mixes we can note, with the Ferro-oxide
peaks (f), the existence of low cristalline
phases in the material, probably because
a silica-lime reaction has taken place,
as in pozzolanic-like materials.

Very similar observations can be made
with regard to the bricks made with fly
ashes, clinker or lime. In this case also,
the structures due to the hydration
reaction, can be attributed mainly to
formation of amorphous phases; on the other
hand some very neat peaks (p) can be
attributed to the formation of crystalline
compounds as plazolite, Massidda (1979),
obtainable at high curing pressures from
materials as the fly ashes.

3.4 Other Tests

The presence of efflorescences is
practically light or negligable, in all
kinds of tested bricks. Specific weights
are neatly higher for the Furnace Slag
bricks, with the same moulding pressure,
20 N/mm^2 of the "green" samples. This
because the microstructure of the
fly-ashes, owned to his origin (a rapid
freezing of kiln exit gases from 300°C
to lower temperature), is practically a
mass of void vitreous bubbles, very
disperded, Sersale (1980).

The water absorbance is strictly related
to specific weights and porosities.

4. CONCLUSIONS

The autoclaved hydration of Coal Ashes
mixes, with an activator, lime or Portland
Cement clinker, in order to obtain

Table 2:Characteristics of bricks.

	I	II
CCS-clinker	2.25	6.40
CCS-lime	2.07	6.85
CPV-clinker	1.24	17.20
CPV-lime	1.10	21.00

I: specific weight, Kg/liter
II: water absorbance, mass %

curing: 20 bar, 4 days;

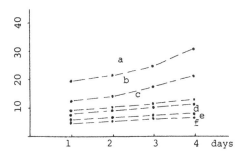

Fig.3:Compressive strenght,N/mm^2,verti-
cal,v/s curing time.a,b,a.s.o.:see Fig.2.

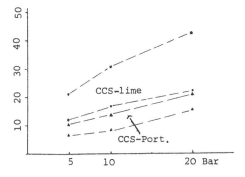

Fig.4:Compressive strenght,N/mm^2,verti-
cal,v/s curing pressure.

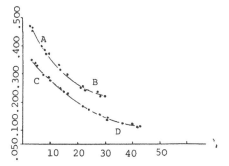

Fig.5:Total porosity,ml of mercury/1 g,v/s
compressive strenght,N/mm^2,abscissas.
A,B:CPV bricks(clinker;lime).
C,D:CCS bricks(clinker;lime).

materials of Brick type, is strictly
dependent on hydration temperature, rather
than the curing time. The Bricks possess
good technical properties, with differences
between the raw materials, fly ashes and
furnace slags. The re-using of Coal Ashes
from the ENEL Power Plants represents,

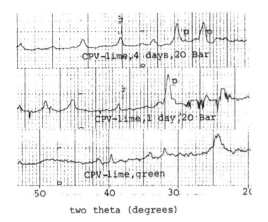

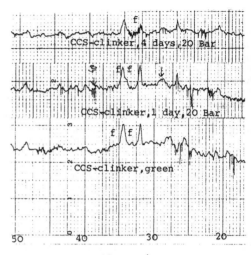

Fig.6 : X-ray diffraction pattern of green
and hydrated CPV-lime

two theta(degrees)

Fig.8 : X-ray diffraction pattern of green
and hydrated CCS-clinker

For the X-ray analysis shown in Figs. 6 to
8, Cu-k-alpha-1 radiation has been used.

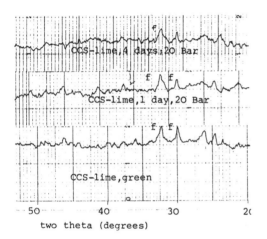

Fig.7 : X-ray diffraction pattern of green
and hydrated CCS-lime

therefore, a valid indication on
alternative raw materials in Ceramic
Industry, and a contribution to solution
of enviromental problem of industrial
wastes disposal.

REFERENCES

Amandola, G. e Terreni, V. 1980, Analisi
Chimica Tecnica e Strumentale Milano:
Tamburini.
Amàt, P. and Usai, G. 1980. Production
of Pozzolanic Cements with Sulcis Coal
Ashes, Rend. Fac. Sci. Univ. Cagliari,
69,7–14.
Collepardi, M. 1980, Scienza e Tecnica
del Calcestruzzo, Milano: Hoepli.

ENEL, 1989. Bollettino Prod. Energia El.
Italia, 2,1–20.
ENEL, 1982. Rapporto Impatto Ambientale
"Gioia Tauro", 10–20.
Giarratano, M. 1935. Carbone ed
Elettricità, Quadrante, 29,10–13.
Joisel, A. 1973. Les Adjuvants du Ciment.
Paris: Joisel.
Mariani, E. 1976. I Leganti Idraulici.
Milano: Casa Editrice Ambrosiana.
Massazza, F. e Cannas, M. 1962. Reattività
Pozzolanica delle Ceneri Volanti, Il
Calore, I,9–15.
Massazza, F. 1974. Sulla Reattività
Pozzol. Il Cemento, 3,131–140.
Massidda, L. and Sanna, U. 1979, Structure
and Characteristic of Compacts, Cement
and Concrete Research; 9,127–134.
Sersale, R. 1980. Structure and
Characterization of Pozzolanas, Proc.
7th Int. Congr. on Cement Chemistry,
Paris, IV,3–18.

Bibliography on autoclaved aerated concrete 1982-1992

Shinsaku Tada

Note: The Second International Conference on AAC was held in Lausanne, Switzerland, in 1982. On this occasion, a bibliography on AAC covering major contributions published before 1982 has been prepared by Y. Houst and F. H. Wittmann. This bibliography has been published in three parts in the proceedings volume of the conference: Autoclaved Aerated Concrete, Moisture and Properties, edited by F. H. Wittmann, 1983, pp 325-369, Elsevier Scientific Publishing Company, Amsterdam, The Netherlands.

The present bibliography covers the period following the second AAC-Conference up to now. It is published in alphabetical order of the first authors. Titles of the papers are indicated in English, French or German. The language in which the original paper is written is given in brackets.

Abbreviations used in this AAC bibliography

ASME HTD.	American Society of Mechanical Engineers, Heat Transfer Division
ASTM Spec Tech Publ.	American Society of Testing and Materials, Special Technical Publication
Autoclaved Aerated Concr Moisture Prop.	Autoclaved Aerated Concrete - Moisture and Properties
Batim. internation.	Batiment Internationale
Batiment Batir.	Batiment-Batir
Bauplanung Bautech.	Bauplanung-Bautechnik
Bautenschutz Bausanierung.	Bautenschutz + Bausanierung
Betonwerk Fertigteil-Tech.	Betonwerk + Fertigteil-Technik.
Build Res Establ Dig.	Buiding Research Establishment Digest
Cem Concr Res.	Cement and Concrete Research
Chem Eng Process.	Chemical Engineering and Processing
Concr Int.	Concrete International
Concr Works Int.	Concrete Works International
DDH.	Das Dachdecker-Handwerk
Dtsch Bauz.	Deutsche Bauzeitschrift
Elem Fertigbau.	Element + Fertigbau
Environ Int.	Environment International
Fiz.-Khim. Mekh.	Fiziko-Khimicheskaya Mekhanika Materialov

Gesund Ing Haustech BauphysUmwelttech.	Gesundheits Ingenieur. Haustechnik, Bauphysik, Umwelttechnik
Health Phys.	Health Physics
Hebezeuge Foerdermittel.	Hebezeuge und Foerdermittel
Indian Concr J.	Indian Concrete Journal
Ing Archit Suisses.	Ingenieurs et Architectes Suisses
Int J Cem Compos Lightweight Concr.	International Journal of Cement Composite and Lightweight Concrete
J. Am. Ceram. Soc.	Journal of American Ceramics Society
J Acoust Soc Am.	Journal of Acoustic Society of America
J Mater Civ Eng.	Journal of Materials in Civil Engineering
J Mater Sci.	Journal of Materials Science
Key Eng Mater.	Key Engineering Materials
Ki.	Ki (Karlsruhe)
Kniznice Odb. Ved. Spisu Vys. Uceni Tech. Brne, B.	Kniznice Odbornych a Vedeckych Spisu Vysokeho Uceni Technickeho v Brne
Kunstst Bau.	Kunststoffe im Bau
Mag Concr Res.	Magazine of Concrete Research
Mater Struct.	Materials and Structures
Numer Methods Therm Problems Part 2	Numerical Methods in Thermal Problems
PB Rep.	PB Reports
Prep. Ann. Meeting Ceram. Soc. Japan.	Preprints for Annual Meeting of Ceramics Society of Japan
Prep. Ann. Meeting CAJ.	Preprints forAnnual Meeting of Cement Association of Japan
Proc. Ann. Meet. JCI.	Proceeding of Annual Meeting of Japan Concrete Institute
Rheol. Fresh Cem. Concr.; Proc. Int. Conf.	Proceedings of International Conference of Rheology of Fresh Cement and Concrete.
Schweiz Ing Archit.	Schweizer Ingenieur und Architekt
Stavebnicky Cas.	Stavebnicky Casopis
Stroit. Mater.	Stroitel'nye Materialy
Struct Eng Part A.	Structural Engineers Part A
Stud. Environ. Sci.	Science Reports. Studies of Fundamental and Environmental Sciences. (Hiroshima Univ. Japan)
Tech Bau.	Technik am Bau
Therm Conductivity.	Thermal Conductivity
Zh. Prikl. Khim.	Zhurnal Prikladnoj Khimii
Ziegelind Int.	Ziegelindustrie International

1. ABD-EL-WAHED, M. G.; EL-DIAMONY, H.; GALAL, A-EL-K. F.; SHATER, M. A.
   Cement dust binder for autoclaved cellular concrete.
   TIZ. 109(4), 265-267 (1985).
   Waste,Manufacture. (English)

2. ABYZOV, A. N.; SERGEEV, S. I.; TYUSHKINA, G. L.
   Heat-resistant phosphate aerated concrete produced from chromium-titanium-corundum
   and zirconium dioxide slurries.
   Issled. i Primenenie Stroit. Materialov na Osnove Mest. 1, 119-124 (1984).
   Fire resistance, Composition. (Russian)

3. ACHTZIGER, J.
   Praktische Untersuchung der Tauwasserbildung im Innern von Bauteilen mit
   Innendaemmung.
   Waermeschutz, Kaelteschutz, Schallschutz, Brandschutz. NO.Sonderausgabe, 9-14 (1985).
   Walls, Moisture. (German)

4. ACHTZIGER,J.; CAMMERER,J.
   Investigation of the moisture influence, as a function of application, on the heat
   transport through insulated external structural members.
   Bauphysik. 12(2), 42-46 (1990).
   Moisture,Thermal properties. (German)

5. AGARWAL,P. D.
   Lightweight concrete from fly ash.
   Congr Rep 13th Congr Int Assoc Bridg Struct Eng., 81-86 (1988).
   Lime, Fly ash. (English)

6. AKERBLOM, G.
   Environmental radon investigations in Sweden.
   U. S. Environ. Prot. Agency, Off. Radiat. Programs. 35(1), 171-188 (1983   ).
   Radio activity. (English)

7. AKIHAMA, S.; TADA, S.
   New building materials and special constructions. Members made of new building
   materials.
   Concrete Technology (Japan). 25(3), 62-66 (1987).
   Fiber, Prefablication, Panels. (Japanese)

8. ALEKSANDROVSKIJ, S .V.
   Fluage et particularités du comportement des bétons cellulaires autoclavés dans les
   conditions de l'Extrème Nord.
   Beton i zelezobeton. (3), 21-23 (1982).
   Durability, Protection. (Russian)

9. AMKHANITSKII, G. YA.; LEVIN, S. N.; SHMIDT, A. M.
   Technology of aerated concrete based on wastes from dressing of Kursk Magnetic
   Anomaly iron ores.
   Povysh. Tekhnol. Snizhenie Mater. Energoemkosti Sbornogo , 119-126 (1982).
   Manufacture, Waste. (Russian)

10. AMKHANITSKII, G. Ya.; LAPIDUS, M. A.; TURKINA, I. A.
    Small wall blocks from unautoclaved aerated ash concrete.
    Stroit. Mater. (7), 14-15 (1991).
    Walls, Fly ash. (Russian)

11. ANON.
The Linford 2 plant of Durox Building Products Ltd.
Concrete plant and production. 8 (6), 231-234 (1990).
Manufacture. (English)

12. ANON.
Foam - lightweight concrete answer?
N Z Concr Constr. 33(May), 2-6 (1989).
Manufacture, Physical properties. (English)

13. ARNDT, H.
Gasbeton - ein Baustoff mit Zukunft.Leistungsfaehigkeit und Einsatzgrenzen.
Betontechnik. 6(3), 93-95 (1985).
Thermal insulation. (German)

14. ARONI, S.
Shear strength of reinforced aerated concrete beams with shear reinforcement.
Mater Struct. 23(135), 217-222 (1990).
Strength, Testing, Panels. (English)

15. ARONI, S.
On energy conservation characteristics of autoclaved aerated concrete.
Mater Struct. 23(133), 68-77 (1990).
Thermal insulation, Standards. (English)

16. ARONI, S.; CIVIDINI, B.
Shear strength of reinforced aerated concrete slabs.
Mater Struct. 22(132), 443-449 (1989).
Slabs, Strength. (English)

17. ATZENI,C.; MASSIDDA.L.; SANNA,U.
Properties of gas concretes containing high proportions of PFA.
Cem Concr Res. 21(4), 455-461 (1991).
Fly ash, Mechanical properties, Rheological properties. (English)

18. BAGDA, E.
Bauphysik der Baustoffoberflaeche.
Bautenschutz Bausanierung. 8(2), 53-55 (1985).
Heat transfer. (German)

19. BAGDA, E.
Der instationaere Temperatur- und Feuchteverlauf an gedaemmten Fassaden.
Kunstst Bau. 19(4), 174-177 (1984).
Environment, Heat transfer. (German)

20. BAGROV, B. O.
Recycled industrial products for the manufacture of cellular concretes.
Beton i zelezobeton. 20(4), 16-18 (1988).
Manufacture, Waste. (Russian)

21. BAGROV, B. O.; VASIL'EVA, T. D.; SADOVSKII, P. A.; FEDYANIN,V. S.; PARASHCHENKO, V. V.
Aerated concretes from industrial wastes.
Beton i zelezobeton. (9), 6-7 (1990).
Manufacture, Waste. (Russian)

22. BARANOV, A. T.
Chemical additives in the production of aerated concretes.
Issled. po Str-vu Stroit. Teplofiz. Dolgovech. Konstruktsii, (43), 28-36 (1989).
Manufacture, Additive. (Russian)

23. BASCOUL, A.; CABRILLAC, R.; MASO, J.C.
Influence de la forme et de l'orientation des pores sur le comportment mécanique
des béton cellulaires.
Proc. 1st Int. RILEM Congr. Vol.1 Pore Struct. Mater. Prop. ed. Maso, J.C., 333-340 (1987)
Pore-size distribution, Modeling, Thermal properties. (French)

24. BAVE, G.
Regional climatic conditions,building physics and economics.
Autoclaved Aerated Concr Moisture Prop. ed. F.H. Wittmann, 1-12 (1983).
Durability, Physical properties. (English)

25. BEERLING, D. J.
The testing of cellular concrete revetment blocks resistant to growth of Reynoutria
japonica houtt (Japanese knotweed).
Water research : (Oxford). 25 (4), 495-498 (1991).
Environment, Protection. (English)

26. BELAU, E.; HARSTEL, G.; GOTTSCHALK, W.
Beitrag zur Automatisierung des Schneidprozesses bei der Gasbetonherstellung im VEB
Metalleichtbaukombinat-Werk Calbe.(Teil I).
Betontechnik. 9(6), 168-172 (1988).
Manufacture. (German)

27. BERETKA,J.; TAYLOR,A.
Use of granite dust for making aerated concrete and ceramics.
Key Eng Mater. 53/55, 512-517 (1991).
Manufacture, Waste. (English)

28. BERNDT, K.
Bauen mit Gasbeton-Bauteilen im Industriebau.
Zentralbl Ind. 29(6), 465-470 (1983).
Fire resistance, Physical properties. (German)

29. BERNDT, K.
Bauen mit Gasbeton-Bauteilen im Industriebau. II.
Zentralbl Ind. 30(1), 47-52 (1984).
Acoustic insulation, Physical properties. (German)

30. BERNDT, K.
Bauen mit Gasbeton-Bauteilen im Industriebau (III).
Zentralbl Ind. 30(3), 204-208 (1984).
Durability, Testing. (German)

31. BERNDT, M.; KELLER, R.; UTZ, W.
Beschichtungen auf Gasbeton.
Betontechnik. 9(5) ,132-134 (1988) .
Finishing, Permeability. (German)

32. BLUMENBERG, J.; STRASSER, A.
Gas- und Schwerbetonabsorber in Kombination mit Sole-Waermepumpen.
Tech Bau. (9) ,705-713 (1987) .
Sandwich panel, Thermal insulation. (German)

33. BOGATYREV, G. M.; MAKAROV, A. B.
Expansion of the raw-material base for the manufacture of aerated concrete.
Stroit. Mater. (2), 4 (1991) .
Manufacture, Composition, Expansion. (Russian)

34. BOLDT, R.; GUSCHMASCH, A.; SCHLEGEL, E.
Ueber die Temperaturabhaengigkeit der Waermeleitfaehigkeit von feuerfesten
Leichtsteinen.
Silikattechnik. 36(4) ,119-121 (1985) .
Modeling, Heat transfer, Thermal properties. (German)

35. BONDI, P.; STEFANIZZI, P.
Rilevamento sperimentale di parametri caratteristici del trasporto di umidite in
materiali porosi:Apparecchiatura e primi risultati.
Termotecnica. 43(6) ,35-39 (1989) .
Heat transfer, Moisture,Testing. (Italian)

36. BOWDEN, G.
Autoclaved aerated concrete: flexible solutions to new regulations.
Concrete (London). 23 (11) ,21 (1989) .
Standards. (English)

37. BRESCH, C-M.
Sanierung von Gasbeton.
Dtsch Bauz. (7) ,919-925 (1989) .
Joints, Sealant, Cracking. (German)

38. BRIESEMANN, D.
Autoclaved aerated concrete structural units in building construction.
Betonwerk Fertigteil-Tech. 53(8) ,565-568 (1987) .
Panels, Mechanical properties. (English)

39. BRIESEMANN, D.
Prefabricated parts made of autoclaved aerated concrete (AAC) for use in
noncombustible buildings.
Betonwerk Fertigteil-Tech. 52(11) ,731-735 (1986) .
Panels, Physical properties. (English)

40. BRIESEMANN, D.
The shrinkage of masonry composed of autoclaved aerated concrete blocks and having
joints of normal or very small width.
Betonwerk Fertigteil-Tech. 50(10) ,705-709 (1984) .
Shrinkage, Joints. (English)

41. BRIESEMANN, D.
Structural elements and masonry of autoclaved aerated concrete.
Autoclaved Aerated Concr Moisture Prop. ed. F.H. Wittmann, 257-266 (1983).
Deformation, Testing, Physical properties. (English)

42. BRIESEMANN, D.
Anmerkungen zur Berechnung bewehrter Bauteile und von Mauerwerk aus
dampfgehaertetem Gasbeton.
Betonwerk Fertigteil-Tech. 48(10), 638-642 (1982).
Panels, Mechanical properties. (English)

43. BRIESEMANN, D.; FREY, E.
Glueing of reinforced autoclaved aerated concrete slabs to form room size wall
units.
Betonwerk Fertigteil-Tech. 53(3), 190-193 (1987).
Slabs, Joints, Prefablication. (English)

44. BRIGHT, N. J.
Aerated concrete blocks - research developments in the United Kingdom.
Concr Works Int. 1(5), 181-185 (1982).
Physical properties. (English)

45. BRIGHT, N. J.; LUCKIN, K. R.
Effects of moisture on the lateral strength of AAC masonry when tested in the form
of wallettes.
Autoclaved Aerated Concr Moisture Prop. ed. F.H. Wittmann, 283-295 (1983).
Moisture, Walls, Mechanical properties. (English)

46. BRUEHWILER, E.; WANG, J.; WITTMANN, F.H.
Fracture of AAC as influenced by specimen dimension and moisture.
J Mater Civ Eng. 2(3), 136-146 (1990).
Strength, Fracture. (English)

47. (Building Research Establishment, Dep. Environment, GBR)
Autoclaved aerated contrete.
Build Res Establ Dig. NO.342, 8p (1989).
Standards. (English)

48. BUSCH, J.; HEIDRICH, K. D.
Laengenaenderungen von Plattenbauten mit Gasbeton-Aussenwaenden.
Bauzeitung. 39(7), 306-308 (1985).
Shrinkage, Walls, Cracking. (German)

49. CABRILLAC, R.; LUHOWIAK, W.; DUVAL, R.
Practical study of non-autoclaved anisotropic aerated concrete blocks.
Mater. Eng. (Modena, Italy). 1(3), 865-872 (1990).
Manufacture. (English)

50. CAMMERER, W. F.
Der Feuchtigkeitseinfluss auf die Waermeleitfaehigkeit von Bau- und
Waermedaemmstoffen.
Bauphysik. 9(6), 259-266 (1987).
Moisture, Thermal properties. (German)

51. CHARIEV, A.; CHRISTOV, YU. D.; VOLZHENSKII, A. V.; LARGINA, O. I.
Use of unautoclaved aerated concrete from Barkhan sand.
Beton Zhelezobeton (Moscow). 62(10), 25-26 (1988).
Composition, Silica. (Russian)

52. CHATTERJI, S.
Freezing of aqueous solutions in a porous medium. Part I. Freezing of
air-entraining agent solutions.
Cem Concr Res. 15(1), 13-20 (1985).
Freezing, Expansion. (English)

53. CHERNOV, A. N.
Béton cellulaire hétérogène de densité cellulaire variable.
Beton i zelezobeton. (3), 25-26 (1982).
Density. (Russian)

54. CHERNYKH, V. F.
Technology of Styropor-modified aerated concrete and prospects of its use in
housing construction.
Asfal't. i Tsement Betony dlya Uslovii Sibiri, Omsk. 63(9), 97-103 (1989).
Manufacture. (Russian)

55. CHITHARANJAN, N.; KALIAPPAN. T. P.
Utilisation of celcrete wastes for low-cost roofing.
Indian Concr J. 62(10), 523-531 (1988).
Roof, Waste, Manufacture. (English)

56. CHITHARANJAN, N.; MANOHARAN, P. D.
Reinforced aerocrete thin joints as a substitute for timber.
Indian Concr J. 63(9), 441-449, 4 (1989).
Fiber, Mechanical properties. (English)

57. CHITHARANJAN, N.; SUNDARARAJAN, R.; MANOHARAN, P. D.
Development and application of Aerocrete with nonmetalic fibres.
Proc Int Symp Fibre Reinf Concr 1987 Vol 2, 7.63-7.74 (1988).
Fiber, Mechanical properties. (English)

58. CHITHARANJAN, N.; SUNDARARAJAN, R.; MANOHARAN, P. D.
Development of Aerocrete: A new lightweight high strength material.
Int J Cem Compos Lightweight Concr. 10(1), 27-38 (1988).
Waste, Fracture, Mechanical properties. (English)

59. CHONAN, S.; KUGO, Y.
Acoustic design of a three-layered plate with high sound interception.
J Acoust Soc Am. 89(2), 792-798 (1991).
Acoustic insulation, Sandwich panel. (English)

60. CHUCHUEV, A. S.
Kinetics of surface tension of model aerated concrete mixes in the early structure
formation period.
Tekhnol. Mekh. Betona, Riga. (12), 74-79 (1987).
Hydration, Rheological properties, Hadening. (Russian)

61. CHUSID, M.
    Autoclaved cellular concrete.
    Prog. Archit. 71 (5) , 43-46 (1990) .
    (English)

62. CIVIDINI, B.
    Investigation of long time deflectionon reinforced aerated concrete slabs.
    Autoclaved Aerated Concr Moisture Prop. ed. F.H. Wittmann , 267-281 (1983) .
    Moisture, Slabs, Creep. (English)

63. CORMON, P.
    Amélioration   des bétons mousse.
    Batiment Batir. 8(11) , 26-32 (1982) .
    Manufacture. (French)

64. CUI, K.; HAIYUN, Y.; CUI, C.; MA, B.
    On the technique of wet mixed-grinding for the production of aerated concrete made
    from fly-ash.
    Proc. 2nd Int. Symp. Cem. Concr. Beijing. vol.3 296-303 (1989)
    Manufacture. (English)

65. DANILOVA, T. A.; LABOZIN, P. G.
    Heat-resistant cellular concrete.
    Refractories (USSR) 24 (1-2) , 38-41 (1983) .
    Thermal properties. (English)

66. DAPKUS, G.; STANKEVICIUS, V.
    Humidity behaviour of cellular concrete compact roofs.
    CIB 86 Vol 6 , 2029-2037 (1986) .
    Roof, Moisture. (English)

67. DAPKUS ,G.; STANKEVICIUS, V.
    Cellular concrete carbonation.
    Batim. internation. 18 (3) , 184-187 (1985) .
    Carbonation, Durability. (English)

68. DE VEKEY, R. C.; ARORA, S. K.; BRIGHT, N. J.; LUCKIN, K. R.
    Research results on autoclaved aerated concrete blockwork.
    Struct Eng Part A. 64(11) , 332-340 (1986) .
    Walls, Standards. (English)

69. DOBRY, O.; DROCHYTKA, R.; VOLEC, J.
    Zur Erforschung von Ursachen der Gasbetonkorrosion.
    Betontechnik. 11(3) , 70-72 (1990) .
    Durability, Chemical   reaction, Testing. (German)

70. DOMBROVSKII, A. V.; SAZHNEV, N. P.; GORYAINOV, K. E.
    Increase in the homogeneity of aerated concrete during impact molding.
    Stroit. Mater. , 17-18 (1982   ) .
    Porosity, Manufacture. (Russian)

71. DOUVEN, E.; VAN KOEVERINGE, L.; RAMAKERS, J.
    Application of fly ash in aerated concrete.
    Report. 18(6),53 pp. (1984   ).
    Manufacture, Fly ash. (Dutch)

72. DROCHYTKA, R.
    A study of the course of reactions between cellular concrete and SO2 and CO2.
    Silikaty. 32(1),85-91,96 (1988).
    Durability, Chemical   reaction, Microstructure. (Czech)

73. DROCHYTKA, R.
    Atmospheric corrosion of concrete and aerated concrete.
    Kniznice Odb. Ved. Spisu Vys. Uceni Tech. Brne, B. 97,209-213 (1984).
    Corrosion, Environment, Protection. (Czech)

74. DUDEROV, YU. G.; SHTARKH, G. S.; TUL'SKII, G. V.; MEL'NIKOV, A. M.
    Testing the industrial use of phosphate aerated concrete of decreased bulk density
    for heat insulation of glass furnaces.
    Pr-vo i Primenenie Fosfat. Materialov v Str-ve, M. ,92-97 (1983).
    Thermal insulation, Density. (Russian)

75. DVORKIN, L. I.; MIRONENKO, A. V.;SHANBAN, I. B.
    Cement-free unautoclaved aerated concrete.
    Stroit. Mater. (11),11-13 (1990).
    Manufacture, Hadening. (Russian)

76. EBENHAN, W.
    Sekundaerenergienutzung bei Autoklavisierungsprozessen.
    Betontechnik. 7(2),43-44 (1986).
    Autoclaving, Manufacture. (German)

77. EINRE, A.
    Energy conservation in the production of aerated concrete.
    Stroit. Mater. (2),22-23 (1986).
    Manufacture. (Russian)

78. EL-HEMALY, S. A. S.; TAHA, A. S.; EL-DIDAMONY, H.
    Influence of slag substitution on some properties of sand-lime aerated concrete.
    J Mater Sci. 21(4),1293-1296 (1986).
    Slag, Autoclaving, Strength. (English)

79. ENGELMANN, H.
    Bewegungsfugen zwischen Bauteilen aus Porenbeton.
    Baugewerbe. (21),8-10,12 (1990).
    Joints, Thermal properties, Sealant. (German)

80. ERNST, L.
    Brueckenkrane mit speziellen Anforderungen fuer Gasbetonwerke.
    Hebezeuge Foerdermittel. 29(11),338-341 (1989).
    Manufacture. (German)

81. ESKUSSON, K.; KILKSON, A.; ESKUSSON, I.
Modeling of the frost resistance of aerated silicate concrete.
Avtoklav. Silikat. Betony, Tallin. 9(4),55-65 (1988).
Modeling, Freezing. (Russian)

82. FEDIN, A. A.
Study of structure formation and structure of aerated concretes.
Issled.po Str-vu Stroit. Teplofiz. Dolgovech. Konst. Tallin.,4-15 (1989).
Chemical reaction, Microstructure. (Russian)

83. FEDYNIN, N. I.
Manufacture of unautoclaved aerated ash concrete with increased strength and
durability.
Stroit. Mater. (11),8-11 (1990).
Manufacture, Fly ash, Durability. (Russian)

84. FENG, N.
Fundamental study of cellular concrete aerated by a gas carrier. Part 1.
Guisuanyan Xuebao. 14(1),63-72 (1986).
Hardening, Aluminium powder. (Chinese)

85. FENG, N.; MUKAI, T.; EBARA, K.
Fundamental study on the use of zeolite powder as admixture for cellular concrete.
ANN. Rep. CAJ. (38),142-145 (1984).
Chemical reaction, Rheological properties. (Japanese)

86. FENG, N.
The fundamental investigation of gas-carrier cellular concrete with natural zeolite
rock.
Proc. Int. Symp. Cem. Concr. Beijing. vol.3 279-291 (1985)
Manufacture. (English)

87. FITZKE, S.; SCHOENEMANN, H
Erarbeitung der Aufgabenstellung fuer das Prozessleitsystem "audatec" zur
Automatisierung technologischer Prozesse.
Betontechnik. 10(3),72-75 (1989).
Manufacture. (German)

88. FLASSENBERG, G.
Building with aerated concrete members for industrial buildings and structures.
Betonwerk Fertigteil-Tech. 57(8),33-42 (1991).
Manufacture. (German)

89. FLASSENBERG, G.
Dimensioning of vertical joints between horizontally assembled wall panels made of
autoclaved aerated concrete.
Betonwerk Fertigteil-Tech. 56(12),39,41-44 (1990).
Joints, Panels, Sealant. (English)

90. FRTUSOVA, S.
Highly active aluminum foaming agents for the manufacture of aerated concrete.
Experience with Czechoslovak paste Albo 542.
Stavivo. 65(4),136-137 (1987).
Aluminium powder, Manufacture. (Czech)

91. FUJIWARA, H.; SHIMOYAMA, Y.; TANAKA, T.
Studies on methods for making high-strength aerated concrete.
Prep. Ann. Meeting CAJ. 44th, 446-451 (1990).
Chemical reaction, Mechanical properties, Waste. (Japanese)

92. FUKUKAWA, Y.; KONDO, K.; YOSHIDA, T.
Development in a CAD system for the mapping of ALC panels.
Onoda Kenkyu Hokoku. 35(1), 74-83 (1983).
Walls, Manufacture. (Japanese)

93. GAAL, M.
Special aerated concrete for building ends by inflation of wet paste.
Magy. Alum. 23(9), 302-304 (1986).
Manufacture. (English)

94. GALIBINA, E. A.; KREMERMAN, T. B.; MEINART, G. O.; DOMBROVSKII, A. V.; SAZHENV, N. P.
Aerated concrete manufactured by impact technology with a mixed binder
Stroit. Mater. (3), 20-22 (1990).
Manufacture, Setting. (Russian)

95. GALIBINA, E. A.; KREMERMAN, T. B.
Feasibility of using the fourth-phase shale ashes of the Baltic State Regional
Electric Power Plant for production of aerated concrete.
Stroit. Mater. iz Poput. Produktov Prom-sti, L. 24(6), 67-70 (1988).
Manufacture, Fly ash. (Russian)

96. GALIBINA, E. A.; REMNEV, V. A.; KREMERMAN, T. B.
Effect of superplasticizers on hydraulic activity of shale ashes and qualitative
and performance indexes of autoclaved aerated concrete
Issled.po Str-vu Stroit. Teplofiz. Dolgovech. Konst. Tallin. 113(8), 84-90 (1989).
Manufacture, Additive. (Russian)

97. GAVRILENKO, V. N.; DAVIDENKO, V. P.; TARASENKO, A. A.
Use of wastes in aerated concrete products.
Stroit. Mater. Konstr. (2), 14-15 (1988).
Manufacture, Waste. (Russian)

98. GELLERSHTEIN, E. M.; OKULOVA, L. I.; POMAZANOV, V. N.; FINTIKTIKOVA, O. I.; SUSHKO, A.
Increase of corrosion resistance of aerated concrete by using powder polymers.
Gidrotekhn. Sooruzh. i Vopr. Ekspluat. Orosit. Sistem. 19(109), 125-127 (1986).
Corrosion, Additive, Protection. (Russian)

99. GEORGIADES, A.; FITIKOS, C.; MARINOS, J.
Effect of micropore structure on autoclaved aerated concrete shrinkage.
Cem Concr Res. 21(4), 655-662 (1991).
Shrinkage, Autoclaving, Pore-size distribution. (English)

100. GLADKOV, D. I.; EROKHINA, L. A.; ZAGORODNYUK, L. KH.; ZAITSEV, N. K.
Unautoclaved aerated concrete based on fly ash from the Kurakhovo State Regional
Electric Power Plant.
Fiz.-khim. Kompozits. Stroit. Mater.; Belgorod. 55(6), 191-193 (1989).
Manufacture, Fly ash. (Russian)

101. GLADYSHEV, B. M.; SHMANDII, M. D.; DIKII, YU. A.; NEMERTSEV V. S.
Aerated concrete from kieselguhr sand with a cement-lime binder.
Beton i zelezobeton. (1), 9-11 (1988).
Manufacture, Composition. (Russian)

102. GLASER, H.
Wasserdampfdiffusion durch hygroskopische Baustoffe im Hinblick auf die DIN 4108.
Gesund Ing Haustech BauphysUmwelttech. 107(2), 85-97 (1986).
Permeability, Moisture. (German)

103. GLASER, H.
Die Feuchtigkeitsaufnahme hygroskopischer Waende durch Dampfdiffusion.
Ki. 14(2), 57-60 (1986).
Moisture, Standards, Thermal insulation. (German)

104. GLAVAS, V.
Quality of cement for aerated concrete production.
Hem. Ind. 37(12), 312-315 (1983).
Manufacture, Hadening. (Serbo-Croatian)

105. GOEBEL, K.
The radioactivity of building materials. Amount, types and effects.
Ziegelind Int. 43(4), 237-240 (1990).
Radio activity. (English)

106. GORYAJNOV, V. EH.; KHOLOMKIN, P. G.
Détermination du taux d'hydratation d'un ciment dans une suspension préparée pour
l'imprégnation et en qualité de composant des bétons cellulaires.
Z. prikl. him. 55 (10), 2170-2174 (1982).
Hydration. (Russian)

107. GRIEBENOW, G.; KOCH, R.; SITKA, R.; VOELKEL, G.
Loadbearing floor plates of autoclaved aerated concrete components. Experimental
investigations and design models.
Betonwerk Fertigteil-Tech. 55(6), 56-61 (1989).
Floors, Slabs, Mechanical properties. (English)

108. GRIEBENOW, G.; KOCH, R.; SITKA, R.; VOELKEL, G.
Loadbearing floor plates of autoclaved aerated concrete components. part 2.
Experimental investigations and design models.
Betonwerk Fertigteil-Tech. 55(8), 62-66 (1989).
Floors, Slabs, Mechanical properties. (English)

109. HANAFI, S.; TAHA, A. S.; EL-FAROUK, O.; ABO-EL-ENEIN, S. A.
Surface area and pore structure of some autoclaved cellular concrete products.
TIZ. 113(8), 659-661 (1989).
Slag, Composition, Pore-size distribution. (English)

110. HAYNES, R. D.; YARBROUGH, D. W.
Apparent thermal conductivity of low density concrete obtained with a
radial-heat-flow apparatus.
Therm Conductivity. 20th, 265-274 (1989).
Thermal properties, Lightweight aggregate concrete. (English)

111. HELLER, W.; KUPSCH, H.; LEONHARDT, J. W.
Optimization of a production line of porous (aerated) concrete.
Isotopenpraxis. 25(4), 166-168 (1989).
Manufacture. (English)

112. HENGST, R. R.; TRESSLER, R. E.
Fracture of foamed portland cements.
Cem Concr Res. 13(1), 127-134 (1983).
Fracture, Pore-size distribution, Mechanical properties. (English)

113. HILDEBRAND, H.
Wissenschaftlich-technische Entwicklungsloesungen beim Vorhaben Gasbetonwerk
Zehdenick.
Betontechnik. 11(1), 6-9 (1990).
Manufacture. (German)

114. HILDEBRAND, H.
Wissenschaftlich-technische Entwicklungsloesungen beim Vorhaben Gasbetonwerk
Zehdenick.
Baustoffindustrie. 32(6), 174-177 (1989).
Manufacture. (German)

115. HILDEBRAND, H.
Anlage fuer den Umschlag und die Verpackung von Daemmbaustoff DBk 300.
Betontechnik. 9(6), 188-189 (1988).
Manufacture. (German)

116. HILDEBRAND, H.; FISCHER, H-P.
Herstellung neuer veredelter Erzeugnisse im VEB Gasbetonwerk Hennersdorf.
Betontechnik. 3(6), 186-189 (1982).
Manufacture. (German)

117. HILDEBRAND, H.; SIELAFF, M.
Reduzierung des Wasserverbrauchs und Wiederverwendung von Abwasser bei der
Gasbetonherstellung.
Betontechnik. 8(4), 118-121 (1987).
Manufacture. (German)

118. HILDINGSON, O.
Radon measurements in 12,000 Swedish homes.
Environ Int. 8(1/6), 67-70 (1982).
Radio activity. (English)

119. HOFFMANN, B.; SEIDEL, H.
Grinding aids for quicklime and their effect on its processability and reactivity.
Zem.-Kalk-Gips, Ed. B. 39(5), 269-273 (1986).
Manufacture, Lime, Chemical reaction. (German)

120. HOFFMANN, O.
Application of high-temperature dilatometry in the study of calcium
hydrosilicates.
Thermochimica Acta. 93, 529-532 (1985).
Expansion, Deformation, Testing. (ENGLISH)

121. HOSHIKA, Y.; MURAYAMA, N.
    A case report of gas chromatographic determination of odorants in ambient air near
    of autoclaved lightweight cellular concrete (ALC) factory.
    Taiki Osen Gakkai-si. 20(2), 89-95 (1985).
    Environment, Waste, Testing. (English)

122. HOUST, Y.; ALOU, F.; WITTMANN, F. H.
    Influence of moisture content on mechanical properties of autoclaved aerated
    concrete.
    Autoclaved Aerated Concr Moisture Prop. ed. F.H. Wittmann, 219-234 (1983).
    Moisture, Modeling, Deformation. (English)

123. HOUST, Y.; ALOU, F.; WITTMANN, F. H.
    Influence de l'humidité sur les proprietés mécaniques du béton cellulaire
    autoclavé.
    Ing Archit Suisses. 109(1), 9-15 (1983).
    Modeling, Moisture, Mechanical properties. (French)

124. HUANG, C. L. D.
    Effects of thickness of cylinders on moisture and heat transfer in porous concrete
    tubes.
    ASME HTD (Am Soc Mech Eng Heat Transf Div) 22, 103-109 (1982).
    Heat transfer, Moisture. (English)

125. HUMS, D.
    Relation between humidity and heat conductivity in aerated concrete.
    Autoclaved Aerated ConcrMoisture Prop. ed. F.H. Wittmann, 143-152 (1983).
    Moisture, Heat transfer. (English)

126. IBARRA, E. E.; HIBINO, H.; MITSUDA, T.
    Electron micrographs of calcium silicate hydrates. II. Electron microscopy of
    tobermorite in autoclaved aerated concrete.
    Yogyo Gijutsu Kenkyu Shisetsu Nenpo (Nagoya Kogyo Daigaku). 16, 13-31 (1989).
    Tobermorite, Composition. (English)

127. IKEDA, Y.; YASUIKE, Y.; TAKASHIMA, Y.
    29Si MAS NMR Study on Structural Change of Silicate Ions in Tobermorite with
    Carbonation of ALC.
    Nippon Seramikkusu Kyokai Gakujutsu Ronbunshi. 99(5), 423-426 (1991).
    Carbonation, Chemical reaction, Tobermorite. (Japanese)

128. IL'ICHEV, G. D.
    New technology for production of aerated concrete.
    Vopr. Atom. Nauki i Tekhn. Ser. Proektir. i Str-vo, Moskva. (3), 37-39 (1989).
    Manufacture. (Russian)

129. INOUE, A.
    Fire resistance characteristic of ALC panel joint. Fire performance of wall with
    joint gap.
    Kenzai Siken Joho. 24(3), 6-10 (1988).
    Joints, Fire resistance, Fiber. (Japanese)

130. ISHIBASHI, K.; RIN, G.
    Ultimate strength of light-weight air-mixing concrete construction members.
    Prep. Ann. Meeting JCI. 9(2), 391-396 (1987).
    Mechanical properties, Walls. (Japanese)

131. ISOME, Y.
     Light-Weight, High-Strength Foamed Concrete. Manufacture and Performance.
     Chemistry and Education. 39(4), 397-400 (1991).
     Manufacture, Prefablication, Water absorption. (Japanese)

132. IWASAKI, M.; TADA, S.
     Carbonation of aerated concrete.
     Proc. Int. Symp. Cem. Concr. Beijing. vol.3, 343-356 (1985).
     Carbonation, Adsorption, Pore-size distribution. (English)

133. JAEGERMANN, C.; PUTERMAN, M.; HAVIV, E.
     Blistering of membranes over foam-concrete roofs.
     J Mater Civ Eng. 1(1), 31-45 (1989).
     Roof, Water absorption. (English)

134. JAMBOR,J.; BAGEL,L.; ZIGO,O.
     Effect of pore structure on the drying of AAC.
     Stavebnicky Cas. 38(5), 341-357 (1990).
     Moisture, Drying, Pore-size distribution. (Czech)

135. JAMBOR, J.; ZIVICA, V.
     Investigation of the relative resistance of hydration products of cement against
     corrosion due to aggressive carbon dioxide water.
     Therm. Anal.; Proc. ICTA, 8th Ed. Blazek, Antonin, 2, 605-608 (1985).
     Durability, Carbonation, Hydration. (English)

136. JEONG, H. D.; TAKAHASHI, H.; TERAMURA, S.
     Low temperature fracture behaviour and the characteristics of autoclaved aerated
     concrete (AAC).
     Cem Concr Res. 17(5), 743-754 (1987).
     Freezing, Fracture. (English)

137. JEONG, H. D.; TAKAHASHI, H.; TERAMURA, S.
     Low temperature fracture behaviour and AE characteristics of autoclaved aerated
     concrete (AAC).
     Prog Acoust Emiss 3. 7(2), 529-537 (1986).
     Freezing, Fiber, Fracture. (English)

138. JUSSEN, D.
     Fertigkeller.
     Elem Fertigbau. 20(1), 32-36 (1983).
     Walls. (German)

139. KAMADA, E.; KOH, Y.; TABATA, M
     Experimental study on testing method for frost resistance of autoclaved lightweight
     concrete.
     Proc. 3rd Int. Conf. Durability of Building Materials and 3, 372-382 (1984).
     Freezing, Testing, Water absorption. (English)

140. KAMADA, E.; TABATA, M.; SENBU, O.; ITO, K.
     Test method for the investigation of mechanism and evaluation of frost damage.
     Cement Concrete. NO.432, 20-27 (1983).
     Durability, Freezing,Testing. (Japanese)

141. KASAI, Y.; MATSUI, I.; OZAWA, M.; SAITO, T.
   Utilization of fly-ash and granulated blast furnace slag for autoclaved lightweight concrete.
   Proc. Int. Symp. Cem. Concr. Beijing. vol. 3 304-317 (1985)
   Fly ash, composition. (English)

142. KAWAKAMI, O.
   Test method of deformation performance of non-bearing wall.
   Kenzai Siken Joho. 25(1) , 29-34 (1989) .
   Strength, Walls. (Japanese)

143. KELLER, R.; KUEHNE, G.
   Neue Gasbetonklassen in der DDR.Erlaeuterungen zur TGL 33416/01 Gasbeton, Technische Bedingungen.
   Betontechnik. 6(5) , 144-146 (1985) .
   Standards. (German)

144. KESLI, E. O.; BUYAL'SKAYA, V. Z.; NIKITIN, M. K.; GUREVICH, N. I.
   Frost resistance of aerated concrete with an addition of organosilicon compounds.
   Konstruktsii iz Yacheist. Betonov dlya Zhil.-grazhd. Str-va 8(4) , 53-60 (1987) .
   Freezing, Chemical reaction, Additive. (Russian)

145. KHOLOPOVA, L .I.; VESELOVA, S. I.
   Matériau granulé et coloré pour la finition des panneaux mureaux en béton cellulaire.
   Beton i zelezobeton. (3) , 29-30 (1982) .
   Finishing. (Russian)

146. KIREEV, YU. N.; ANISIMOV, YU. P.; ORLOVA, I. G.; SAZHNEV, N. P.; SAVEL'EVA, N. P.
   Study of the effect of an iminodisuccinic acid addition on hydration and swelling of an aerated concrete mix.
   Avtoklav. Silikat. Betony, Tallin. Vol 2 , 77-85 (1988) .
   Hydration, Hadening, Additive. (Russian)

147. KLOSE, D.
   Verpackungslastaufnahmemittel fuer Gasbetonstreifenelemente als Grundlage einer geschlossenen Transportkette.
   Bauplanung Bautech. 36(9) , 404-406 (1982) .
   Manufacture. (German)

148. KNIGINA, G. I.; ZAVADSKII, G. V.; BELOZEROVA, N. G.
   Organomineral lubricants for molds used in autoclaved aerated concrete production.
   Stroit. Mater. , 13-14 (1983   ) .
   Manufacture, Moulding. (Russian)

149. KOEHLER, D.
   Rationalisierung des Energieeinsatzes bei der Produktion von Silikatbetonen in Autoklaven.
   Energieanwendung. 33(3) , 87-90 (1984) .
   Manufacture. (German)

150. KOEHNE, J. H.
   Gascon gereed voor de woning bouwmarkt.
   Cement. 39(6) , 22-24 (1987) .
   Fly ash, Aluminium powder. (Dutch)

151. KOGA, K.
Effect of surface layers of ALC panels on salt attack.
Haseko Giho. (7) , 74-81 (1991) .
Finishing, Polymer, Protection. (Japanese)

152. KOHLER, N.
Global energetic budget of aerated concrete.
Autoclaved Aerated Concr Moisture Prop. ed. F.H. Wittmann , 13-26 (1983) .
Thermal insulation, Manufacture. (English)

153. KORINCHENKO, I. V.
Leaching corrosion of aerated concrete as a material for drainage pipes.
Metody Rascheta Protsessov Massoperenosa v Gidrogeol. 5(4) , 33-36 (1984) .
Corrosion, Protection. (Russian)

154. KOTH, W.; SCHNEIDER, K.
Gasbetonaussenwand fuer mehrgeschossige Mehrzweckgebaeude in der SKBS 75.
Bauplanung Bautech. 39(1) , 10-12 (1985) .
Thermal properties, Walls. (German)

155. KRASHENINNIKOV, A .N.; SHMYGLYA, T. A.
Les bétons cellulaires avec des adjuvants polymères.
Beton i zelezobeton. (7) , 36-38 (1982) .
Polymer. (Russian)

156. KRASHENINNIKOV, O.N.; ZOTOVA, K. V.; ZHURBENKO, G. V.; MEOS, M. A.; NEPEINA, O. V.
Vermiculite aerated concrete for thermal insulation of coatings.
Fiz.-khim. Osnovy Pererab. i Primeneniya Mineral. Syr'ya, 38(5) , 27-31 (1990) .
Finishing, Thermal insulation. (Russian)

157. KRITOV, V. A.; SKATYNSKY, V. I.; CHIKOTA, E. I.; UDACHKIN, I. B.
Building enclosing structures made of energy saving autoclave concrete.
Proc. 9th CIB Congr.; Stockholm 4 , 143-152 (1983) .
Manufacture. (English)

158. KRUML, F.
Influence of saturation degree of autoclaved aerated concretes on their creep.
Autoclaved Aerated Concr Moisture Prop. ed. F.H. Wittmann , 249-256 (1983) .
Moisture, Creep, Density. (English)

159. KUENZEL, H.
Der Feuchteschutz von Gasbeton-Aussenbauteilen.
Schweiz Ing Archit. 106(39) , 1075-1079 (1988) .
Moisture, Thermal properties. (German)

160. KUENZEL, H.
Keine Dampfsperre zwischen Daemmstoff und Gasbeton. Untersuchung der
Feuchteverhaeltnisse in Gasbetonflachdaechern mit Zusatzdaemmung.
DDH. 108(19) , 14,16,18 (1987) .
Moisture, Roof. (German)

161. KUENZEL, H. M.; KIESSL, K.
Bestimmung des Wasserdampfdiffusionswiderstandes von mineralischen Baustoffen aus Sorptionsversuchen.
Bauphysik. 12(5),140-144(1990).
Water absorption, Standards, Moisture. (German)

162. KUPSCH, H.; HELLER, W.
An accelerator-produced short-lived radionuclide for the tracer technique of gypsum in a large-scale production plant of gas concrete.
3rd Proc. Radioisot. Appl. Radiat. Process. Ind. Ed. Luther, 2,1003-1008(1986).
Radio activity. (English)

163. KURZWEIL, M.; GOTTSCHALK, W.; BERTRAM, M.
Entwicklung und Rationalisierung der Gasbetonproduktion im VEB Metalleichtbaukombinat - Werk Calbe.
Betontechnik. 7(4),99-101(1986).
Manufacture. (Czech)

164. KUZNETSOV, YU. B.; MAKARICHEV ,V. V.; UKHOVA T. A.
Une homogénéite élevée dans la qualité. C'est une réserve d'économie de liants dans les bétons cellulaires.
Beton i zelezobeton. (5),19-20(1984).
Density. (Russian)

165. LANE, S. N.
Evaluation of absorptive sound barrier samples in freeze/thaw environments.
Public Roads. 53(2),46-48(1989).
Acoustic insulation, Freezing, Environment. (English)

166. LANG, I.
Applications of AAC by-products.
Stud. Environ. Sci. 48,647-648(1991).
Manufacture, Waste. (English)

167. LAPSA, V.; SPACA, L.; BRENARD, A.;STEINERT, A.
Sulfur addition in nonautoclaved aerated concrete.
LLA Raksti. 209,,3-8(1984).
Hydration, Manufacture. (Russian)

168. LASYS, A.
Thermal and sound insulating building material on the base of gas concrete.
TIZ. 109 (5),350-353(1985).
Manufacture. (German)

169. LAURENT, J-P.
La conductivité thermique 'à sec' des bétons cellulaires autoclavés: un modèle conceptuel.
Mater Struct. 24(141),221-226(1991).
Thermal properties, Porosity, Modeling. (French)

170. LEIMBOECK, R.
Designing with gas concrete components.
Betonwerk Fertigteil-Tech. 52(9),585-588(1986).
Walls. (English)

171. LINDNER, W. O.
Ist Gasbeton ein aussergewoehnlicher Baustoff?
Dtsch Bauz. (9) , 1167-1169 (1985) .
Finishing, Protection. (German)

172. LOBANOV, .I A.; PUKHARENKO ,YU .V.; MORGUN, L. V.
Les bétons cellulaires non autoclavés avec armatures en fibres synthétiques.
Beton i zelezobeton. (8) , 28-30 (1983) .
Fiber, Reinforcement. (Russian)

173. LOUDON, A. G.
The effect of moisture content on thermal conductivity.
Autoclaved Aerated Concr Moisture Prop. ed. F.H. Wittmann , 131-142 (1983) .
Moisture, Thermal insulation. (English)

174. LYSEK, N.
Manufacture of quicklime for the aerated concrete industry in the Silesian Lime
Plant in Tarnow Opolski.
Cem.-Wapno-Gips. 6(1) , 100-103 (1985   ) .
Lime. (Polish)

175. MAI, J.; KORTHALS, H.
Das Reihenhaus als ein Beitrag zum energieoekonomischen Bauen unter dem Aspekt der
Anwendung von Gasbeton.
Bauzeitung. 39(4) , 164-166 (1985) .
Thermal insulation. (German)

176. MALI, E.; SUSTERSIC, J.; ZAJC, A.
Use of fine-grained and highly microaerated concrete (foamed lightweight concrete)
in hydraulic engineering.
Zem. Beton (Vienna) 34(2) , 63-65 (1989) .
Manufacture, Composition. (German)

177. MANNONEN, M. O.
Thermal insulation of lightweight concrete.
Proc 9th Congress Fed Int Precontrainte 1982 Vol 3 , 32-38 (1982) .
Thermal insulation. (English)

178. MANTOKU, A.; SAITO, T.
Study in the vapor proofing and frost damage to exterior components in the cold
district V. Water proofing and frost damage of ALC panels.
Hokkaido Kanchi Kentiku Kenkyuusho Kenkyu Hokoku-shu. 1982 , 129-132 (1982) .
Freezing, Finishing. (Japanese)

179. MARKAN, I. F.; ZAVOLOKA, M. V.; POBOKIN, A. A.; NECHITALIO, L. A.
Wastes from superphosphate manufacture - effective raw material for aerated
concrete.
Stroit. Mater. Konstr. (2) , 14-15 (1990) .
Manufacture, Waste. (Russian)

180. MATHEY, R. G, ROSSITER, W. J. JR
A summary of the manufacture, uses, and properties of autoclaved aerated concrete.
ASTM Spec Tech Publ. (1030) , 15-37 (1990) .
Manufacture, Durability, Physical properties. (English)

181. MATIDA, K.; OHUCHI, T.; FURUHIRA, A.
Experimental research on thermal constant measurement of refractory material.
Kenzai Siken Joho. 24(4), 6-13 (1988).
Thermal properties, Fire resistance. (Japanese)

182. MATSUMOTO, M.; SATO, M.
A periodic solution of moisture condensation and re-evaporation process in the building wall.
Numer Methods Therm Problems Part 2 2, 819-829 (1985).
Modeling, Moisture, Adsorption. (English)

183. MATSUMURA, A.
Shear strength and behavior of reinforced Autoclaved Lightweight Cellular Concrete members. Study on strength and behavior of reinforced ALC members. Part 1.
Nihon Kentiku Gakkai Ronbun Hokoku-shu. (336), 42-52 (1984).
Panels, Mechanical properties. (Japanese)

184. MATSUMURA, A.
Shear strength and behavior of reinforced Autoclaved Lightweight Cellular Concrete members. Study on strength and behavior of reinforced ALC members. Part 2.
Nihon Kentiku Gakkai Ronbun Hokoku-shu. (343), 13-23 (1984).
Walls, Reinforcement, Mechanical properties. (Japanese)

185. MERKIN, A. P.; UDACHKIN, I. B.; ZAKHARCHENKO, P. V.; SEMIDID'KO, A. S.; GRYUNER, G. F.
Active silica-containing industrial wastes - raw material for autoclaved building materials.
Stroit. Mater. (5), 23-24 (1987).
Composition, Silica. (Russian)

186. MERLET, J. D.
Le béton cellulaire autoclavé.
CSTB Magazine. (23), 2-7 (1984).
(French)

187. MICHALIK, H.; PIPPIG, R.; ERBRING, M.; HILDEBRAND, H.
Technische Loesung zur Verpackung und zum TUL-Prozess von Daemmbaustoff 300.
Betontechnik. 7(3), 93-96 (1986).
Manufacture. (German)

188. MICHEL, W.
Reinigungsfaehige Beschichtungen -Voraussetzungen und Einsatzmoeglichkeiten eines Systems.
Bautenschutz Bausanierung. 14(6), 42, 47-48 (1991).
Walls, Finishing, Chemical reaction. (German)

189. MICHEL, W.
Reinigungsfaehige Beschichtungen.
Bauplanung Bautech. 45(7), 305-308 (1991).
Manufacture, Finishing, Protection. (German)

190. MIMORI, T.; TAIRAKU, T.
An experiment on frost damage of autoclaved lightweight concrete.
Kusiro Kogyo Senmon Gakkou Kiyo. (20), 99-108 (1986).
Freezing, Testing. (Japanese)

191. MIRZA, W. H, AL-NOURY, S. I.
Utilisation of Saudi sands for aerated concrete production.
Int J Cem Compos Lightweight Concr. 8(2) , 81-85 (1986) .
Silica, Composition. (English)

192. MITSUDA, T.
Advances in hydrated calcium silicate materials.
Gypsum and Lime. (229) , 464-470 (1990) .
Tobermorite, Chemical reaction. (Japanese)

193. MITSUDA, T.
Autoclaved calcium silicate products.
Ceramics. 23(8) , 748-752 (1988) .
Hydration, Composition, Pore-size distribution. (Japanese)

194. MITSUDA, T.; IBARRA, E. E.
Analytical electron microscopy of tobermorite in ALC.
Prep. Ann. Meeting Ceram. Soc. Japan. , 71 (1989) .
Tobermorite, Hydration, Chemical reaction. (Japanese)

195. MITSUDA, T.; SASAKI, K.; ISHIDA, H.
Phase evolution during autoclaving process of aerated concrete.
J. Am. Ceram. Soc. 75 , 1858-1863 (1992) .
Autoclaving, Tobermorite, Pore-size distribution. (English)

196. MOYNE,C.; BATSALE,J. C.; DEGIOVANNI,A.; MAILLET, D.
Thermal conductivity of wet porous media: Theoretical analysis and experimental measurements.
Therm Conductivity. 21st , 109-120 (1990) .
Moisture,Testing,Thermal properties. (English)

197. MUNTEAN, M.; PAUL, F.; DINU, M.; IVAN, E.; GHINDESCU, C.
Hydrated phases in sample of cellular concrete.
Cemento. 81(3) , 119-126 (1984) .
Hydration, Microstructure. (English)

198. MUROMSKIJ, K. P.
Evaluation du béton cellulaire en tant que milieu élastique.
Beton i zelezobeton. (5) , 24-25 (1984) .
Rheological properties. (Russian)

199. MUTO, T.; ICHIYA, N.
Elucidation of crack occurrence of steel reinforced light-weight concrete under autoclave cure.
Ann. Rep. CAJ. (43) , 198-203 (1989) .
Autoclaving, Bond, Cracking. (Japanese)

200. MYASNIKOV, O. E.; PUGAEV, A. V.; ANISIMOV, YU. P.
Study of the effect of iminodisuccinic acid on hydration of a lime binder.
Khimiya Kompleksonov i ikh Primenenie, Kalinin. , 135-140 (1986) .
Hydration, Additive, Lime. (Russian)

201. NAGATA, M.; YONEDA, S.; MATSUNAGA, A.
Experimental studies on super lightweight cellular concrete using special inorganic expanded aggregate.
Ann. Rep. CAJ. (43), 346-351 (1989).
Manufacture, Hadening, Admixture. (English)

202. NAKAMURA, M.; HATA, Y.; YAMAOKA, S.
Inorganic surface modification of autoclaved light weight concrete.
Nihon Seramikkusu Kyokai-si. 96(11), 1093-1097 (1988).
Freezing, Finishing, Pore-size distribution. (Japanese)

203. NAKAMURA, M.; KETANI, K.
Surface modification of ALC by low temperature plasma treatment.
Nippon seramikkusu kyokai gakujutsu Ronbun-shu. 98(10), 1164-1168 (1990).
Finishing. (Japanese)

204. NAKAMURA, M.; KITANO, S.; ARAKAWA, M.
Surface modification of autoclaved lightweight concrete.
Nihon Seramikkusu Kyokai-si. 95(10), 1007-1011 (1987).
Finishing, Adsorption, Protection. (Japanese)

205. NAKAMURA, M.; TSUJII, S.
Improvement of frost durability on ALC by nonflammable inorganic coating.
Nihon Seramikkusu Kyokai-si. 97(12), 1471-1477 (1989).
Finishing, Chemical reaction, Freezing. (Japanese)

206. NAKANO, S.
Autoclaved aerated concrete - History and technical development.
Ceramics. 17(7), 507-512 (1982).
Manufacture, Physical properties. (Japanese)

207. NEUE, J.
Kapazitive Feuchtemessung an Gasbeton.
Bauplanung Bautech. 40(5), 215-218 (1986).
Moisture, Testing. (German)

208. NICKOL, D.; RUECKERT, H.
Unterschungen zur Verbesserung des Schwindverhaltens von vorgefertigten Bauelementen aus Schaumbeton.
Betontechnik. 12(2), 73-77 (1991).
Shrinkage, Manufacture, Strength. (German)

209. NICKOL, D.; RUECKERT, H.
Stehende Fertigung von Wandelementen aus Schaumbeton.
Betontechnik. 10(2), 47-50 (1989).
Manufacture. (German)

210. NIELSEN, A.
Shrinkage and creep - Deformation parameters of aerated, autoclaved concrete.
Autoclaved Aerated Concr Moisture Prop. ed. F.H. Wittmann, 189-206 (1983).
Moisture, Shrinkage, Creep. (English)

211. NIELSEN, A. F.
Gamma-ray-attenuation used on free-water intake tests.
Autoclaved Aerated Concr Moisture Prop. ed. F.H. Wittmann, 43-53 (1983).
Moisture, Testing, Water absorption. (English)

212. Nihon Cement Co.
Repairing method using polymer cement-based material for ALC building. AR-ALC
method.
Cement Industry. (209), 19-24 (1988).
Durability, Protection, Finishing. (Japanese)

213. NIKEZIC, D.; VASILJEVIC, L.
Indoor radon concentration measurements by using SSNTD.
Proc Int Workshop Radon Monit Radioprot Environ Radioact (9), 484-487 (1990).
Radio activity, Testing. (English)

214. NISCHER, P.
Foamed concrete.
Betonwerk Fertigteil-Tech. 49(3), 148-151 (1983).
Carbonation, Shrinkage. (German)

215. OB'EDKOV, V. A.
Effect of sodium chloride on the thermal conductivity of aerated concrete.
Comments.
Beton i zelezobeton. (12), 9-10 (1986).
Thermal properties. (Russian)

216. ORLOVA, I. G.; KUZ'MICHEVA, M. B.
Effect of plasticizer S-3 on characteristics of expansion and early structure
formation at a decreased water-solid ratio.
Tekhnol. Silikat. Mater.; Tallin., 28-39 (1989).
Microstructure, Composition, Expansion. (Russian)

217. ORLOVA, I. G.; SAZHNEV, N. P.; SAVEL'EVA, N. P.; BATRAKOV, V. G.; FALIKMAN, V. YA.
Study of the applicability of new plasticizing additives in the impact technology
for molding of aerated concrete articles.
Avtoklav. Silikat. Betony, Tallin. 24(3), 66-76 (1988).
Hydration, Additive, Manufacture. (Russian)

218. OSTRAT, L. I.; ORLOVA, I. G.; ESKUSSON, K. K.
Increase of the impact strength of fly ash aerated concrete.
Beton i zelezobeton., 24 (1984).
Impact resistance, Mechanical properties. (Russian)

219. PAK, A. A.; SUKHORUKOVA, R. N.; KRASNOVA, G. G.
Thermal analysis and x-ray diffraction studies of siliceous waste material
containing aerated silicate concrete.
Fiz.-khim.Osnovy Pererab.i Primeneniya Mineral.Syr'ya Apat. (11), 35-38 (1990).
Waste, Testing. (Russian)

220. PAVLOV, V. I.; GEVORKYAN, A. A.
Effect of sodium chloride on the thermal conductivity of aerated concrete.
Beton i zelezobeton. 13(4), 26-27 (1985).
Thermal properties. (Russian)

221. PETROV, I. A.; ROZENBLYUM, A. YA.; KAN, L. A.; KUTYRINA ,T. M.; DMITRIEV, YU. V.
Ossatures allégées de batiments industriels sans étage.
Beton i zelezobeton. (5),15-16 (1982).
(Russian)

222. PETTERSSON, H.; HILDINGSON, O.; SAMUELSSON, C.; HEDVALL, R.
Radon release from building material.
PB Rep. NO.PB-83-159665,59p (1982).
Radio activity, Testing. (English)

223. PIAZZA, J. L.; DIAZ, J. S. V.
Utilization of mineral coal ash for production of building materials.
Low-Cost Energy Sav Constr Mater. Vol 1,331-338 (1984).
Manufacture, Fly ash. (English)

224. PILZ, E.
Korrosionsschutzsystem fuer autoklavgehaerteten Gasbeton - Internationaler Stand und
Perspektive.
Betontechnik. 9(1),27-28 (1988).
Reinforcement, Protection, Corrosion. (German)

225. POSPISIL, F.
Carbon dioxide influence on setting mixture for cellular concrete.
Silikaty. 30 (1),41-48 (1986).
Carbonation, Setting. (Czech)

226. PRAETSCH, J.
Zur energieoekonomischen Wirksamkeit von Gasbeton im innerstaedtischen Bauen.
Energieanwendung. 36(6),208-209 (1987).
Thermal insulation, Heat transfer. (German)

227. PRIM, P.; WITTMANN, F. H.
Structure and water absorption of aerated concrete.
Autoclaved Aerated Concr Moisture Prop. ed. F.H. Wittmann,55-69 (1983).
Moisture, Water absorption, Pore-size distribution. (English)

228. Ravindrarajah, R. S.
Properties of cellular concrete.
Proc. Int. Symp. Cem. Concr. Beijing. vol.3 318-330 (1985)
(English)

229. REICHVERGER, Z.
Using an impact device with sliding drop collar for in situ evaluation of
compressive strength of insulating cellular concrete.
Journal of Testing and Evaluation. 14 (6),298-302 (1986).
Impact resistance, Testing. (English)

230. REIHER, S.; BOETHIN, K.
Instandsetzung von Gasbetonaussenwaenden.
Bauzeitung. 39(2),78-79 (1985).
Durability, Protection. (German)

231. RIEDER, M.
Silicon-Massehydrophobierung von Gasbeton und Gips.
Baugewerbe. (20) ,47-49 (1985) .
Finishing, Water absorption, Chemical reaction. (German)

232. ROBERTSON-DUNN, D. J.
Autoclaved aerated concrete.
Concrete (Leatherhead) 16(11) ,40-42 (1982) .
Walls, Thermal insulation. (English)

233. ROELFSTRA, P. E.; WITTMANN, F. H.
Numerical analysis of drying and shrinkage.
Autoclaved Aerated Concr Moisture Prop. ed. F.H. Wittmann ,235-248 (1983) .
Moisture, Shrinkage, Modeling. (English)

234. ROTHWELL, G. W.
Effects of external coatings on moisture contents of autoclaved aerated concrete
walls.
Autoclaved Aerated Concr Moisture Prop. ed. F.H. Wittmann ,101-117 (1983) .
Finishing, Permeability, Moisture. (English)

235. ROULET, C-A.
Expansion of aerated concrete due to frost - Determination of critical saturation.
Autoclaved Aerated Concr Moisture Prop. ed. F.H. Wittmann ,157-170 (1983) .
Moisture, Freezing, Testing. (English)

236. RUECKWARD, W.
Luftschalldaemmung von Waermedaemmverbundsystemen - Leichte und schwere
Putzschichten im Vergleich.
Bauphysik. 4(5) ,161-165 (1982) .
Acoustic insulation. (German)

237. RUST,W. W.; ROBERTS,A.S.JR
Porous media heat transfer experiment and theoretical model analysis.
ASME HTD. 152 ,61-68 (1990) .
Panels, Heat transfer, Modeling. (English)

238. RYUMIN, K. I.
Study of the possibility of producing autoclaved aerated concrete from nepheline
sludge.
Beton i zelezobeton. 209 ,56-62 (1984) .
Manufacture, Waste. (Russian)

239. SAGAE, A.
Initial moisture content of AAC can cause condensation problems.
Kentiku Gijutsu. (484) ,128-132 (1991) .
Panels, Moisture, Roof. (Japanese)

240. SAKAEV, R. V.; SHCHERBACHENKO, V. L.
Fine-grained slag-ash concrete aerated with cellular polystyrene.
Issled. Svoistv i Tekhnol. Poluch. Effektiv. Stroit. Mater. ,93-98 (1989) .
Manufacture, Slag, Fly ash. (Russian)

241. SAKAMOTO, I.
Fitting method of autoclave lightweight concrete panel to exterior wall.
Kentiku Gijutu. (433), 105-112 (1987).
Joints, Walls, Deformation. (Japanese)

242. SAKHAROV, G .P.; LOGINOV, EH. A.
Résistance du béton cellulaire.
Beton i zelezobeton. (6), 10-12 (1982).
Durability, Protection. (Russian)

243. SANDBERG, P. I.; ERLANDSSON, B.
Investigations of moisture conditions in a cellular concrete roof.
Autoclaved Aerated Concr Moisture Prop. ed. F.H. Wittmann, 307-312 (1983).
Moisture, Modeling, Roof. (English)

244. SANDBERG, P. I.; ERLANDSSON, B.
Computer calculations of internally insulated aerated concrete roofs.
Autoclaved Aerated Concr Moisture Prop. ed. F.H. Wittmann, 297-306 (1983).
Moisture, Modeling, Roof. (English)

245. SAUMAN, Z.
Temperature reduction possibilities in the autoclave process of cellular concretes technology.
Silikaty. 29(4), 335-342 (1985).
Manufacture, Autoclaving, Composition. (Czech)

246. SAUMANN, Z.; SEVCIK, F.; JEDLICKA, A.
Preparation and properties of refractory aerated concrete.
Stavivo. 61(4), 144-147 (1983).
Manufacture, Fire resistance. (Czech)

247. SCHLEGEL, E.
Makroporoese Kalziumsilikat-Waermedaemmstoffe fuer Anwendungstemperaturen bis zu 900°C.
Silikattechnik. 38(4), 140-142 (1987).
Manufacture, Creep, Physical properties. (German)

248. SCHLEGEL, E.; BOLDT, R.
Die Waermeleitfaehigkeit hitzebestaendiger Gasbetone SILTONTHERM in Abhaengigkeit von der Rohdichte und Porengroesse.
Silikattechnik. 41(12), 424-427 (1990).
Thermal properties, Porosity, Testing. (German)

249. SCHLEGEL, E.; BOLDT, R.
Dependence of specific thermal conductivity of the SILTONTHERM cellular concrete on its apparent density and size.
Silikaty. 31 (3), 227-236 (1987).
Thermal properties, Density. (Czech)

250. SCHLEGEL, E.; UNVERRICHT, M.
Hitzebestaendiger, makroporoeser Kalziumsilikat-Waermedaemmstoff mit einer Rohdichte von 300kg/m3.
Silikattechnik. 37(3), 81-83 (1986).
Porosity, Thermal properties. (German)

251. SCHMIDT-REINHOLZ, CH.; SCHMIDT, H.
Physikalische und chemische Untersuchungsverfahren in der Grobkeramik. XXV.
Bestimmung derFeuchtigkeitsdehnung.
Sprechsaal. 121(11) ,1085-1088 (1988) .
Moisture, Expansion. (German)

252. SCHOBER, E.; POLSTER, H.
Erfahrungen beim Einsatz bewehrter Gasbetonelemente unter hoher
Immissionsbeanspruchung.
Bauplanung Bautech. 39(8) ,358-361 (1985) .
Durability, Reinforcement, Corrosion. (German)

253. SCHOBER, E.; UHLIG, B.
Quecksilber-Druck-Porosimetrie. Eine Methode zur Ermittlung der
Porengroessenverteilungen bei silikatischen Baustoffen.
Betontechnik. 5(4) ,104-107 (1984) .
Pore-size distribution, Testing. (German)

254. SCHRAMM, R.
Autoclaved aerated concrete: a building material in a Common Market.
Betonwerk und Fertigteil-technik. 56 (12) ,36-38 (1990) .
(German)

255. SCHUBERT, P.
On the shrinkage behaviour of aerated concrete.
Autoclaved Aerated Concr Moisture Prop ed. F.H. Wittmann ,207-218 (1983) .
Moisture, Shrinkage, Cracking. (English)

256. SCHUBERT, P.; BACKES, H-P.
Carbonation tests on expanded polystyrene lightweight concrete.
Betonwerk Fertigteil-Tech. 49(5) ,306-311 (1983) .
Carbonation. (English)

257. SCHULZE, D.; HELLER, W.; KUPSCH, H.
Photon activation analysis on building materials.
Isotopenpraxis. 24(6) ,224-228 (1988) .
Radio activity, Testing. (English)

258. SCHWARZ,P.
Building with reinforced wall slabs of aerated concrete.
Betonwerk Fertigteil-Tech. 56(12) ,45-48 (1990) .
Walls, Protection, Standards. (German)

259. SCHWERM, D.
Cladding - construction and design in precast concrete.
Betonwerk Fertigteil-Tech. 57(1) ,76-81 (1991) .
Thermal insulation, Sandwich panel, Fire resistance. (English)

260. SENBU, O.; KAMADA, E.; TAIRAKU, T.; MIMORI, T.
Evaluation method of frost damage for external surface finishings of ALC buildings.
Nihon Kentiku Gakkai Ron-bun Hokoku-shu. (396) ,17-26 (1989) .
Durability, Freezing, Testing. (Japanese)

261. SENBU, O.; KAMADA, E.; TAIRAKU, T.; MIMORI, T.
Experimental study on testing method for frost resistance of autoclaved lightweight concrete.
Proc. Workshop on Low Tempe. Effects on Conc., Sapporo., 384-401 (1988).
Freezing, Testing, Water absorption. (English)

262. SENBU, O.; KAMADA, E.; KOH, E.; TABATA, M.
Frost damage without repetition of freezing and thawing and evaluation method of porous inorganic materials.
Nihon Kentiku Gakkai Ron-bun Hokoku-shu. (367), 23-29 (1986).
Freezing, Water absorption, Testing. (Japanese)

263. SENBU, O.; KAMADA, E.; KO, Y.; TABATA, M.
Evaluation method for frost damage of porous inorganic materials.
Nihon Kentiku Gakkai Ron-bun Hokoku-shu. 357, 1-7 (1985).
Freezing, Testing. (Japanese)

264. SHARIFOV, A.
Preparation of aerated lightweight concretes using a foaming agent from scrubber paste.
Izv. Vyssh. Uchebn. Zaved.; Stroit. Arkhit. (4), 48-52 (1990).
Setting, Rheological properties. (Russian)

265. SHIKHNENKO, I. V.; KRUGOLOV, V. A.
Aerated concrete from industrial wastes.
Stroit. Mater. Konstr. (4), 14-15 (1990).
Manufacture, Waste. (Russian)

266. SHIKHNENKO, I. V.; KRUGLOV, V. A.
Naturally hardened aerated concrete.
Stroit. Mater. Konstr. (4), 18 (1986).
Hardening. (Russian)

267. SHUSTOVA, L. L.
Unautoclaved monolithic heat-insulating aerated concrete from power-plant ashes.
Issled. Mestnykh Stroit. Mater.: Sb. Nauch. Tr.; Ufa. (2), 33-37 (1990).
Manufacture, Fly ash. (Russian)

268. SIELAFF, M.
Moeglichkeiten zur Aufbereitung und Wiederverwendung von Gasbetonbruch.
Betontechnik. 5(5), 139-141 (1984).
Manufacture. (German)

269. SIELAFF, M.; BEHNE, J.
Automatische Messung der Abgussviskositaet in Gasbetonwerken.
Betontechnik. 9(4), 102-104 (1988).
Manufacture, Setting, Rheological properties. (German)

270. SIGEKURA, H.
Durability of ALC panel and construction work using it.
Sekou. (270), 27-29 (1988).
Durability, Panels. (Japanese)

271. SILAENKOV, E. S.
Life of Articles Produced from Aerated Concrete (Dolgovechnost' Izdelii iz Yacheistykh Betonov).
PUBLISHER:Stroiizdat,Moscow, USSR BOOK, 176 pp. (1986).
(Russian)

272. SOLIS,M. M.
Autoclaved cellular concrete for residential construction.
Concr Int. 12(9), 41-44 (1990).
Manufacture, Standards. (English)

273. SOUTHERN, J. R.
Moisture in solid blockwork walls at Glenrothes.
Autoclaved Aerated Concr Moisture Prop. ed. F.H. Wittmann, 313-323 (1983).
Moisture, Walls, Drying. (English)

274. STAERKER, K.
Anforderungen an die Ausbildung von Aussenwandkonstruktionen mit Gasbeton.
Bauplanung Bautech. 38(9), 396-398 (1984).
Walls, Thermal insulation. (German)

275. SUDELAINEN, N. N.; VERETEVSKAYA, I. A.
Study of properites of ash from two-stage combustion of shale for production of aerated autoclaved concrete.
Tekhnol. Silikat. Mater.; Tallin., 20-28 (1989).
Manufacture, Waste. (Russian)

276. SUN, B.; LI, G.; JIA, C.
Relation between hydrates and the strength and shrinkage of autoclaved aerated concrete.
Guisuanyan Xuebao. 11(1), 77-84 (1983).
Microstructure, Shrinkage, Strength. (Chinese)

277. SUN, B.; LI, G.; JIA, C.
Quantitative analysis of hydrates in autoclaved aerated concrete.
Guisuanyan Xuebao. 10(3), 357-363 (1982).
Tobermorite, X-ray diffraction, Hydration. (Chinese)

278. SUN BAOZHEN, SU ERDA
Relation between properties of aerated concrete and its porosity and hydrates.
Proc.1st Int.RILEM Congr.Vol.1, Pore Struct. Mater. Prop. ed. Maso, J.C., 232-237 (1987).
Carbonation, Porosity, Mechanical properties. (English)

279. SUN G., TANG,D., ZHAO,Y.
Carbonation of autoclaved aerated concrete and its hydration products.
Guisuanyan Xuebao. 13(4), 414-423 (1985).
Carbonation, Microstructure. (Chinese)

280. SUN G.; TANG, D.; ZHAO, Y.
The carbonation of autoclaved aerated concrete and its hydration products.
Proc. Int. Symp. Cem. Concr. Beijing. vol.3, 331-342 (1985).
Carbonation, Hydration, Composition. (English)

281. SVANHOLM, G.
Measurement of thermal conductivity of moist specimen.
Autoclaved Aerated Concr Moisture Prop. ed. F.H. Wittmann, 153-156 (1983).
Thermal properties, Heat transfer, Moisture. (English)

282. SVANHOLM, G.
Influence of water content on properties.
Autoclaved Aerated Concr Moisture Prop. ed. F.H. Wittmann, 119-130 (1983).
Moisture, Physical properties. (English)

283. TADA, S.
Sorption isotherm and the freezing point depression of autoclaved aerated concrete.
Prep. Ann. Meet. JCI. 14(1), 627-630 (1992).
Adsorption, Freezing, Moisture. (Japanese)

284. TADA, S.
Pore structure and freezing behavior of autoclaved lightweight concrete.
Concrete Research and Technology (JCI). 2(1), 95-103 (1991).
Porosity, Moisture, Freezing. (Japanese)

285. TADA, S.
Pore structure and moisture characteristics of autoclaved lightweight concrete.
Concrete Research and Technology (JCI). 1(1), 155-164 (1990).
Porosity, Moisture, Physical properties. (Japanese)

286. TADA, S.
Autoclaved aerated concrete.
Gypsum and Lime. (222), 269-276 (1989).
Durability, Microstructure, Physical properties. (Japanese)

287. TADA, S.
Material design of aerated concrete -An optimum performance design.
Mater Struct. 19(109), 21-26 (1986).
Modeling, Physical properties, Porosity. (English)

288. TADA, S.; TANAKA, T.; MATSUNAGA, Y.
Measurement of pore structure of aerated concrete.
Proc. Int. Symp. Cem. Concr. Beijing. vol.3, 384-393 (1985).
Pore-size distribution, Testing. (English)

289. TADA, S.; NAKANO, S.
Microstructural approach to properties of moist cellular concrete.
Autoclaved Aerated Concr Moisture Prop. ed. F.H. Wittmann, 71-89 (1983).
Moisture, Water absorption, Microstructure. (English)

290. TAM, C. T.; LIM, T. Y.; RAVINDRARAJAH, R. S.; LEE, S. L.
Relationship between strength and volumetric composition of moist-cured cellular concrete.
Mag Concr Res. 39(138), 12-18 (1987).
Microstructure, Mechanical properties, Composition. (English)

291. TERAMURA, S.
Fracture toughness and acoustic emission spectroscopy of fiber reinforced autoclaved lightweight concrete.
Nihon Kentiku Gakkai Ron-bun Hokoku-shu. (381), 26-34 (1987).
Fiber, Fracture, Microstructure. (Japanese)

292. TERAMURA, S.; TSUKIYAMA, K.; WATANABE, T.; TAKAHASHI, H.
Relationship between silica rock and fracture toughness of autoclaved calcium silicate.
Gypsum and Lime. (215), 203-213 (1988).
Silica, Microstructure, Fracture. (Japanese)

293. TERENT'EV, A.; KUNNOS, G.; RUDZINSKI, L.
Rheological model of structurized swelling viscoelastic medium under compressive deformation.
Rheol. Fresh Cem. Concr.; Proc. Int. Conf. Ed. Banfill, P. F., 103-112 (1991).
Modeling, Rheological properties. (English)

294. TERENT'EV, A.; KUNNOS, G.
Method of determining rheological parameters for linear viscoelastic medium under four kinds of uniform loading.
Rheol. Fresh Cem. Concr.; Proc. Int. Conf. Ed.Banfill, P. F., 93-102 (1991).
Modeling, Rheological properties. (English)

295. THOMAS-HESSELBARTH, J.; SCHARLIPP, K-P.
Entwicklung von Kraftschlusszangen fuer den Transport von Gasbetonelementen im VEB Betonprojekt Dessau.
Betontechnik. 10(1), 9-11 (1989).
Manufacture, Prefablication. (German)

296. THURNER, F.; STIETZ, M.
Bestimmung der Sorptionsisothermen loesungsmittelfeuchter Sorbentien nach der Durchstroemungsmethode.
Chem Eng Process. 18(6), 333-340 (1984).
Adsorption, Testing. (German)

297. TITOV, V. A.; VASCHENKO,T. P.; VARLAMOV, V. P.; SUDINA, N. K.
Estimation of type and magnitude of macrostructure defects of porous concretes.
Int. Symp. Brittle Matrix Composites (2nd) Cedzyna. 34(2), 536-541 (1989).
Microstructure, Mechanical properties, Manufacture. (English)

298. TRAETNER, A.
Experimentelle Untersuchungen zur Querkrafttragfaehigkeit von Stahlzellenverbundkonstruktionen.
Bauplanung Bautech. 38(6), 252-257 (1984).
Walls, Reinforcement, Mechanical properties. (German)

299. UNVERRICHT, M.
Entwicklung neuer Gasbetongueten mit hohen Daemmeigenschaften.
Betontechnik. 4(2), 35-36 (1983).
Thermal properties, Porosity. (German)

300. URABE, K.
Structural changes of calcium silicate hydrate on heating.
Prep. Ann. Meeting CAJ. 44th, 38-43 (1990).
Tobermorite, Thermal properties. (Japanese)

301. VAN DIJK, W.; DE JONG, P.
Determining the 222Rn exhalation rate of building materials using liquid scintillation counting.
Health Phys. 61(4),501-509 (1991).
Testing, Radio activity. (English)

302. VERMA, C. L.; TEHRI, S. P.; RAI, M.
Techno-economic feasibility study for the manufacture of lime-fly ash cellular concrete blocks.
Indian Concr J. 57(3),67-70 (1983).
Manufacture, Fly ash. (English)

303. VESNIN, B. G.; MIRONOV, V. S.; BOKOVA, L. I.; KOZLOV, A. D.; SHURABINA, L. N.
Electric heating of monolithic walls from aerated concrete.
Beton i zelezobeton. ,26-27 (1988).
Thermal insulation. (Russian)

304. VOLZHENSKII, A. V.; CHISTOV, YU. D.; KARPOVA, T. A.; ISKHAKOVA, A. A.
Manufacture and properties of products from unautoclaved aerated concrete with standardized moisture content and thermal conductivity.
Stroit. Mater. (11),7-8 (1990).
Manufacture, Moisture, Thermal properties. (Russian)

305. VOROB'EV, KH. S.; GONTAR, YU. V.; CHALOVA, A. I.; SERGEJKINA, E .M.
Procédés de finissage d'éléments muraux de grandes dimensions en béton cellulaire.
Beton i zelezobeton. (6),16-17 (1982).
Panels. (Russian)

306. VOZNESENSKII, V. A.; BAROVSKII, N. D.;VIROVOI, V. I.; SHINKEVICH, E. S.
Pore-structure formation and use of its parameters for prediction of the properties of cellular composites.
Fiz.-Khim. Mekh. 17,3-10 (1990).
Pore-size distribution, Physical properties. (Russian)

307. WACHI, E
Lime and ALC.
Lime. (334),6-14 (1983).
Lime, Composition. (Japanese)

308. WALTHER, K.
Stationaere Feuchteverteilung in einer monolithischen Aussenwandecke.
Gesund Ing Haustech Bauphys Umwelttech. 110(4),197-200 (1989).
Walls, Moisture, Adsorption. (English)

309. WATANABE, H.; KINOSITA, M.; TSUCHIKABE, K.
Autoclaved lightweight precast concrete curtain wall with the fire resistant joint and the drained joint, "Wall 21".
Kawasaki Seitetsu Giho. 20(4),360-361 (1988).
Walls, Prefablication, Joints. (Japanese)

310. WEISE, J.
Zur Berechnung von Konstruktionen aus Gasbeton nach Grenzzustaenden.
Betontechnik. 6(1),9-11 (1985).
Strength. (German)

311. WELLENSTEIN, R.
Der Einfluss der Dichte mineralischer Baustoffe auf ihre Waermeleitfaehigkeit.
Betonwerk Fertigteil-Tech. 48(10),598-601 (1982).
Thermal properties, Density. (German)

312. WENZEL, H. P.
Jointing solutions for front wall parts of composite type made of reinforced
concrete and gas concrete.
Wiss. Zeit. der Hoch. Arch. Bau. Weimar. 33 (5-6),254-258 (1987).
Walls, Joints. (English)

313. WENZEL, H-P.
Vorstellungen zum Einsatz von Gasbeton in mehrschichtigen Aussenwaenden aus
bauphysikalischer Sicht.
Betontechnik. 8(4),110-113 (1987).
Sandwich panel, Walls. (German)

314. WIDMANN, H.; ENOEKL, V.
Foam Concrete-Properties and Production.
Betonwerk Fertigteil-Tech. 57(6),38-44 (1991).
Physical properties, Durability,Manufacture. (English)

315. WITTMANN, F. H.
Effect of waterproofing on the properties of concrete.
Bautenschutz Bausanierung. 10(4),151-155 (1987).
Finishing, Physical properties. (German)

316. WITTMANN, F. H.; GHEORGHITA, I.
Fracture toughness of autoclaved aerated concrete.
Cem Concr Res. 14(3),369-374 (1984).
Fracture, Mechanical properties, Porosity. (English)

317. WOLEK, W.; DROZDZ, M.
Aerated refractory concretes.
Mater. Ogniotrwale. 38(4),95-98 (1986).
Thermal properties. (Polish)

318. XIAO, Y.
Fine pore structure and its optimum design of fly-ash silicate.
Proc. 2nd Int. Symp. Cem. Concr. Beijing. vol.3 313-319 (1989)
Modeling, Pore-size distribution. (English)

319. XIAO, Y.
An approach to the mathematical model of casting technological process of
autoclaved aerated concrete.
Proc. Int. Symp. Cem. Concr. Beijing. vol.3,359-374 (1985).
Manufacture, Modeling. (English)

320. XIE, Y.; XIA, G.
Study on the wet blend-milling technique of cement-lime-sand for making aerated
concrete.
Proc. 2nd Int. Symp. Cem. Concr. Beijing. vol.3 304-312 (1989)
Manufacture. (English)

321. XIE Y.; XIA, G.; WAMG, Q.
Research on compatibility of gas forming and thickening of fresh aerated concrete.
Proc. Int. Symp. Cem. Concr. Beijing. vol.3, 357 (1985).
Manufacture, Hydration. (English)

322. XU, E.; ZHAO, Z.
The new technology on the wet mixing grinding mill manufacturing mortar of aerated concrete.
Proc. Int. Symp. Cem. Concr. Beijing. vol.3, 358 (1985).
Manufacture, Strength. (English)

323. YAMAMOTO, M.; MORIMITSU, M.; AKIYAMA, T.
Pore-structure and sound absorption characteristics of porous sound absorbing materials.
Asahi Glass Kenkyu Hokoku. 39(2), 211-226 (1989).
Pore-size distribution, Acoustic insulation. (Japanese)

324. YAMAMOTO, Y.; TOMATURI, K.; YAMADA, H.; NAGASE, K.
High quality foamed light weight concrete (FLC).
Taisei Kensetu Gijutu Kenkyuusho-ho. (20), 109-113 (1987).
Hydration, Manufacture. (Japanese)

325. YONEDA, S.; MATSUNAGA, A.; NANJO, K.
Studies on high-strength cellular concrete using special inorganic expanded aggregate.
Cement Concrete. (506), 30-36 (1989).
Water absorption, Lightweight aggregate, Strength (Japanese)

326. YSSELDYKE, D. A.
Testing structural foams.
Adv Mater Processes. 134(1), 37-40 (1988).
Modeling, Strength, Porosity. (English)

327. YURINA, N. M.
Corrosion-preventing properties of aerated concrete produced from wastes from beneficiation of ferruginous quartzites.
Fiz.-khim. Osnovy Pr-va Stroit. Mater.; M. 30 (1), 36-40 (1986).
Corrosion, Waste. (Russian)

328. ZACHARIAS B, VENZMER H ; CERNY R
Zum Einfluss des hydraulischen Druckes auf den Feuchtetransport in Kapillarporoesen Stoffen.
Bauphysik. 12(5), 133-136 (1990).
Water absorption, Permeability. (German)

329. ZAJICEK, M.; SISKA, L.
Radioactivity of aerated concrete and raw materials used for its manufacture.
Stavivo. 68 (7-8), 244-250 (1990).
Radio activity. (Czech)

330. ZHAO, Y.; CHEN, Z.
Effect of Alkaline compound on aerated calcium silicate hydrates and their product strength.
Proc. 2nd Int. Symp. Cem. Concr. Beijing. vol.3 320-328 (1989)
Composition, Strength. (English)

331. ZHAO, Y.; CHEN, Z.
   Effect of Alkaline compound on aerated calcium silicate hydrates and their product strength.
   Proc. 2nd Int. Symp. Cem. Concr. Beijing. vol.3 320-328 (1989)
   Composition, Strength. (English)

332. ZHENG, W.
   The pore structure of aerated concrete and its effect on strength and drying shrinkage.
   Proc. Int. Symp. Cem. Concr. Beijing. vol.3 , 394-405 (1985) .
   Shrinkage, Strength, Pore-size distribution. (English)

Advances in Autoclaved Aerated Concrete, Wittmann (ed.) © 1992 Taylor & Francis. ISBN 90 5410 086 9

Author index

Note: Numbers following a name indicate the first page of a contribution of the corresponding author to this volume. Numbers of papers included in the AAC bibliography 1982-1992 included in this volume are given in brackets.

Printed and bound by CPI Group (UK) Ltd, Croydon, CR0 4YY

23/10/2024

01777686-0011